Introduction to
Plant Population Biology

For Rissa, Eva, Alfred, Brian and Jane

Introduction to Plant Population Biology

Jonathan Silvertown

Department of Biological Sciences
The Open University
Milton Keynes

Deborah Charlesworth

Institute of Cell, Animal and Population Biology
University of Edinburgh
Edinburgh

FOURTH EDITION

Blackwell
Science

© 2001 by J.W. Silvertown &
D. Charlesworth
Editorial Offices:
Osney Mead, Oxford OX2 0EL
25 John Street, London WC1N 2BS
23 Ainslie Place, Edinburgh EH3 6AJ
350 Main Street, Malden
    MA 02148–5018, USA
54 University Street, Carlton
    Victoria 3053, Australia
10, rue Casimir Delavigne
    75006 Paris, France

Other Editorial Offices:
Blackwell Wissenschafts-Verlag GmbH
Kurfürstendamm 57
10707 Berlin, Germany

Blackwell Science KK
MG Kodenmacho Building
7–10 Kodenmacho Nihombashi
Chuo-ku, Tokyo 104, Japan

Iowa State University Press
A Blackwell Science Company
2121 S. State Avenue
Ames, Iowa 50014–8300, USA

First published 1982
by Longman Scientific & Technical
Reprinted 1984
Second edition 1987
Third edition published 1993
by Blackwell Scientific Publications
Reprinted 1994, 1997
Fourth edition 2001
by Blackwell Science Ltd

Set by Kolam Information
    Services Pvt. Ltd.,
    Pondicherry, India
Printed and bound in Great Britain

The Blackwell Science logo is a
trade mark of Blackwell Science
Ltd, registered at the United
Kingdom Trade Marks Registry

DISTRIBUTORS

Marston Book Services Ltd
PO Box 269
Abingdon, Oxon OX14 4YN
(*Orders*: Tel: 01235 465500
        Fax: 01235 465555)

The Americas
    Blackwell Publishing
    c/o AIDC
    PO Box 20
    50 Winter Sport Lane
    Williston, VT 05495-0020
    (*Orders*: Tel: 800 216 2522
            Fax: 802 864 7626)

Australia
    Blackwell Science Pty Ltd
    54 University Street
    Carlton, Victoria 3053
    (*Orders*: Tel: 3 9347 0300
            Fax: 3 9347 5001)

A catalogue record for this title
is available from the British Library

ISBN 0-632-04991-X

Library of Congress
Cataloging-in-Publication Data

Silvertown, Jonathan W.
    Introduction to plant
    population biology / Jonathan
    Silvertown, Deborah
    Charlesworth.– –4th ed.
        p. cm.
    Includes bibliographical
    references.
    ISBN 0-632-04991-X (pbk.)
    1. Plant populations. 2.
    Vegetation dynamics. I.
    Charlesworth, Deborah. II. Title.

QK910 .S54 2001
581.7′88– –dc21

                    2001035131

For further information on
Blackwell Science, visit our website:
www.blackwell-science.com

Contents

CONTENTS

Acknowledgements

We thank Brian Charlesworth, Greg Cheplick, David Gibson and Jeff Ollerton for their comments on drafts of various chapters of the present edition and, in particular, Richard Abbott who read the entire manuscript. These acknowledgements should not be taken to imply agreement with what we have written: responsibility for errors of omission and comission remains ours.

We are particularly grateful to Rissa de la Paz and Brian Charlesworth for their help and support throughout.

1 Introduction

1.1 Plants

Plants and their products are all around and inside us. This book you are holding was once a plant. The oxygen you are breathing was freed to the air by photosynthesis. The clothes you are wearing may be made from plant fibres; if they are not, the chances are that they are spun from polymers derived from oil, which was once ancient chlorophyll. Perhaps you have a cup of coffee or tea by your side? If so, the secondary compounds which defend the seeds of *Coffea arabica* or the leaves of *Camellia sinensis* from their natural enemies will help you stay awake while you read this book! There is nothing dull about plants, but it does require some subtlety to appreciate their peculiar ways. There is enough promiscuous sex and sudden death in the plant kingdom to furnish a hundred cable TV channels, but it all happens with the sound turned off.

This book is an *introduction* to the population biology of plants. In it we seek to understand how plant populations are structured and how they change from an ecological and evolutionary perspective. The fundamental ecological and evolutionary principles of the subject are not exclusive to plants; indeed plants are ideal model organisms in which to study those fundamentals. However, being sessile autotrophs means that plants experience and respond to their environments in ways that are sometimes difficult for mobile heterotrophs, such as ourselves, to fathom intuitively. Therein lies the fascination of the plant world.

1.2 Population biology

A **population** is a collection of individuals belonging to the same species, living in the same area—the water hyacinth *Eichhornia crassipes* in a ditch, the grass *Lolium perenne* in a lawn, or Norway spruce *Picea abies* in a forest are examples. This definition has two components, a genetic one (individuals belong to the same species) and a spatial one (individuals live in the same area), but populations are neither genetically nor spatially homogeneous. Populations have several kinds of structure. The **genetic structure** describes the patchiness of gene frequency and genotypes, and the **spatial structure** describes the variation in density within a population. Populations also have an **age structure** that describes the relative numbers of young and old individuals, and a size structure describing the relative numbers of large and small individuals. Population biology attempts to explain the origin of these different kinds of structure, to understand how they influence each other, and how and why they change with time. Changes over time in the genetic make-up of a population are the subject of **evolution**; change in numbers with time is the subject of **population dynamics**. These are the two principal organizing themes of this book. We deal with the special characteristics of plants that affect their population biology in the second half of this chapter.

1.2.1 Demography

The essence of population biology is captured by a simple equation that relates the numbers per unit area of an organism N_t at some time t to the numbers N_{t+1} one time unit (e.g. year) later:

$$N_{t+1} = N_t + B - D + I - E, \qquad (1.1)$$

where B is the number of births, D the number of deaths, and I and E are, respectively, immigrants into the population and emigrants from it. B, D, I and E are known as **demographic parameters** and are central to both population dynamics and evolution. The dynamics of a population may be summarized by the ratio N_{t+1}/N_t, which is called the **annual** or the **finite rate of increase**, and is given the Greek symbol λ (lambda). The balance between the two demographic parameters which increase N_t (B & I) and those which decrease it (D & E) determines whether the population remains stable ($N_{t+1} = N_t$ and λ = 1), increases ($N_{t+1} > N_t$ and λ > 1) or decreases ($N_{t+1} < N_t$ and λ < 1). Among other things, the values of B, D, I and E in natural plant populations are variously influenced by pollinators, herbivores, diseases, animals that disperse seeds, soil, climate, by the density of the population itself and by that of other plant species.

Furthermore, the influence of these factors on the demographic parameters often has a genetic component, because there is often genetic variation in natural populations: some individuals are more susceptible to disease than others, for instance, and some are more distasteful to herbivores, and some more tolerant of climatic extremes. A consequence is that one genotype may be favoured in one locality, and another somewhere else, and this may produce local differences in allele frequency that correlate with environmental conditions. For example, allele frequencies at an acid phosphatase (APH) locus in *Picea abies* correlate with altitude in the Seetaler Alps of Austria (Fig. 1.1e), and on a larger scale with latitude in northern Europe (Fig. 1.1d). Polymorphism at this locus appears to be an indication of a more general correlation between the genetic composition and geographical location of *P. abies* populations in Europe. Using 22 enzyme loci to characterize genotypes, Lagercrantz and Ryman (1990) found a similar geographical correlation stretching across populations sampled from most of the species' natural range. Note that *correlations* between allele frequencies and environment do not mean that the particular loci studied are necessarily the direct *cause* of

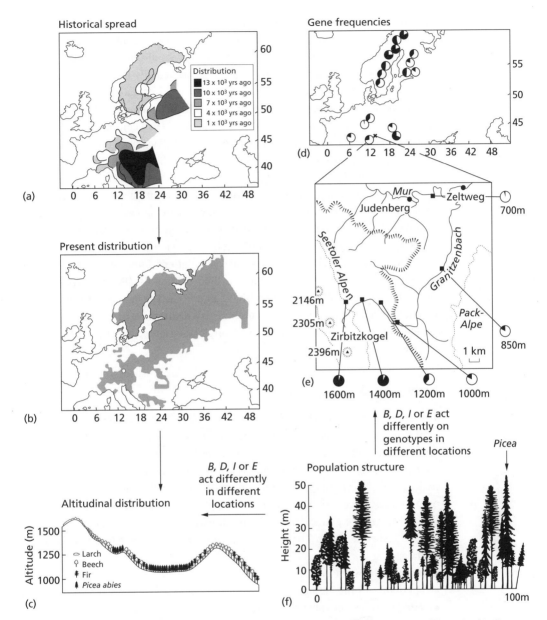

Fig. 1.1 How demographic parameters in populations of Norway spruce (*Picea abies*) in Europe underlie its (a) historical spread, (b) present geographical distribution, (c) local, altitudinal distribution, (d & e) APH allele frequencies, and (f) population structure (from Silvertown 1987a).

ecological differences between genotypes. Because the causative relationship between loci such as APH and the genes responsible for ecological differences is unknown, such loci are referred to as **genetic markers** (see next paragraph and Chapter 2).

Demographic processes underlie the distribution of species as well as the distribution of genotypes within species. At a local scale, differences in B, D, I and E must explain why *Picea abies* forms a distinct altitudinal belt in the Alps (Fig. 1.1c). Distributions can be investigated experimentally by sowing or planting the species along transects that cross the boundary of its natural range, to assess why boundaries lie where they do (Chapter 7). Climate changes and, in response, species' boundaries are, to varying degrees, dynamic. On a continental scale this is easily seen by comparing the historical spread of *P. abies* since the last glaciation (Fig. 1.1a) with its present geographical distribution (Fig. 1.1b). If rates of migration are slower than climatic change, boundaries may reflect historical limits rather than demographic ones. This is true of the distribution and genetic structure of Norway spruce in Europe today. The pollen record shows that the postglacial recolonization of western Europe by *P. abies* occurred from two refugia: one south of the Alps, in a region including northern Italy, and another on the eastern side of the Baltic Sea (Fig. 1.1a). Norway spruce populations found in these two areas today have distinctive genetic markers in their chloroplast DNA that reflect their different historical origins (Vendramin *et al.* 2000). The migratory history of a population can often be read in the present-day distribution of selectively neutral genetic markers. This is the science of **phylogeography**, which is discussed in Chapter 7.

1.2.2 Fitness and natural selection

The genetic structure of a population is subject to a number of forces that may bring about evolutionary change. The two principal forces are **gene flow** and **natural selection**. Gene flow simply describes the changes in gene frequency brought about by the migration of individuals, their seeds and their pollen. Special attention is given to natural selection because this is the only known process that can produce **adaptive** evolutionary change. This was the great, enduring insight in Charles Darwin's book *The Origin of Species*, first published in 1859. Three conditions are necessary for natural selection to operate:

- Variation between individuals;
- Inheritance of the variation;
- Differences in fitness between variants.

Fitness (usually denoted by the symbol w or W) is defined in population genetics terms in Section 3.5.2.1. For our purposes, it is usually accurate to define the fitnesses of different genotypes as the relative success with which they transmit their genes to the next generation. This definition also makes it possible to define fitnesses of phenotypes, as we shall see when we discuss evolution of plant breeding systems in Chapter 9. However, we do not always need to be so formal in our use of the word fitness, particularly as the strict definition of fitness is rarely something we can measure in the real world. The term is also used more generally, for instance in the term 'components of fitness' to refer to characteristics that we expect to correlate with increased fitness (although it may often be better to be explicit, and speak directly of survival or fertility).

Ecologists sometimes use the term fitness in an even more general way, and ignore the fact that, in sexually reproducing populations, individuals with different genotypes breed with one another. The concept of phenotypic fitnesses can sometimes justify use of the simplifying assumption that different genotypes do not interbreed. This assumption can, for instance, lead to approximately correct inferences when a new phenotype enters a population (e.g. by a new mutation), and we are interested in asking whether it will increase in frequency. In that case, we may use λ to estimate fitnesses by applying Eqn 1.1 separately to different phenotypes. For example, in *Picea abies* we could compare λ_H for the high altitude phenotype with λ_L for trees with the low altitude phenotype in different parts of the Seetaler Alps. At low altitudes we might expect $\lambda_L > \lambda_H$, and at high altitudes we might expect $\lambda_H > \lambda_L$. By convention, the phenotype with the highest fitness has a value $W = 1$, and the fitness of other phenotypes is given as a proportion of this. Fitnesses usually change with environmental conditions. So, *if* the APH phenotypes are adapted to the respective environments in which they are most frequent (and this has *not* yet been tested experimentally), at low altitudes the fitness of the low phenotype W_L would be 1, and the fitness of the high phenotype $W_H < 1$, while at high altitudes $W_H = 1$ and $W_L < 1$.

Given heritable variation and fitness differences between variants (i.e. certain genotypes have phenotypes that cause their genes to multiply faster than those in others), there is the potential for adaptive evolutionary change to take place. Whether change actually occurs, and what direction of changes occur, may depend upon the fitnesses in other populations of a species, and on the rate of migration of seeds and pollen from neighbouring areas, where selection may have a different direction. Field observations have shown that natural selection may be strong enough to produce marked genetic differences between adjacent populations of the same species and to produce adaptation to local conditions. For example, the annual *Veronica peregrina* grows in and around temporary pools, called vernal pools, that form in early springtime in the Central Valley of California, USA. Although pools are only a few metres wide, plants in the centre are genetically different from those around the periphery (Keeler 1978). Phenotypically, they are more tolerant of flooding (Linhart & Baker 1973), and differ in a number of other ways that adapt them better to the conditions of intense intraspecific competition that occur in the centre than to competition with grasses that affects those genotypes growing at the pool edge (Fig. 1.2) (Linhart 1988). We shall explore natural selection and other evolutionary forces in more detail in Chapter 3.

1.2.3 Life tables and age dependence

The probability that an individual dies or reproduces is often related to its age; thus, the age structure of a population is likely to affect its future. The proportion

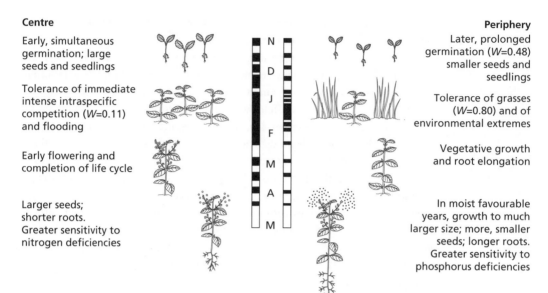

Centre

Early, simultaneous germination; large seeds and seedlings

Tolerance of immediate intense intraspecific competition (*W*=0.11) and flooding

Early flowering and completion of life cycle

Larger seeds; shorter roots. Greater sensitivity to nitrogen deficiencies

Periphery

Later, prolonged germination (*W*=0.48) smaller seeds and seedlings

Tolerance of grasses (*W*=0.80) and of environmental extremes

Vegetative growth and root elongation

In moist favourable years, growth to much larger size; more, smaller seeds; longer roots. Greater sensitivity to phosphorus deficiencies

Fig. 1.2 Genetically determined differences between plants of *Veronica peregrina* sampled from the centre and periphery of a vernal pool. Bars show the periods between November and May of adequate soil moisture in the two microhabitats. '*W*' indicates the relative fitness, based upon seed production, of plants from the other microhabitat when grown experimentally in the conditions indicated (after Linhart 1988).

of youngsters is an indication of the likely future, but this does depend upon the chances of surviving to adulthood. This information is conventionally summarized in a **life table** (Table 1.1).

Acacia suaveolens is a small shrub that grows in fire-prone arid habitats of SW Australia. Figure 1.3 shows a **survivorship curve** for *A. suaveolens*, based upon the data in the life table. The life table and survivorship curve for *A. suaveolens* indicate that substantial mortality occurs throughout the lifespan of the population. In fact, the mortality rates affecting this population of *A. suaveolens* are quite modest by comparison with many plant populations that have been studied, for which 90% mortality in the first year of life is not unusual. Such mortality may alter the genetic structure of a population if some genotypes are more susceptible than others.

Life tables were originally devised as a means of studying the human population, and are used by actuaries to calculate the risk of insuring the life of their clients (Hutchinson 1978). A person's age is, of course, not the only factor that influences their longevity; nor is this so in plants. Although life tables are used for plant populations, in many ways they are inappropriate to the peculiarities of plants. The rate of growth in plants is highly dependent upon the local environment and consequently is very variable, so that two genetically identical plants of the same age may be quite different in size. This is an example of **phenotypic**

Table 1.1 Life table and fecundity schedule for a population of the shrub *Acacia suaveolens*, in Australia (data from T. Auld & D. Morrison pers. comm.).

Age (yr) x	Number N_x	Survival l_x	Mortality d_x	Mortality rate, d_x/l_x q_x	Survival rate, $1-(d_x/l_x)$ p_x	Seeds / plant m_x
0	1000	1	0.174	0.174	0.826	0
1	826	0.826	0.145	0.176	0.824	41
2	681	0.681	0.159	0.233	0.767	33
3	522	0.522	0.122	0.234	0.766	31
4	400	0.4	0.093	0.233	0.768	31
5	307	0.307	0.076	0.248	0.752	18
6	231	0.231	0.057	0.247	0.753	9
7	174	0.174	0.043	0.247	0.753	9
8	131	0.131	0.015	0.115	0.885	9
9	116	0.116	0.013	0.112	0.888	7
10	103	0.103	0.012	0.117	0.883	5
11	91	0.091	0.011	0.121	0.879	3
12	80	0.08	0.009	0.113	0.888	6
13	71	0.071	0.009	0.127	0.873	–
14	62	0.062	0.007	0.113	0.887	–
15	55	0.055	0.007	0.127	0.873	2
16	48	0.048	0.005	0.104	0.896	4
17	43	0.043	–	–	–	3

– indicates missing values

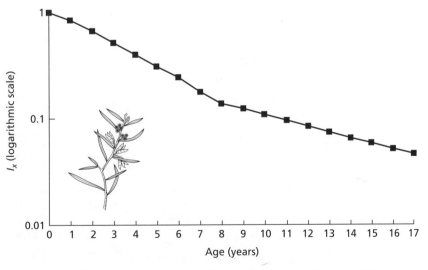

Fig. 1.3 A survivorship curve for *A. suaveolens* (pictured), based upon the data in Table1.1.

plasticity, a feature in which plants excel (Chapter 2). Genetically identical plants may have quite different phenotypes if they have been exposed to different environments.

1.2.4 Life cycle graphs and stage dependence

Size is a major influence on the fate of individual plants and on their fecundity (Fig. 1.4). Because of phenotypic plasticity, size and age tend to be only loosely correlated. The dependence of plant fate on age is further weakened by the existence of dormant stages – notably seeds, but also tubers and stunted seedlings – that may persist for many years before entering a phase of active growth. Thus, while it is clearly possible to draw up plant life tables, and for some purposes it is useful to do so, this approach leaves much demographic variation in plant populations unaccounted for.

A simple alternative to the life table that uses age or plant **stage** to classify individuals in the population is the **life cycle graph** (Hubbell & Werner 1979). Each age-class or stage in the life cycle is represented by a node, and transitions between nodes are shown by arrows joining them (Fig. 1.5a,b). Appropriate rates of transition between stages and seed production are obtained from field studies. For example, Werner (1975) studied a population of the herb *Dipsacus sylvestris* in an abandoned field in Michigan, and classified plants into several stages: (1) first-year dormant seeds, (2) second-year dormant seeds, (3) small rosettes, (4) medium rosettes, (5) large rosettes and (6) flowering plants. A life cycle graph based on the annual rates of transition between these stages is shown in Fig. 1.5(c). *D. sylvestris* dies after flowering, so the arrows from (6) to the other stages all represent numbers of individuals in those stages that were produced from seed that were produced that year.

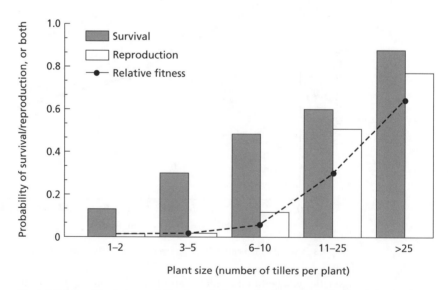

Fig. 1.4 Relationships between size and survival, reproduction and relative fitness (survival × reproduction) of the grass *Bouteloua rigidiseta* (data from Fowler 1986).

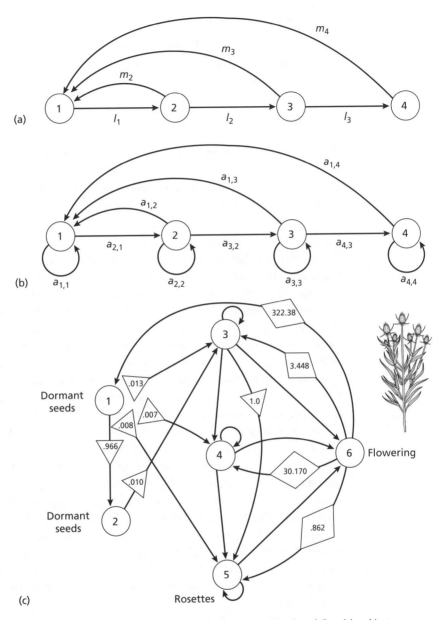

Fig. 1.5 Examples of life cycle graphs. Diamonds represent numbers (of seeds) making a transition, and triangles show rates of transition between nodes. Graphs for: (a) an age-classified life cycle, (b) a stage-classified life cycle, and (c) the life cycle graph for a population of the herb *Dipsacus sylvestris* (pictured; Caswell 1989).

1.3 Some consequences of being a plant

Two features shared by all plants deserve special mention because of their fundamental consequences for the population biology of these organisms: plants' growth and development from meristems, and the fact that they are sessile.

1.3.1 Meristems

Not all cells in a plant are capable of dividing. Dividing cells occur in discrete regions called **meristems**, predominantly found in the apices of shoots, in the buds in the axils of leaves, near the tips of roots, and in the cambial layer beneath the surface of the stem in dicotyledonous plants and gymnosperms.

1.3.1.1 *Structure and life history*

The number and distribution of meristems on a plant, which of them develop and when, determine how a plant grows and its overall structure (Fig. 1.6). A bud that develops into a shoot usually multiplies the number of meristems on a plant, because each shoot has its own meristems. This produces the **modular construction** typical of plants. A bud that develops into a flower or inflorescence consumes the meristem and ends its career, so the plant must use other meristems for vegetative development if it is to continue to grow and survive. For example, when the shoot of a grass, called a **tiller**, flowers its apical meristem differentiates, causing the death of the tiller after reproduction. If all the individual tillers on a grass plant flower simultaneously, the whole plant will die.

The **life history** of a plant describes how long it typically lives, how long it usually takes to reach reproductive size, how often it reproduces and a number of other attributes that have consequences for demography and fitness. Because of the manner in which plants develop from meristems, there is a close relationship between the structure of a plant and its life history. Torstensson and Telenius (1986) found that the major difference in life history between the annual herb *Spergularia marina* and the related perennial *Spergularia media* was the result of a difference in how their axilliary meristems behaved. Axilliary meristems were committed to flowering at the 6th or 7th node in the annual, but in the perennial these meristems produced shoots instead and flowering began much later in development.

As well as allowing plants to grow and branch vertically, meristems allow plants to proliferate and spread horizontally. This is **clonal growth** (Fig. 1.6d). The whole plant comprises the **genet**, which is defined as an individual arising from seed, which (barring mutation) will be a single genotype, however large and fragmented it may later become through clonal growth. A clonally produced part of a plant, with its own roots and a potentially independent existence, is known as a **ramet**. The tillers produced from lateral meristems in grasses are an example. The branching pattern of growth produced by meristem activity results in a hierarchical structure: a genet is composed of ramets, ramets are composed of one or more

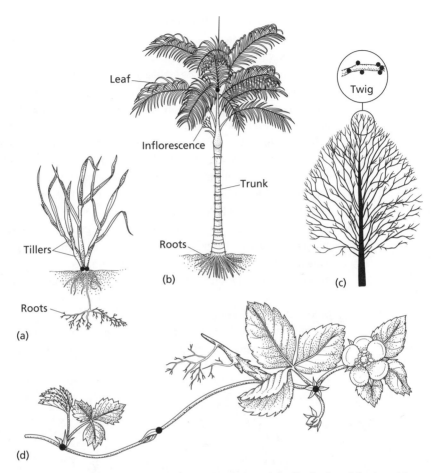

Fig. 1.6 Schematic diagrams showing plant construction and the distribution of shoot meristems, indicated by dots, in (a) a grass, (b) a palm, (c) a dicot tree, and (d) a dicot clonal herb.

branches, these bear inflorescences, inflorescences bear flowers, and flowers contain ovules and pollen. This hierarchical structure affects life history because the survival and reproduction of the genet depends upon the behaviour of ramets, and ramets depend upon the behaviour of their parts (Silvertown 1989). Flowering terminates the life of a meristem, so if all meristems on a branch produce flowers the branch will die, if all branches flower the ramet dies, and if all ramets flower the genet dies. Plants in which the genet dies after flowering are termed **semelparous** (or monocarpic), and those which can flower more than once are **iteroparous** (or polycarpic). An extreme example of semelparity occurs in some bamboos. The plant is clonal, forming many, perennial ramets, but all of them flower at the same time and then the whole genet dies. In the giant semelparous bamboo *Phyllostachys bambusoides* the prereproductive period is 120 years! (Janzen 1976).

There is a continuum of types of plant structure between trees, whose branches all depend upon a common trunk, and clonal plants, in which the 'trunk' is reduced to no more than the crown of a rootstock and the 'branches' develop roots of their own. Both extremes can been seen by comparing *Salix pentandra*, which is a small tree, with *Salix herbacea*, which is a dwarf, creeping shrub (Fig. 1.7a). There are similar contrasting examples among birches (*Betula* spp.) and dogwoods (*Cornus* spp.). Among clonal plants there is also great diversity in the method of spread, including stolons (above-ground creeping stems), rhizomes (below-ground creeping stems), tubers like the potato, bulbs such as the onion, or corms as in crocus.

Once a ramet has its own roots, it is usually self-sufficient in carbon, although water and minerals may still be imported from other parts of the clone if the connections between ramets persist (see Section 10.5). Plant species vary a good deal in how long the connections between ramets last. In white clover the stolons that connect ramets together are short-lived and the parts of a genet can wander far from each other, so that a single successful genet may be represented in many parts of a field. At the other extreme is what is probably the biggest tree on earth: a banyan *Ficus benghalensis* growing in Calcutta, India. It is a clone over 200 years in age, with a thousand connected tree trunks covering more than 1.5 ha.

1.3.1.2 *Genetics and evolution*

Meristematic growth and modular construction have a number of genetic and evolutionary consequences for plants. First, the number of flowers a plant is

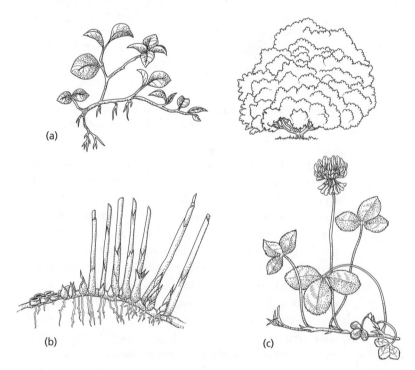

Fig. 1.7 (a) The growth form of *Salix herbaceae* (left) and *S. pentandra*; (b) rhizome of *Cyperus alternifolius*, and (c) *Trifolum repens*.

capable of producing is limited by the number of meristems, so its size is likely to be a major determinant of its total reproductive success and fitness. Secondly, meristems are like perpetually embryonic tissue—they are 'forever young', so it is possible for a very old and large genet to be entirely composed of much younger ramets that perpetually replace themselves by clonal growth. Huge clones in which the genet is inferred to be thousands of years old are known in many species. Probably the record is held by creosote bush *Larrea tridentata*, which has some clones in SW USA that, judged by their size and rate of growth, may be 11 000 years old (Vasek 1980). This is as ancient as the deserts in which they live.

1.3.1.3 *Behaviour*

Textbooks of ethology are strangely reluctant to define 'behaviour'; although ethologists clearly know what they mean by the word, some mistakenly believe plants to be incapable of any activities that deserve the term. To encompass all the *animal* activities that ethologists study, behaviour must be defined broadly, for example as a response to some event or change in the organism's environment (Silvertown & Gordon 1989). Most plants are capable of some kind of movement, for example in the orientation of leaves, and *Mimosa pudica*, the 'sensitive plant', is able immediately to fold up its leaves, exposing a spiny stem, when touched by a herbivore. However, rapid movement is rare and the chief method of behavioural response in plants depends upon their modular pattern of growth and the ability to alter the size, type and location of new organs to match an environmental change. Thus, for example the tropical rainforest liana *Ipomoea phillomega* produces several types of shoot, depending upon the nature of the light environment. In shaded conditions, stolons with long internodes and rudimentary leaves extend rapidly over the ground (Fig. 1.8). When a gap is reached, twining shoots with large leaves are produced and these ascend to the tree canopy where the liana forms a crown of its own (Peñalosa 1983).

The parallel between the behaviour of *Ipomoea phillomega* and animal foraging is difficult to resist. Such behaviour is especially common in climbers which, because of their extensive shoot systems, traverse several types of microenvironment during

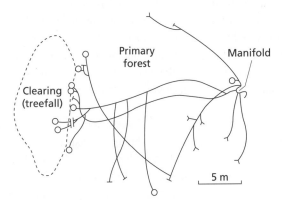

Primary forest

Manifold

Clearing (treefall)

5 m

Fig. 1.8 A map of the shoot system of *Ipomoea phillomega* on the floor of a tropical rainforest in Veracruz, Mexico. This plant originated at the 'manifold', which has an ascending shoot and a crown in the canopy. Circles represent ascending shoots with crowns. Stolons that have lost their tips end in a 'T', and those that are still growing are shown with a 'Y' (Peñalosa 1983).

growth and often have different types of leaf or shoot to match. Perhaps the simplest behaviour of which virtually all plants are capable is etiolation in response to shade. Most plant responses to shade from other plants are mediated by the substance **phytochrome**, which has two forms that interconvert with each other. The Pr form absorbs red light and is then converted to the Pfr form, which absorbs far-red light. The equilbrium between the two forms is determined by the *ratio* of red to far-red light. Leaves absorb red and blue light (this is why they are green), but are relatively transparent to far-red light, so a plant can detect the presence of leaf shade by the effect this has on the Pr : Pfr ratio. In fact, the system is so sensitive that plants can detect the quality of reflected light and respond to the presence of neighbours that do not shade them directly (Ballaré *et al.* 1990; Novoplansky *et al.* 1990). Using the phytochrome system, seeds of many species can discriminate between shade caused by other plants and darkness caused by burial. They will germinate in the dark but not when shaded by leaves (Silvertown 1980a).

In *Arabidopsis thaliana* and many other broad-leaved flowering plants there are five phytochrome genes (*PHYA–PHYE*) with partly complementary but often overlapping functions (Mathews & Sharrock 1997; Smith 2000). Genotypes with mutations that disable one or more of these genes can be used to test the adaptive significance of plant behavioural responses to shading by neighbours (Schmitt *et al.* 1999). Cucumbers with a mutation in the *PHYB* gene, which in the wild type modulates response to shade, were unable to find light gaps in a canopy of maize, while normal cucumbers could do so (Ballaré *et al.* 1995).

1.3.2 Sessileness

1.3.2.1 *Defence, dormancy and dispersal*

The fact that plants are rooted to the spot has a number of important consequences. Plants are unable to escape their herbivorous enemies by flight, but they are often packed with toxins that make them unpalatable. Others resist grazing with spines or, as grasses and sedges do, by hiding their meristems in the very base of the plant. Plants cannot migrate in unfavourable seasons of the year, but they have a range of protective strategies that include leaf deciduousness, seed dormancy and retreating to underground storage organs from which they may regrow in the next season. Some movement is of course possible, but creeping stems and roots grow too slowly to form any effective means of escape from most hazards. However, by forming an extensive network, stolons and roots can reduce the risk that local damage will kill the entire plant, and it has been suggested that this is a major advantage of the clonal growth habit (Eriksson & Jerling 1990).

Although plants are sessile, their genes are mobile and are transported in seeds and pollen, both equipped with a variety of dispersal aids. Pollen grains are produced in vast numbers, and wind-dispersed grains are light and sometimes, as in pines, have buoyancy aids attached to them. Animal-pollinated plants have colourful, fragrant flowers that attract insects, bats and birds as pollinators that are often rewarded with nectar (Chapter 2). It is worth remembering that the fruit and flowers so enjoyed by humans originate from organs evolved by sessile organisms that rely on animals for transport.

Virtually all plants disperse most of their seeds. Seed dispersal is often a two-stage process involving a primary stage in which seeds leave the mother and reach the ground and a secondary stage when seeds that reach the ground are moved again by animals or water. A wide range of mechanisms is involved in primary seed dispersal: ballistic expulsion of seeds by exploding fruits such as in the cucumber *Echballium elaterium*; wind transport aided by a wing or pappus such as in the dandelion *Taraxacum officinale*; transport of seeds in hooked structures that become trapped in animal fur; and carriage in the guts of animals that have eaten berries and which defecate viable seeds (Fig. 1.9). Each mechanism produces a characteristic distribution of seeds around the mother, or **seed shadow**. Wind dispersal mostly carries seeds further than primary dispersal by animals (Willson 1993; Clark *et al.* 1999), but in all cases dispersed seeds are most concentrated near the parent (Fig. 1.10).

Harvester ants in deserts and various vertebrate seed predators (particularly rodents) in forests are important agents of secondary seed dispersal. Both cache their food but fail to consume all of it before some seeds germinate (Brown *et al.* 1979; Forget 1996). In several species of woodland violet *Viola* spp., seeds are expelled from dehiscent fruit and then they are collected from the ground by ants that carry them to their nests (Beattie 1985). There, ants remove a fatty appendage from the seed, called the elaiosome, and deposit the seed, still viable, on their refuse heap. In a study of the ant-transported herb *Corydalis aurea* growing in the Rocky Mountains, USA, Hanzawa *et al.* (1988) compared the fitness of seeds transported by ants with that of seeds not carried away, and found that ant transport increased λ by 38%.

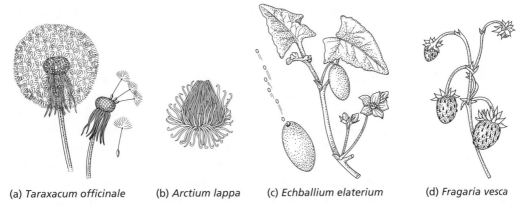

(a) *Taraxacum officinale*   (b) *Arctium lappa*   (c) *Echballium elaterium*   (d) *Fragaria vesca*

Fig. 1.9 Examples of modes of seed dispersal (a) Wind dispersed seeds of *Taraxacum officinale*, (b) hooked, animal dispersed seed head of *Arctium lappa*, (c) seeds explosively expelled from the fruits of *Echballium elaterium*, and (d) fleshy, animal-dispersed 'berry' of *Fragaria vesca*.

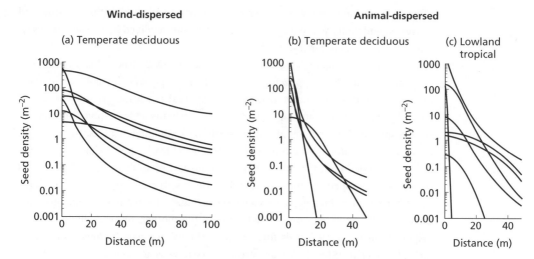

Fig. 1.10 Primary seed dispersal curves estimated by inverse modelling for (a) seven wind-dispersed deciduous tree species and (b) five animal-dispersed deciduous tree species of temperate North American forest; (c) seven animal-dispersed species of lowland tropical forest (from Clark *et al.* 1999).

The shape and length of the tail of seed dispersal curves is hard to quantify because it is difficult to identify the source of a seed when it has travelled a long way. However, two techniques offer a solution to this problem. One is the use of genetic markers such as microsatellites to identify the parents of seedlings (Chapter 3.3.2). The other employs a statistical method known as inverse modelling to estimate the entire seed shadows of individual forest trees from the distribution of dispersed seeds around the mapped locations of adults. This method predicts long, 'fat' tails for many tree species (Fig. 1.10; see Clark *et al.* 1999). Such distributions account for observed rates of spread of trees into the northern temperate zone in postglacial times (e.g. Fig. 1.1a) much better than earlier methods of estimation have (Clark 1998), suggesting that the dispersal curves produced by inverse modelling are biologically realistic.

1.3.2.2 *Mates and neighbours*

Because they are sessile, most of the ecological and genetic interactions between plants are with their immediate neighbours. Despite the agency of wind and animals in moving pollen, most plants mate with their neighbours and are frequently surrounded by relatives, creating genetic patchiness, even within continuous populations (Chapter 3). The genetic and spatial structures of populations interact with one another. Spatial patchiness in the density of populations reinforces the genetic isolation of patches, and this can lead to genetic divergence.

If below-ground resources such as water or nutrients limit plant growth, competing neighbours will interact chiefly by the depletion of resources in the common pool. However, when plants grow in a closed stand and light limits growth, a plant that is taller than its neighbours will not only receive a disproportionate share, but is also likely to *suppress* the neighbours in its shade. As a result of this inherent asymmetry among plants competing for light, even a slight difference in height or canopy area can be decisive in determining which plant wins (Chapter 4). This introduces a further complication, because the plant that is initially the largest among its neighbours is often merely the one that germinated first. The early plant catches the light. So, because plants are sessile, not only *where* a plant is but also precisely *when* it appeared there may be a matter of life and death.

1.4 Summary

A **population** is a collection of individuals belonging to the same species, living in the same area. Population biology aims to explain **spatial structure**, **age structure** and **size structure** in terms of the four basic **demographic parameters** which measure birth, death, immigration and emigration rates. Changes in numbers of individuals over time are the subject of **population dynamics**. When rates affect genotypes differently they may produce **evolutionary change** in the frequencies of alleles at genetic loci. Fitness differences can often be quantified in models of evolutionary change in terms of differences in numbers of genes transmitted from parental to progeny generations. The principal force leading to **adaptative** evolutionary change is **natural selection**. Natural selection can operate only when there is variation between individuals in heritable characters that affect fitness. The dynamics of a population may be summarized by the ratio N_{t+1}/N_t, which is called the **annual** or the **finite rate of increase** (λ). The finite rate of increase for a genotype can sometimes be used as a measure of its Darwinian **fitness**.

Birth and death rates are influenced by the age and size of individuals. Plant size is particularly sensitive to local environmental conditions and shows great **phenotypic plasticity**. A **life table** may be used to describe the age-dependence of demographic parameters and a **life cycle graph** to describe a stage- (or size-) classified population. The **life history** of plants tends to be tied up closely with their **modular construction**, which arises from **meristematic** development. **Clonal growth** is common in plants. The individually rooted parts of a clone are called **ramets**, and the whole clone is called a **genet**. Unlike in unitary animals where gametes and somatic tissues derive from separate cell lines, in plants both cell types derive from meristems. This has the important consequence that clonal plants may be potentially immortal.

Modular growth and the ability to produce structures that match a change in the environment give plants a **behavioural** repertoire that enables them to forage for light and avoid competitors. Because plants are unable to escape their animal enemies, they often rely on **chemical defences**. Seeds and pollen transport plants' genes, but plants still usually end up mating with their **neighbours**.

1.5 Further reading

Bell, A. D. (1991) *Plant Form*. Oxford University Press, Oxford.

Harper, J. L. (1977) *Population Biology of Plants*. Academic Press, London.

Hutchinson, G. E. (1978) *An Introduction to Population Ecology.* Yale University Press, New Haven, CT.

Raven, P. H., Evert, R. F. & Eichorn, S. E. (1998) *Biology of Plants.* Worth, New York.

1.6 Questions

1 Program a spreadsheet to calculate all the terms of the life table shown in Table 1.1 from the values of N_x given.

2 What is meant by *fitness* and why is the concept so important in population biology?

3 Why is age usually a less useful predictor of a plants' fitness than its size?

4 How do modularity, sessileness, phenotypic plasticity and behaviour relate to one another in plants?

5 Explain why neighbours are so important in plant population biology.

2 Variation and its inheritance in plant populations

2.1 Introduction

The two foundation stones of modern biology are Darwinian evolution and Mendelian genetics. Both are based on an understanding of variation within populations, the subject of this and the next chapter. Darwin's theory of evolution by natural selection is central to all biology (Maynard Smith 1993). It is founded on the realization that heritable variation among individuals in a population could be the raw material for evolutionary change. Darwin's work on inheritance in plants and animals is, however, little known today because he failed to understand how inheritance worked. Darwin did several experiments with cabbages and antirrhinums. With hindsight, we can now make sense of his results in terms of Mendelian genetics. He crossed a peloric snapdragon (*Antirrhinum majus*), which has narrow tubular flowers, with pollen of plants with the normal asymmetrical snapdragon flowers, and the latter reciprocally with pollen of peloric plants (see Fig. 2.1a). Among the many seedlings produced, none was peloric. When these progeny plants were allowed to self-fertilize, 88 seedlings out of 127 had normal snapdragon flowers, two had flowers of an intermediate condition, and 37 were perfectly peloric (Darwin 1868). He also reported that, when seeds are obtained from the short-styled floral form of *Primula auricula* (a heterostyled species, see Fig. 2.8b, in which this flower morph is a heterozygote) 'about one-fourth of the seedings appear long-styled' (Darwin 1877, p. 223) but, lacking a satisfactory theory of inheritance, he failed to appreciate the significance of this 3:1 ratio.

Mendel realized that a 3:1 ratio in the second generation of such a cross implies that the traits are inherited as particles, which we now call genes. If each plant has two variants of a gene (in modern terminology, two 'alleles', say A and a), and one allele (A) is dominant to the other, then the initial cross can be symbolized by $AA \times aa$, and the first generation (F_1) progeny will be Aa. If the F_1 plants are self-fertilized, half of the ovules or pollen will receive the A allele, and the other half will receive a. The combinations of progeny genotypes in the progeny, or F_2, generation, will therefore be the familiar 'Mendelian single factor ratio' 1 AA: 2 Aa: 1 aa, or one quarter of recessive homozygotes (3 of the dominant to 1 of the recessive phenotype). The particulate basis of inheritance could be discerned by Mendel because he chose (perhaps deliberately) to study character differences which were each determined by a single gene.

By the 1930s, Darwin's theory of evolution had been combined with Mendelian genetics (Maynard Smith 1993). This is called the Neo-Darwinian synthesis, or the 'Modern Synthesis' (Huxley 1942). In this chapter, we review heritable variation, and also describe the diversity of breeding systems of plant populations, which affect how genes are transmitted from generation to generation. The following chapter outlines some simple population genetics concepts that will be used throughout the rest of the book, particularly in Chapters 7, 9 and 10. We will see how genetic variability is measured and how the genetic structure of plant populations can be described, based on frequencies of alleles and genotypes in populations. We will also show how various factors, including breeding systems and natural selection, affect these frequencies.

2.2 Types of trait

Individuals of a plant species or population are often as different from one another as individuals in human populations. Morphological and physiological variants range from discontinuous differences to continuous variability (Fig. 2.1). **Discontinuous variation** between plants, such as the difference between peloric and normal *Antirrhinum* flowers (Fig. 2.1a) or the phenotypic differences studied by Mendel, is variation that falls into 'discrete' phenotypic classes. These often correspond to different genotypes with a simple underlying genetic basis, such as single gene inheritance. This is often called **major gene inheritance**. Such traits show less variability within genotypes (caused by environmental influences) than between different genotypes. Flower colour is an example of a phenotype that is often controlled by major genes, and is variable in some natural plant populations (e.g. *Ipomoea purpurea*; Section 3.5.6). Another well-known example of polymorphism occurs in populations of white clover *Trifolium repens*, in which some individuals have the ability to produce cyanogenic glucosides and to release cyanide from them when leaves are damaged (an antipredator defence), while others (acyanogenic) do not (Ennos 1981). The ability to produce cyanogenic glucoside is controlled by the presence of an allele, *Ac*, at a single locus and cyanide release requires an active enzyme, linamarase. Plants of the genotype *li/li* at a second locus have no active linamarase, and only plants with the *Li* allele can release cyanide. The variability in phenotypes is thus caused by variability at two loci. A special category of genes with clear-cut, discrete inheritance patterns are molecular genetic variants, including **marker genes**. These will be described in more detail in Section 2.5.1.

Quantitative variability, on the other hand, is the situation when variability is more or less **continuous**, without distinctly demarcated phenotypic classes (e.g. Figure 2.1b). Such characters are often subject to strong environmental effects. Plant size, and size of plant parts, for example, are often heritable, but because many genes, as well as many environmental influences, can affect size, its genetic basis is difficult to discern. The genes underlying such **quantitative characters** are

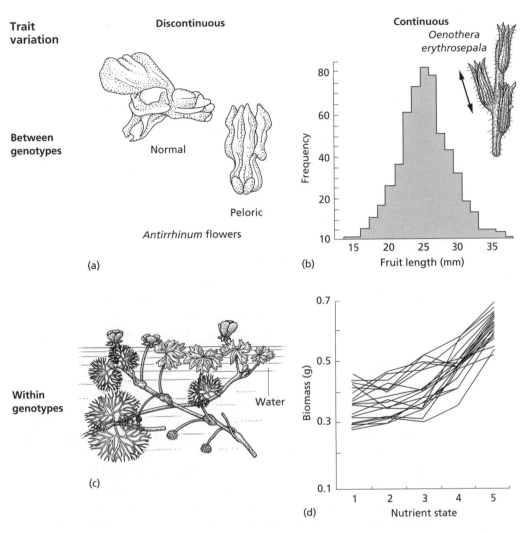

Trait variation

Discontinuous

Continuous

Between genotypes

Normal

Peloric

Antirrhinum flowers

(a)

Oenothera erythrosepala

Fruit length (mm)

Frequency

(b)

Within genotypes

Water

(c)

Biomass (g)

Nutrient state

(d)

Fig. 2.1 Types of plant trait. (a) Segregation of peloric snapdragon (*Antirrhinum majus*), a trait controlled by major gene variants. In this case the normal corolla is dominant to peloric. (b) Quantitative variation in length of fruit of the evening primrose *Oenothera erythrosepala*. (c) Heteromorphism in leaves of *Ranunculus heterophyllus*. (d) The reaction norm of plant size in response to a nutrient gradient for the progeny of 20 maternal familes of *Abutilon theophrasti*. Each line represents the mean sizes for a family grown at different points along the gradient.

sometimes called **QTLs** (Quantitative Trait Loci). They have also been called polygenes, but this has misled people into thinking that there is a special kind of gene involved in the control of such characters. There is a large body of evidence suggesting that QTLs are normal Mendelian genes, that are carried on the chromosomes and segregate just like other genes, which can be mapped with respect to other loci in the plants' chromosomes (see, e.g., Falconer & Mackay 1996). Individual gene differences at QTLs do not affect the phenotype enough to stand out above the environmental 'noise' in the character, and can be detected only by statistical approaches.

2.3 Genotype and phenotype

The phenotype expressed by a given genotype depends, to some degree, on its environment. A change in phenotype in response to the environment is called **phenotypic plasticity**. Most plant characters are phenotypically plastic to a greater or lesser extent. Discontinuous variation in a trait, for example between the floating and submerged leaves of aquatic plants (Fig. 2.1c) or the two kinds of flowers of cleistogamous species (Section 2.7.2.4) is known as **heteromorphism**, whereas a more or less continuous set of phenotypes expressed by a single genotype across an environmental range is called a **reaction norm** (Fig. 2.1d). MacDonald and Chinnappa (1989) investigated plasticity in five populations of *Stellaria longipes*. Means for 11 traits were generally different between populations, most of which proved to be the result of phenotypic plasticity. The phenotypic plasticity of a trait may itself have a heritable component and can be selected independently of the trait itself (Bradshaw 1965; Schlichting & Levin 1986; Sultan 1987). For example, the phenotypic response of plants to changes in light quality caused by shading from competitors is controlled by a family of phytochrome genes whose function and evolution is now well understood at the molecular level (Pigliucci *et al.* 1999; Mathews *et al.* 1995).

When phenotypic differences between genotypes vary between environments, this is called a **genotype × environment interaction**. For example, in *Arabidopsis thaliana*, cold treatment of seeds or rosettes for about 4 weeks dramatically reduces the interval between germination and flowering (under a given light regime), for plants from some parents, but not others (Figure 2.2a, Nordborg & Bergelson 1999). Figure 2.2b illustrates one way to investigate genotype × environment interactions. Twenty inbred lines of the poppy *Papaver dubium* were studied in 16 soils with different nutrient status (Zuberi & Gale 1976). For each trait, some environments yielded low means over all inbred lines, while others gave high overall means. The regression was calculated for each genotype's mean in the different environments against the means for all genotypes. When regression slopes differ significantly from one another, as in this example, this tells us that genotypes respond in different ways to environmental variation, i.e. genotype × environment interaction occurs. When the regression lines cross each other, the 'best' genotype depends upon the environmental conditions. For one character, leaf number at 10 weeks, the genotype that produced the largest plants in the treatment with the best soil, produced almost the smallest plants on the poorest soil. All 11 quantitative traits studied showed significant genotype × environment interactions.

Phenotypic variation can be separated into environmental and genetic variability by growing plants in a uniform environment such as a 'common garden'. Phenotypic differences between the plants should then be largely genetic, provided that precautions are taken to prevent the carry-over of environmental effects from the

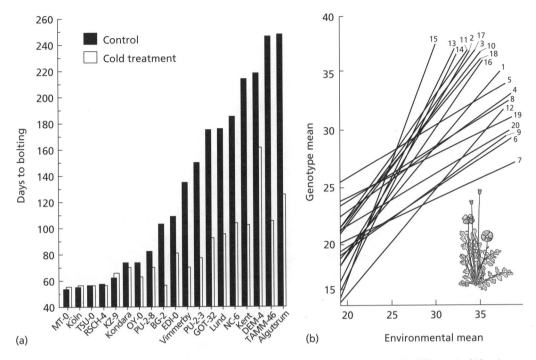

(a)

(b)

Fig. 2.2 (a) Number of days from germination to start of flowering for different *Arabidopsis thaliana* families (horizontal axis), showing the different responses to cold treatment at the rosette stage of growth. After the treatment, all plants were grown under the same conditions (Nordborg & Bergelson 1999). (b) Regressions of the mean number of leaves at ten weeks for each of 20 genotypes of *Papaver dubium* grown in 16 environments, on the mean for all genotypes in each environment. Regression lines are numbered for genotypes 1–20 (from Zuberi & Gale 1976).

wild (for example in the soil), or nongenetic maternal effects (such as plants from good environments having larger seeds and their progeny consequently growing faster). In a classic study (Clausen *et al.* 1948), this method was used to demonstrate genetic differences between populations of *Achillea lanulosa* along a transect across an altitude gradient in the Sierra Nevada of California (Fig. 2.3a). However, the garden is not the natural habitat for any of the genotypes planted in it. Another approach is therefore to transplant individuals between field sites (including reciprocal transplants, and, as controls for the possible effects of transplanting, replanting some plants in their native site). This is illustrated in Fig. 2.3b.

2.4 Quantitative inheritance

2.4.1 Heritability and environmental variation

The phenotypic value (P) of an individual under specified environmental conditions can be expressed in terms of two components:

$$P = \quad G \quad + \quad E$$

genotypic value   environmental deviation

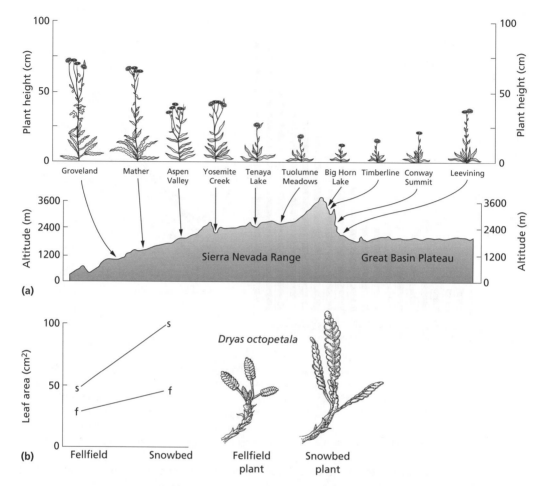

Fig. 2.3 (a) Results of a common garden experiment with *Achillea lanulosa* (Clausen *et al.* 1948). Plants collected along the transect were raised from seed and grown in a common garden at Stanford. (b) Reciprocal transplant experiment between fellfield and snowbed populations of *Dryas octopetala* in Alaska, showing the effects on leaf area (McGraw & Antonovics 1983).

In studying quantitative inheritance one would like to be able to measure how much of the phenotypic variation is genetic, as opposed to the amounts resulting from differences in the environments to which different individuals are exposed or to random perturbations arising during development. However, it is impossible to tell how much of the phenotypic value of any *individual* is genetic vs. environmental (i.e. to partition P into G and E). We can, however, meaningfully ask what proportion of the *population's* variability in phenotypic values is genetic, and what proportion is the result of nongenetic (environmental) differences. The total phenotypic variance is the sum of contributions from genetic and environmental causes (Falconer & Mackay 1996). The phenotypic variance (V_P) can be broken down into genetic (V_G) and environmental (V_E) components:

$$V_P = V_G + V_E + 2COV_{GE}$$

where the term COV_{GE}, the covariance of genotype and environment, allows for the possibility that particular genotypes may be concentrated in particular environments, rather than randomly distributed. Analysis of variance allows one to test whether there is statistically significant genetic variation, and to quantify the extent to which genetic effects, as opposed to environmental differences, cause the phenotypic variability. This can be done by calculating the ratio of genetic to phenotypic variance V_G/V_P (the **broad-sense heritability** of the trait). V_G can be estimated in plant species that can be cloned (e.g. from runners or cuttings), so that we can replicate individual genotypes, and eliminate genetic variability. Alternatively, in highly self-fertilizing plants, individuals can be assumed to be homozygous at all loci, so that their progeny will all have the same genotype as their parent. Finally, in self-compatible, but not naturally highly inbreeding plants, one can grow highly inbred lines by inbreeding for several generations. Individuals in such a line should have identical genotypes, whereas different lines may differ genetically. If the replicates of each clone or genotype are grown over a range of natural environments, or in a uniform environment (which can sometimes be managed in a greenhouse) the variance between genotypes estimates the genetic variance (while that within genotypes is entirely environmental).

Scheiner and Goodnight (1984) measured environmental variability values (V_E), genetic variances (V_G) and broad-sense heritabilities of 12 traits in five populations of the grass *Danthonia spicata*. Individuals were clonally replicated and then grown in six different environments: there were highly significant differences between replicates of the same genotype in phenotypic variance (V_P). It is interesting to note that genetic variances did not differ significantly between populations, and that the genetic variance of a population and its phenotypic plasticity were not significantly correlated.

2.4.2 Estimating genetic variance and the number of loci affecting a quantitative character

How can one find out the numbers of loci controlling a quantitative character? A rough estimate of how many loci differ between two strains of plants that differ in a phenotypic character can be obtained as follows. Consider a cross between two inbred lines of plant which each have genotypes with different alleles at some loci affecting a character (in the example of Fig. 2.4, a character affected by four unlinked loci). For a quantitative character, the variances of the parental, F_1 and F_2 generations will be different. If the two parental strains are each uniform genetically, then all variation within each strain must be environmental. The same will also be true in the F_1 generation, although these plants have a heterozygous

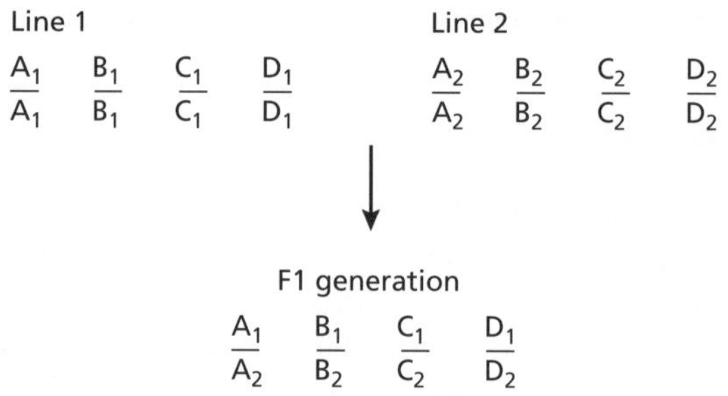

Line 1   Line 2

$\dfrac{A_1}{A_1}$ $\dfrac{B_1}{B_1}$ $\dfrac{C_1}{C_1}$ $\dfrac{D_1}{D_1}$ $\dfrac{A_2}{A_2}$ $\dfrac{B_2}{B_2}$ $\dfrac{C_2}{C_2}$ $\dfrac{D_2}{D_2}$

F1 generation

$\dfrac{A_1}{A_2}$ $\dfrac{B_1}{B_2}$ $\dfrac{C_1}{C_2}$ $\dfrac{D_1}{D_2}$

Fig. 2.4 A cross between two lines, each fixed for different alleles at the loci affecting a quantitative character. A–D are four loci with two alleles (1 or 2) each.

genotype, different from the parents. The variance in the character in these generations is therefore simply V_E. The F_2 generation, however, will be a mixture of the genotypes at all the loci affecting the character, so this generation will have genetic in addition to environmental variability. If four unlinked loci affect the character, as in the cross in Fig. 2.4, the F_2 would be expected to contain all three genotypes at each locus: a total of 81 genotypes.

The genetic variance, V_G, can therefore be estimated from the difference between the variances in the F_1 and F_2 generations. If we assume n genetic factors, each with the same effect on the phenotype, a minimum estimate of the number of factors is:

$$n = \frac{(\text{Mean of line 1} - \text{Mean of line 2})^2}{4V_G}$$

The numbers of genes affecting several floral characters were estimated to range from 3 to 6 in a cross between *Mimulus guttatus* and a closely related copper-tolerant species, *M. cupriphilus* (Macnair & Cumbes 1990).

2.4.3   Additive genetic variance and heritability

A particularly interesting partition of phenotypic variance is one that helps us to understand how the phenotypes of individuals and their progeny are expected to be related to one another. Such an understanding would go beyond merely saying that a trait is heritable, and enables one to predict quantitatively how the character would respond to artificial or natural selection. This is a difficult problem because only genes and not genotypes are passed on; genotypes are created afresh each generation. We thus want to know how much of the heritable variation is transmissible via the genes. We can do this in terms of the phenotypes in the parent and progeny generations. The **breeding value**, A, of an individual is a way to measure the part of the genetic variation that leads to most of the resemblance between relatives. It is measured as the deviation of the mean of an individual's progeny from the population mean, if the individual were mated to a large set of others in the population. In the simplest case, the genes determining a character have intermediate dominance, and there are no interactions between the genes (i.e. all the genetic factors act completely additively). The genotypic value G then depends only on the additive effects of genes, and the genetic variance V_G equals the variance of breeding values, or **additive genetic variance**, V_A. In general, however, V_G includes contributions from dominance and interactions between genes, as well as the variance of breeding values. The phenotypic variance can thus be partitioned as follows (see Falconer & Mackay 1996):

$$V_P \quad = \quad V_A \quad + \quad V_D \quad + \quad V_I \quad + \quad V_E$$

	↑	↑	↑	
	variance of breeding value (additive genetic variance)	dominance term (between alleles at same locus)	interaction term (between alleles at different loci)	

The additive genetic variance is particularly useful because the additive effects of genes are important in causing the resemblance between relatives, and thus determine responses to selection. The additive genetic variance is often expressed as the ratio V_A/V_P or 'heritability' (by convention, symbolized by h^2). Heritability (sometimes called 'narrow-sense heritability') can be estimated from the resemblance between relatives. For example, the regression of offspring values of a character on the mid-parent value is an estimate of $V_A/V_P = h^2$, and in a randomly mating non-inbred population the correlation between half-sibs is an estimate of $h^2/4$ (see Falconer & Mackay 1996). Many plant characters show significant heritability values in natural populations (Lawrence 1984; Charlesworth & Charlesworth 1995; Waldmann & Andersson 1998).

It is important to understand that such estimates assume that individuals occur randomly over environments, so that there are no environmentally caused similarities between relatives, but that any similarity in their phenotypes is purely genetic. A further important point is that, because heritability is a ratio of V_A over the sum of several variances, its value depends on all the variances, not just on V_A. High heritability may be the consequence of high V_A, but can also arise because the denominator in the ratio is low, thus it may merely indicate that V_E is low. Consider the following example. Suppose a plant is grown in uniform conditions in a greenhouse, and families are studied for some quantitative trait, such as plant height. If the heritability of the trait is high, this means that there is some additive genetic variation and that most of the variance in plant height is not environmental, which is reasonable because the environment has been standardized. If the same set of plants was grown in two environments, one of which is good and the other bad for plant growth, the heritability would be much lower because height would be determined mainly by the growing conditions. This shows that heritability is not a constant for a given character in a given species, but depends on environmental conditions. This creates problems if the aim is to compare heritabilities for different characters, or different species; thus, it is sometimes better to measure V_A, and use the additive genetic coefficient of variation, defined as $\sqrt{V_A}$ divided by the mean of the trait of interest; this provides a measure of heritable variability that is independent of the scale of measurement (Houle 1992).

2.4.4 Heritability and the response to selection

Heritability is useful to help predict the effect of artificial selection in plant and animal breeding. Natural or artificial selection changes the frequencies of alleles at loci that have effects on the trait selected. We usually do not know the individual genes concerned, but simply observe changes in the mean value of the trait in the population. The process of selection can be summarized in terms of three means:

1 the mean of the generation before selection;
2 the mean of the group of individuals that contribute progeny to the next generation;
3 the mean of the offspring.

The form of artificial or natural selection in which the 'best' set of parents, in terms of the phenotype of interest, are used for breeding, while others are discarded, is known as truncation selection; it is a simple form of directional selection (see Section 3.5.2). The strength of selection is calculated easily for truncation selection. It can be measured as the difference between the first and second means listed above, the **selection differential**, S. The difference between the parental and offspring generations is the **response**, R (Fig. 2.5). Plant or animal breeders need to know the relationship between the selection differential that they employ and the response that will be obtained. This is evidently related to the problem, treated in the previous section, of determining the extent to which variation in a character is genetic and transmissible to the offspring. Remembering the definition of breeding value, it is obvious that if V_A is high relative to the total phenotypic variance, then a large response to selection will be realized. It follows simply from the definition of heritability that under truncation selection, the selection response in the next generation (and, to a good approximation, often for several more generations) depends on the strength of selection applied and the heritability (Falconer and Mackay 1996):

$$R = Sh^2$$

As an example of the use of this theory, the distance between the tips of the opposite corolla lobes (corolla flare) in a perennial herb, *Polemonium viscosum*, found in the Rocky Mountains, USA, was shown both to be heritable, and to be under selection by pollinators (Galen 1997). In this species, corolla flare was 12% larger in a high-altitude population in which flowers were mainly bumblebee-pollinated, than in another population 500 m lower down the same mountain, in which other pollinators were also present. Heritability was estimated from the regression of offspring values on those of 54 maternal parents, corrected for age differences in corolla flare (Fig. 2.5). The population studied had a high heritability, suggesting that only a few generations would be needed to

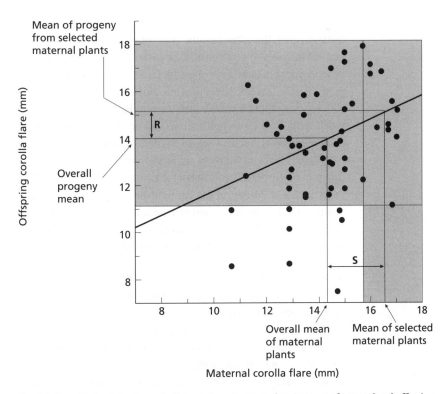

Fig. 2.5 Quantitative inheritance in *Polemonium viscosum*, showing a set of parental and offspring plants' flower width (Galen 1997). The thick line shows the regression of offspring on maternal parent values. The selection differential and response are illustrated by showing (grey area, and mean value of these plants on the *X*-axis) the top 20% of maternal plants, and (on the *Y*-axis) the offspring mean if these plants were used as parents.

produce the observed difference between the populations, if the selection differential exerted by bumblebees is a few percent. An experiment was performed by covering plants and allowing only bumblebees to pollinate them, together with a control in which flowers were randomly hand-pollinated. The differences in offspring production in the two treatments provides data to estimate the selection differential exerted by the bees themselves. This yielded an *S*-value between 7 and 17%. The progeny of plants pollinated by bumblebees had 9% wider flowers than the progeny of control plants, a result close to that which these h^2 and *S*-values predict.

2.5 Discrete genetic variation

2.5.1 Morphological and molecular markers

The term **marker gene** simply means a gene whose inheritance is so easy to follow that it can be used in genetic experiments, including making genetic maps. In other words, the genotypes must be easily and unambiguously scored, as in Mendel's work. In Mendel's peas, the phenotypes of the different genotypes such as *RR* (round) and *rr* (wrinkled) were nonoverlapping (although *Rr* and *RR* could not be distinguished; this is the familiar concept of dominance, the *R* allele being dominant and *r* recessive). White flowers are often caused by recessive alleles. Such major genetic markers have been used to detect seeds sired by pollen from plants other than the seed parent. Given a white flowered *ww* seed parent, in a population in which other plants have the dominant allele for coloured flowers (the *WW* genotype), seeds from cross-pollinations will be *Ww*, and will grow up to be plants with coloured flowers. We will see an example of this method in Section 2.7.3.3.

Different types of markers are useful for different purposes (Table 2.1; see also Ouborg *et al.* 1999). The most useful kind of marker allows all three genotypes at a locus with two alleles to be distinguished, i.e. A_1/A_1, A_1/A_2, A_2/A_2 can all be distinguished from one another (**codominance**). The development of **molecular markers** since 1966 has given us several kinds of markers that have this useful property, and the next sections will decribe some of these.

2.5.1.1 *Allozyme markers*

Variant enzymes with changes in their amino acid sequences that yield bands of different mobilities on electrophoretic gels are one important kind of marker. Different protein sequence variants of an enzyme can be stained on such gels, using standard histological stains that detect the presence of active enzyme (Fig. 2.6a). Variants of enzymes detected by staining in this way appear as two or more different bands or zones of enzyme activity on the gels, and are called **isozymes**. Plants often have two isozymes for many enzymes, such as the glycolytic enzyme phosphoglucose isomerase, one in the cytosol and another in the plastids. Each of these forms will be encoded by a different gene in the nuclear genome of the plant. In addition, each of these loci may have different alleles that encode variant forms of the enzyme with different charges on the protein molecules.

Table 2.1 Different types of genetic markers.

Type of marker	Dominance	Examples	Most important uses
Morphological	Usually dominant/recessive	White flowers in *Ipomoea*	Estimation of selfing rates
Molecular allozymes	Codominant	See Fig.2.6a	Genetic differentiation, genetic distances between populations and species, genetic maps
DNA			
RFLP	Codominant	See Fig.2.6b	Genetic variability and differentiation, genetic maps
RAPD, AFLP	Dominant/recessive		Genetic maps,
microsatellite	Codominant	See Fig. 2.6c	Genetic maps, paternity exclusion analysis

The allelic forms can be detected on gels that separate proteins according to their charge, and are termed **allozymes**. For instance, there may be fast (*F*) and slow (*S*) variants of cytosolic phosphoglucose isomerase: *PgiC*, and sometimes more than two alternative alleles are seen. If different forms are found in a population, and the frequency of the commonest allele is less than 99% (or some other chosen cut-off), the gene is said to be **polymorphic**. A polymorphic gene in a population will occur in homozygous genotypes (*F*/*F* and *S*/*S*) and heterozygotes (*F*/*S*), and these all look different on the gels (Fig. 2.6a). In 16 British populations of *Arabidopsis thaliana*, for example, seven loci were polymorphic out of 17 such loci studied, based on samples of on average 30 plants from each population (Abbott & Gomes 1988). Only two populations had no variants at any of the loci, but most populations were polymorphic for only some of the loci, so that the mean percentage of polymorphic loci in a population was about 17% (compared with $7/17 = 41\%$ for Britain overall).

There may also be differences between populations in the alleles present, so that one population consists of homozygotes for one allele while plants from a different population all have a different allele. Less extreme differences, i.e. differences in allele frequency, are also common. Such differences can be used in measures of genetic distance between populations and differentiation between sets of populations (see Section 3.3.7 and Nei 1987).

2.5.1.2 *RFLP markers and other DNA markers*

There are now also several kinds of markers based on differences in the DNA sequences of genes or other stretches of DNA. We cannot describe these in great detail as this is beyond the scope of this book, but will mention briefly some that are frequently used in plant population biology. Isozyme variants include largely amino acid replacement polymorphisms within coding sequences. They thus ignore 'silent variants' in coding sequences that do not change the amino acid (see Section 2.5.1.4), intron sequence variability, and variants in flanking and intervening sequences, all of which may be detected by DNA-based methods. In this sense, isozyme studies underestimate diversity, although this is not necessarily a problem unless variants are needed (for instance to study the mating system), and are hard to obtain. It is important to understand that some markers are variants within nuclear DNA sequences, while others are in the DNA of the organelles (chloroplasts and mitochondria). Chloroplast DNA variants have been very useful for estimating phylogenetic relationships between plant species. The chloroplasts (and mitochondria) of flowering plants are largely or wholly maternally inherited (Reboud & Zeyl 1994). This makes them useful for tracing maternal lineages, for instance to discover the maternal parent of a hybrid (Soltis *et al.* 1992), or to discover whether a present-day population is descended from one particular ancestral population or another (Dumolin-Lapegue *et al.* 1998; McCauley 1997), although it is important to be aware that many plastid sequences are also found in the nuclear genomes of higher plants (Ayliffe *et al.* 1997). In Section 3.2.4, we shall see how use of both maternally inherited and biparentally

(a) Allozyme marker variants

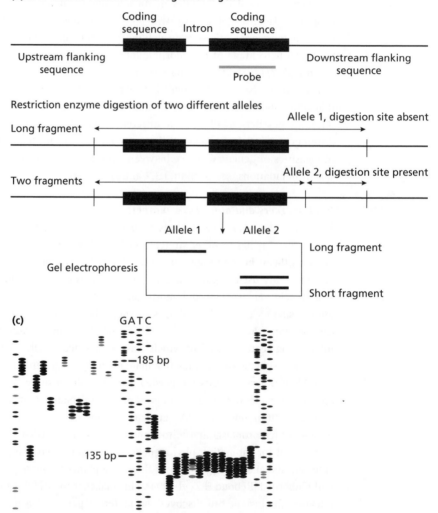

(b) RFLP marker variants in the region of a gene

(c)

Fig. 2.6 Patterns of genotypes that can be detected on gels. (a) Allozymes and an example of how they are detected by staining (note the lower intensity of staining when the enzyme is distributed into more than a single band, in heterozygous genotypes). (b) RFLP variants (note that the short fragment formed by digesting allele type 2 is not visible on the gel because the probe does not include any part of this sequence). (c) Microsatellite variants in *A. thaliana* (Todokoro *et al.* 1995).

inherited nuclear markers can help understand gene flow between populations, and Chapter 7 discusses the use of such markers for inferring population history.

RFLP (**restriction fragment length polymorphism**) markers are variants in DNA sequences that are detected by using a cloned piece of DNA that may be either a gene or an 'anonymous' piece of DNA (whose function, if any, is unknown). This cloned **probe** can be labelled in such a way that it can be used to hybridize with, and thus stain, DNA fragments separated by gel electrophoresis according to differences in their sizes. The probe will stain just the fragments that contain sequences similar to the probe; other DNA remains invisible. If the DNA fragments were produced by digesting a genome with a restriction enzyme that cuts the DNA at particular recognition sequences, then a variant with such a recognition sequence missing from the region being studied will produce a larger fragment than one with it present. Gels that separate fragments according to their sizes allow us to detect these differences and use them as markers (Fig. 2.6). RFLP variants occur in both the nuclear genomes and the organelle (chloroplast and mitochondria) genomes of plants. Like allozymes, heterozygotes for nuclear RFLP variants can often be distinguished from the two alternative homozygotes, i.e. they are codominant.

It is evident that the differences revealed in RFLP studies are the result of differences in the DNA sequences of different individuals, i.e. nucleotide sequence polymorphisms. When molecular studies, including sequencing of particular genes of interest, reveal differences between individuals, it may be possible to find a restriction enzyme that cuts the DNA sequence of a given gene of some individuals, but not others. With the invention of the polymerase chain reaction (PCR) technique, it has become possible to combine PCR amplification with restriction enzyme digestion of the products. The PCR primers are chosen to amplify a particular gene, or part of a known gene, and will produce a fragment of a defined size. If such a fragment contains a restriction enzyme digestion sequence, digestion will produce smaller bands than if the digestion sequence is absent, and thus variants can be scored relatively simply by electrophoretic analysis of the fragments sizes (see Konieczny & Ausubel 1993).

In addition, new PCR-based DNA markers have been developed. Among them are RAPD markers (random amplified polymorphic DNA). PCR amplifications are carried out using random nucleotide sequences as the primers, so that the markers are 'anonymous', rather than sequences from known genes. A given amplification usually yields bands of various sizes, which can be separated by electrophoresis on agarose gels. Comparisons of different individuals often shows that some of these bands are variable, with some individuals having a band that is absent in others. AFLP is another, somewhat similar, PCR-based approach, that yields very large numbers of bands on a single gel (Vos *et al.* 1995). The variability for AFLP, and even for RAPD bands, can be so great that the band pattern can effectively provide a 'fingerprint' of the genotype, distinguishing it from others of the same species, and this can be useful in identifying cultivated varieties of plants.

33

Although these presence/absence polymorphisms are usually inherited as simple Mendelian variants, they are generally not codominant, because PCR amplification can yield a band, whether the sequence is present in a heterozygote or a homozygote. Thus, the presence of a band is generally a dominant character and, because recessive sequence types are not detected, variants may be missed. RAPD and AFLP markers can therefore be useful to detect variability, and give us rough measures of diversity, such as the proportion of bands that individuals share, which can be used to compare different populations, and to estimate the proportion of total variability between vs. within populations. They do not provide very good measures of genetic diversity within, or genetic differences between, populations, for several reasons. First, the lack of codominance is a disadvantage if we want to measure genetic diversity (see Section 3.2.1.2), because the numbers of all variants cannot be counted, and the allele frequencies cannot be estimated directly from population samples. Second, we merely observe the presence or absence of bands, not whether the individuals compared are similar or very different in sequence. Finally, the different kinds of markers cannot readily be compared between different studies, because, as already explained, some reveal much more variability than others, even in the same population. The RAPD approach is widely used in genetic mapping, as many markers can be discovered, and a map developed relatively quickly, based on a cross between two genetically different strains. A drawback, however, is that there is no guarantee that the same variants will exist in a different family; thus, such maps cannot be used for other crosses in the same species without a great deal more work.

2.5.1.3   *Microsatellites*

Microsatellites are another kind of codominant marker. These are sequences with small repeated DNA motifs, for instance GA repeated many times. Individuals often differ in their numbers of repeats (as a result of a high mutation rate in repeat number), and these variants can be detected on gels that separate DNA fragments according to size (like the different fragments shown in Fig. 2.6c). ISSR variants are rather similar (Wolfe *et al.* 1998). It is common for many different alleles to be found at these loci, which are often referred to as 'hypervariable'. This property makes them ideal when one wants to identify individuals or particular strains of plants (e.g. Fang *et al.* 1997). They are the best kind of marker for detecting or excluding paternity, as we will see when we discuss pollen migration (Section 3.2.4.2).

Although microsatellite markers can be laborious to develop, in comparison with some of the other kinds of markers just mentioned, they are usually useful in any population of the species in which they were developed, and even in related species (Kelly & Willis 1998). They are thus very good for making genetic maps. Of all the DNA markers described here, only microsatellites provide any clues about which alleles are most closely related. For the others, we have no way of knowing whether one is a minor variant of another (i.e. closely related to, and therefore probably recently originated from, another variant). This information on related-ness provides rigorous ways to compare levels of variability between and within populations (Michalakis & Excoffier 1996) or to estimate genetic distances between populations (Goldstein *et al.* 1994), so these markers are better than most of the others for these purposes. Because these loci have high mutation rates, they are most useful for comparing populations that are closely related.

2.5.1.4   *Important properties of molecular markers: polymorphism and selective neutrality*

One important property of the kinds of genetic variants we have just described is their polymorphism in natural populations. The availability of markers opens up the possibility of carrying out many different kinds of studies (Table 2.1) in many species, including plants, that are of interest to ecologists, not just those few species that have been well studied genetically. A second important property of most marker variants is that they are often likely to make little or no difference to the phenotype, and can be assumed to be **selectively neutral**. Some of the variants detected by DNA-based methods are within repetitive or other noncoding parts of the DNA, and may thus have only slight phenotypic effects, and the same is true for DNA sequence variants that are in coding sequences but do not change the amino acid sequence (**silent changes**). Even RFLP markers that are based on cloned genes use the genes merely as probes, and detect variants caused by DNA sequences that lie outside the coding region of the gene used, or in its introns (noncoding regions within the gene). Evidently, for studying gene flow and migration, it is best to have markers that we can assume to be neutral, because if natural selection were acting on the genes whose genotypes we were scoring we would be able only to study the more complex situation of migration plus selection, not migration alone.

2.6   Mutation

Gene replication occurs with high fidelity, cell division after cell division, and generation after generation. Any rare, uncorrected mistakes are the ultimate source of polymorphisms and all genetic variation. De Vries (1905) first proposed the idea of heritable changes, which he termed mutations. One type of mutation is the substitution of one base pair in the DNA for another. DNA can also change by the loss or insertion of a few base pairs, or repeat motifs (such as the changes in microsatellite repeat numbers outlined in the previous section), or of larger segments, including transposable elements (transposons, see below). Mutations are rare and apparently random events. Stadler (1930) estimated mutation rates in maize by a direct method, using strains with known recessive marker alleles, such as the allele for shrunken kernels. If a wild-type plant with nonshrunken kernels (Sh^+/Sh^+) is crossed to the mutant homozygote (sh/sh), all the progeny should be Sh^+/sh, and will have the wild-type phenotype unless a Sh^+/sh mutation occurs. Mutation will produce sh/sh progeny, and their rate of occurrence tells us the mutation rate, which is about one in a million for this locus, based on about 2.5 million gametes tested. This is a fairly typical rate. The number of genes is not very certain for any plant, but now that we have sequences of most of the genome of the plant *Arabidopsis thaliana* (The *Arabidopsis* Initiative 2000), we can estimate that the coding sequence of a typical gene in this species may be much more than 1000 base pairs long, so there are many bases that can mutate in most genes. The data suggest that there are about 27 000 genes in the genome of this plant (many of them duplicates, and thus perhaps 15 000 different genes), and similar numbers in other angiosperms. The rate measured by tests such as Stadler's is the rate at which mutations at a given locus arise with a certain visible phenotype. There may of course be many other mutations that we cannot detect. Some of these may be detectable as RFLP variants, and of course all would be detected if we knew the DNA sequence.

Most spontaneous mutations are thought to occur by copying errors during DNA replication, but some are the result of chromosome breakage or rearrangement. Mutations can also be induced by radiation (including X- and gamma rays), by mutagenesis with chemicals that damage DNA or interfere with its accurate replication, and by insertions of transposons into or near the coding sequences of genes. **Transposons**, or 'jumping genes', discovered in maize by Barbara McClintock (reviewed by Federoff 1984), are sequences that are able to move from one part of the genome and insert into the DNA elsewhere. They seem to be present in all organisms that have been examined, including plants, usually in multiple copies, and can sometimes be a substantial fraction of a species' total DNA (Flavell *et al.* 1994; Kumar & Bennetzen 1999). For instance, one kind of transposon is estimated to form about 25% of the total DNA of *Vicia faba*. The wrinkled-seed allele of Mendel's peas is a mutation caused by insertion of a transposon into a gene for an enzyme involved in starch metabolism (Bhattacharya *et al.* 1990). Stadler's mutation rate estimates certainly include some such mutations.

Although mutation rates at individual loci are usually low, mutations have important consequences for natural populations, as we shall see when we discuss asexual reproduction (Section 9.2) and inbreeding depression (Section 2.8.2.1). In these contexts, the relevant mutation rate is often the overall rate at which deleterious mutations occur in the whole genome. We can get some idea of this for plants by multiplying the average mutation rate per gene by the number of genes a typical plant might have, but this is very crude, for several reasons. First, the number of genes is not very certain for any plant (though a value of between 10^4 and 10^5 loci in the genome of a plant appears reasonable), and secondly, we do not know the deleterious mutation rate for any locus. All that we know is the mutation rate to visible mutations that can be scored, as in Stadler's experiments. Only indirect mutation estimates can give the information we need. We can roughly estimate rates of mutation to the kind of mutant alleles that may be responsible for inbreeding depression (see Section 2.8.2.1 below), from the effects of making outcrosses between plants in very inbreeding populations. A large increase in fitness implies a high mutation rate. Estimates from plants are remarkably similar to ones from quite different organisms, such as *Drosophila*, even though they were obtained in a completely different way (Charlesworth *et al.* 1990; Johnston & Schoen 1995). Both insects and plants seem to have an average of about one new deleterious mutation per diploid genome per generation. If there are 25 000 loci in the genome, the mutation rate would be 0.5×10^{-5} per locus per generation, which fits quite well with the direct estimate mentioned above.

2.7 Plant breeding systems and genetic variability: introduction

Sexual reproduction involves three steps: (i) **meiosis** to form haploid cells, and during which genetic **recombination** occurs; (ii) formation of gametes; and (iii) the **fusion of gametes** to form a new zygote (syngamy). One important genetic consequence of sexual reproduction is that it produces offspring with new genotypes, and novel combinations of genes. New combinations of genes can arise during the first meiotic division when crossing-over may occur between homologous chromosomes, and recombination between genes on different chromosomes also occurs when the chromosomes segregate independently at meiosis (for any particular chromosome, half the gametes will contain a copy derived from the maternal parent and half will contain a copy derived from the paternal parent of the individual whose cells are dividing). Thus, an individual produces a great range of gamete genotypes, and at syngamy a huge variety of diploid genotypes can arise. This is true for progeny derived from genetically different parents, and also for progeny of self-fertilization of genotypes in which some genes are heterozygous. If the parents are closely related and highly homozygous (extreme cases being self-fertilization of a homozygous plant, or crossing two inbred lines with one another) the novelty produced by a mating may be low.

We must therefore be aware that, as a consequence of differences in their breeding systems, these features of plant populations may differ from the animal populations with which we are most familiar. Most familiar vertebrate animals and insects reproduce by finding mates and crossing with them. Some plants reproduce in this way too, with individuals of two sexes that can mate only with members of the opposite sex. Holly is one such plant. This is why one should grow a female plant if one wants berries at Christmas (and even then there will be no berries unless a male plant is growing near enough that some of its pollen can reach the female). Horticulturalists have recently developed hermaphrodite strains of holly, which have both male and female parts in each flower, and can self-fertilize and produce fruits by themselves. This hermaphrodite condition is by far the commonest situation in flowering plants (angiosperms), and many (but not all) of them are self-fertile and frequently reproduce in nature by self-fertilization. Because reproduction is central to the evolutionary process, and to thinking about the evolution of any organisms, plant population biologists must pay attention to their species' breeding systems.

2.7.2 Description and analysis of plant breeding systems

There are three important criteria that are used to describe the breeding systems of plants:
- Is the population sexually reproducing or asexual?

- If it is sexually reproducing, a description of the sexes of flowers (the sex system) is necessary
- Are there other features of the flowers that affect whether individuals will self-fertilize or outcross?

2.7.2.1 Plant reproduction

The main features of plant reproduction that are needed for an understanding of plant population biology were outlined in Chapter 1 (Section 1.3). Here, we will concentrate on flowering plant sex and breeding systems. The terminology for plant sex systems is more complex than that used for animals, because modularity allows plants to combine different sex possibilities within an individual (including changing sex). Table 2.2 gives definitions of the terminology we shall use.

Table 2.2 A classification of plant sex systems. Percentages refer to the sex systems recorded as occurring in the best available compilation, a survey of a large number of angiosperm species by Yampolsky & Yampolsky (1922). Gymnosperms are mostly monoecious or dioecious.

Description	Botanical term and definition	Occurrence in plants and examples
Asexual	**Apomictic**: Seeds have the same genotype as their mother. In angiosperms pollination, and fertilization of the endosperm (**pseudogamy**) is common	Many *Taraxacum* spp.
Sexually monomorphic	**Hermaphrodite**: Flowers have both male and female functions	(72%) e.g. most rose cultivars, *Rosa* spp.
	Monoecious: Separate sex flowers on the same individual plants.	(5%) e.g. cucumbers
	Gynomonoecious: Both female (male-sterile) and hermaphrodite flowers occur on the same individuals.	(2.8%) many Asteraceae including *Bellis* and *Solidago*.
	Andromonoecious: Both male (female-sterile) and hermaphrodite flowers occur on the same individuals.	(1.7%) Most Apiaceae, e.g. carrot *Daucus carota*.
Sexually polymorphic	**Dioecious**: Separate sex (male and female) individuals.	(4%) e.g. hollies, *Ilex* spp.
	Gynodioecious: Individuals either female or hermaphrodite.	(7%) many Lamiaceae, e.g. *Glechoma*, *Thymus* spp.
	Androdioecious: Individuals either male or hermaphrodite.	Very rare. e.g. *Mercurialis annua*.

Most species of flowering plants are hermaphroditic, although hermaphroditism is not, of course, confined to plants. Some animals (for instance, some slugs and snails, and even some fish) are hermaphrodites. Botanists also have the useful term **cosexual** to refer to the situation when the plant performs both sex functions, but without specifying whether these are done within each flower (hermaphrodite), or by separate male and female flowers (monoecious). Because animals have testes and ovaries, and there is no equivalent of flowers, zoologists do not need to distinguish between hermaphrodite and monoecious species, and the terms are used interchangeably for animals. An important distinction is between populations in which all individuals are essentially alike (i.e. **sexually monomorphic**, as in hermaphroditism), and those in which there are different kinds of individuals (i.e. **sexually polymorphic**, for instance, dioecious species like ourselves). In addition, one should be aware that some plants differ from one population to another, for instance the frequency of females differs between different gynodioecious populations (Chapter 9).

2.7.2.2 *Determining a plant's breeding system*

The most obvious way to discover the breeding system of a plant population starts with morphological examination of the flowers. If a population consists entirely of females, it is clear that it must be asexual. If the plants appear to be cosexual, and if isolated individuals produce seeds, the population could either be apomictic or selfing. If isolated plants whose anthers are removed still produce seeds, it is clear that the population must be apomictic. If not, the plant may be apomictic but pseudogamous (see Table 2.2), or it may be self-fertilizing (see below), so further study is required to tell which of these possibilities is correct. The only certain way of showing that a plant reproduces sexually, rather than being pseudogamous, is to pollinate emasculated flowers and produce seeds by a cross to another individual, using genetic markers (see above), proving that the seeds are really progeny of the pollen donor, rather than apomictically generated. For example, this is how it was discovered that *Potentilla argentea* is not asexual, as was once thought (Holm *et al.* 1997), and that *Arabis holboellii* is pseudogamous (crosses between dissimilar genotypes yielded only the maternal genotype; see Roy 1995).

Sometimes, it is immediately apparent that plants are males and females (dioecy), or that there are females present in addition to hermaphrodites (gynodioecy). In dioecious plants, however, females often have rudimentary anthers, and males may have quite well developed ovaries, as in spindle trees (*Euonymus europaeus*) (Darwin 1877). Unless functional studies are done, it can therefore be difficult to be sure whether the species or population consists entirely of hermaphrodites, or whether male-steriles (females) are present (reviewed in Mayer & Charlesworth 1991). It is relatively simple to get information on female fertility, by tagging plants at flowering, and counting their seeds or fruits before they detach from the maternal plants. Male fertility studies are less simple. Even if pollen is produced, one cannot be sure that a plant is male fertile. Some plants produce nonfunctional pollen (probably attractive to pollinators, many of which collect pollen as food; see Vogel 1978;

Anderson & Symon 1989; Mayer & Charlesworth 1991). In order to demonstrate male fertility, the pollen must be tested for its ability to fertilize recipient flowers and to produce seeds. Measuring male fertility is even harder: one must estimate how many of a plant's pollen grains have successfully fertilized seeds. Genetic markers are again very useful for this purpose (see Section 3.3.3).

2.7.2.3 Asexual reproduction in plants

Some plants reproduce asexually. Given the well-known ability of plants to take root from cuttings, it is perhaps not very surprising that some plants can produce entire new individuals starting from diploid cells. Among the angiosperms, many variant systems of asexual reproduction are known (see Richards 1997). Botanists often refer to such systems by the general term **apomixis**. The essential point is that flowers are formed, in which certain cells with the same diploid genotype as their maternal parent can form an embryo sac that develops without fertilization into an embryo, in a manner otherwise similar to normal sexual reproduction in which fertilization occurs. Seeds are formed, and progeny seeds and plants with the same genotype as the maternal plant are produced. One of the best known apomictic species is the dandelion, *Taraxacum officinale.*

In many apomictic flowering plants, fertilization of the endosperm nucleus is necessary. This is called **pseudogamy**. The fact that many asexual plants are pseudogamous suggests the interesting conclusion that asexuality evolved relatively recently from sexual ancestral species. It also implies that pollination, by wind or insect agents, is as necessary for the fertility of these plants as it is for sexually reproducing species. Thus, the most obvious interpretation of apomixis as an evolved response to low density of the plant species, or low availability of pollinating insects, cannot be correct, for these cases.

2.7.3 Mechanisms that prevent selfing

Among the sex systems listed in Table 2.2, those in which individual plants have both sex functions imply the possibility that they could reproduce by self-fertilization. In fact, however, plants have several ways of avoiding self-fertilization (Darwin 1862, 1876, 1877), and it seems that few, if any, species reproduce entirely by self-fertilization, although some species, such as the well-known weed *Arabidopsis thaliana,* have small flowers and self-pollination readily occurs (Fig. 2.7b-i). The closest approach to complete selfing is probably in species that have **cleistogamous** flowers, in which self-pollination occurs without the flower opening (Lord 1981). Most, if not all, such species also have open **chasmogamous** flowers as well (Fig. 2.7b-ii and iii), so individual plants do not reproduce entirely by self-fertilization. For example, many species of violets have two kinds of flower on the same plant: one that never opens, but that produces self-fertilized seeds by cleistogamy, and also flowers that open in the normal way and are pollinated by insects or wind.

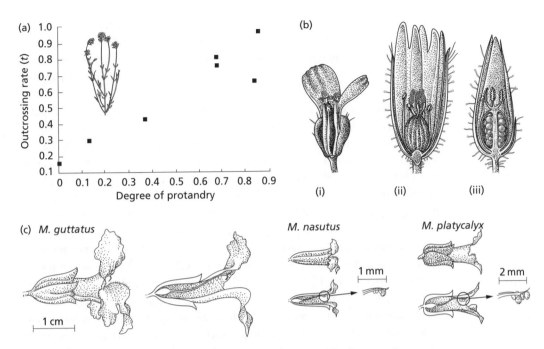

Fig. 2.7 Some floral systems in cosexual plants. (a) Spatial separation of anthers and stigma and protandry and their relationship to selfing rates in populations of *Gilia achillefolia* (pictured; Schoen 1982a). (b) (i) The small, self-fertilizing flowers of *Arabidopsis thaliana*; (ii) and (iii) chasmogamous and cleistogamous flowers of *Cerastium glomeratum* (Proctor & Yeo 1973). (c) Flower sizes of *Mimulus* species, showing small flowers in species with a high frequency of self-fertilization (Dole 1992).

Among species that are predominantly outcrossing, some have flowers with anthers far away from the receptive surface of the stigma, so that pollen is unlikely to get onto the stigma of the same flower (although in some of these plants the anthers or stigmas bend and make contact, if no pollen is received; see Lloyd 1979b). The floral morphology must interact with the pollinators so that pollen that gets onto a particular part of the pollinator is deposited on the stigmas of the flowers visited afterwards. The placement of pollen can be quite precise, as we shall see in the next section when we describe heterostyled flowers, which have different floral morphs in the same species. Other plants release male gametes at different times from the times when female functions are being performed (**dichogamy**). Many such plants are **protandrous**, releasing pollen before they are receptive as females (e.g. the bladder campion *Silene vulgaris*), but some do things the other way round (**protogynous**). Temporal separation between male and female functions is known in mosses and ferns too (Klekowski & Lloyd 1968).

2.7.3.1 *Self-incompatibility*

A particularly interesting mechanism of inbreeding avoidance is **self-incompatibility**. Self-incompatible plants, such as apples, reject their own pollen and so will not produce seeds if self-fertilized. This is why garden catalogues list which varieties of apples go well together—these are cross-compatible varieties that can pollinate one another and yield fruits. Populations of self-incompatible plants sometimes contain rare self-compatible individuals, and there are many known cases in which self-incompatibility seems to have broken down, giving rise to a close relative of an incompatible species that is able to self-fertilize (self-fertile varieties of apple, for example). Rejection is controlled by a self-incompatibility locus (the **S-locus**), or sometimes by more than one locus. In self-incompatible populations these loci have many different alleles (Nasrallah & Nasrallah 1986). Pollen is rejected when it comes from a plant that has the same allele as that of the plant being pollinated. Thus self-pollen, and pollen from other plants that have the same allele, is rejected. Rejected pollen either does not germinate on the stigma surface, or else it germinates and begins to grow in the transmitting tract of the stigma, but does not get as far as the ovary and ovules (Fig. 2.8a).

Two kinds of self-incompatibility systems occur in plants. In **gametophytic** systems, the pollen types are controlled by the haploid genotypes of the pollen grains themselves (i.e. by the gametophyte genotypes). A plant heterozygous for two alleles at the S-locus produces pollen of two incompatibility types, i.e. S_1/S_2 plants produce S_1 and S_2 pollen (Fig. 2.8a). The S_1 type is rejected by stigmas of plants that have the S_1 allele (genotypes such as S_1/S_2, S_1/S_3), but accepted by genotypes without this allele such as S_2/S_3 (although the S_2 half of S_1/S_2 plants' pollen would be rejected). Apples, some poppies (*Papaver rhoeas*) and clovers (e.g. *Trifolium repens*) are examples of species with gametophytic self-incompatibility.

In **sporophytic** systems, the incompatibility type of the pollen is controlled by the diploid S-locus genotype of the plant that produces it. Thus, any given plant produces pollen of just one incompatibility type. If the S_2 allele is dominant to S_1, pollen of S_1/S_2 plants will all be type S_2, whether the grain carries the S_1 or the S_2 allele (Fig. 2.8a). The pollen of such a plant can therefore fertilize S_1/S_3 plants (which do not have incompatibility of type 2), but not plants with the S_2 allele, such as S_2/S_3 (or, of course, their own genotype, S_1/S_2). Sporophytic incompatibility systems are known in many plants of the Brassica (cabbage) family, including edible cabbages, kales, etc., and in the Asteraceae (daisy family).

In most self-incompatible plants, the only way to know the incompatibility types is by laborious genetic testing of family members. The plants and the flowers of all types are indistinguishable by eye, and these are therefore called 'homomorphic' self-incompatibility systems. In one type of sporophytic self-incompatibility system, however, the flowers of the different incompatibility types are morphologically different (Fig. 2.8b). These are called **heterosytyled** species. Unlike species with homomorphic self-incompatibility, there are only a few different incompatibility types, controlled by just one or two genes with two alleles each. Some of these species have two flower morphs (**distyly**), as in primroses (*Primula vulgaris*). Others have three (**tristyly**), as in purple loosetrife (*Lythrum salicaria*; see Darwin 1877). Each 'morph' can fertilize the other two. This is the closest that any organism comes to having three sexes, though of course these plants are all hermaphrodites.

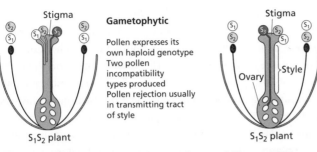

Gametophytic

Pollen expresses its
own haploid genotype
Two pollen
incompatibility
types produced
Pollen rejection usually
in transmitting tract
of style

S_1S_2 plant

Sporophytic

Pollen incompatibility determined
by diploid genotype
Dominance in pollen is possible
One incompatibility type produced,
regardless of pollen grain's
genotype, e.g. S_2 dominant in
pollen: all pollen of S_1/S_2 plants is
S_2 type
Pollen rejection usually on stigma
surface

S_1S_2 plant

(a) The two types of genetic control of pollen self-compatibility specificities

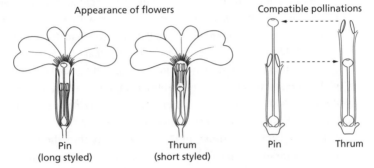

Appearance of flowers

Compatible pollinations

Pin
(long styled)

Thrum
(short styled)

Pin

Thrum

(b) Distyly in the primrose, *Primula vulgaris*

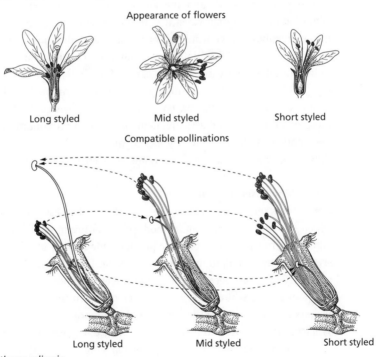

Appearance of flowers

Long styled

Mid styled

Short styled

Compatible pollinations

Long styled

Mid styled

Short styled

(c) Tristyly in *Lythrum salicaria*

Fig. 2.8 Self-incompatibility: (a) homomorphic self-incompatibility; (b and c) Examples of the
two types of heterostyly (b) distyly in P*rimula vulgaris* and (c) tristyly in *Lythrum salicaria*.

43

2.7.3.2 Self-fertilization rates of self-compatible plants

If a population is hermaphrodite or monoecious, one can test plants' self-compatibility by self-pollination (with control cross-pollinations to check on the male and female fertility of the individuals used). The hardest aspect of a plant's breeding system to establish is the self-fertilization rate of self-compatible plants. Observations of pollinators, together with morphological information on flowers, such as their size and apparent attractiveness to pollinators, and the degree of temporal overlap in male and female functions, can help to suggest a population's breeding system. Selfing plants sometimes set fruits without pollinator visits (**autogamous self-fertilization**). In such cases, as with cleistogamous species, selfing is readily recognized by studies in the greenhouse, or with flowers covered to exclude pollinators. Selfers often have smaller flowers than their outcrossing relatives (Lloyd 1965), with anthers closer to the stigmas, lesser temporal separation of male and females floral phases (Dole 1992), and typically produce lower ratios of pollen grains to numbers of ovules compared with more outcrossing species (Cruden 1977). Flower size differences are strikingly seen in monkey flowers (*Mimulus*); inbreeding species have flowers many times smaller than their more outcrossing relatives (Dole 1992, Fig. 2.7c). Within the single species *Leavenworthia crassa*, similar differences are seen between different populations with different levels of autogamous fruit set (Lloyd 1965).

If flowers are small and the anthers are seen to dehisce so that pollen is deposited on the stigma when it is receptive, even in the absence of insects (e.g. in the greenhouse), we would suspect that self-pollination is common. In insect-free conditions, a high ratio of fruits initiating development per flower (fruit per flower ratio) strongly supports a conclusion that the plant is either asexually reproducing (see above) or highly selfing (**autogamous**). A low frequency of fruit development per flower, however, merely shows that self-pollination under these conditions is not very frequent; under natural conditions, the pollinators may transfer pollen to the stigmas of the same flower or to the plant's other flowers, so that the self-pollination and self-fertilization rates might be higher.

2.7.3.3 Estimation of selfing rates

Genetic markers, such as allozymes, can be used to estimate the frequency with which nonautogamous plants reproduce by selfing in nature, by scoring mothers and their progeny at one or more polymorphic loci, to identify the proportion of seeds that have been produced by outcrossing. The idea is a simple one. If a plant is a homozygote for a recessive allele that gives a visible phenotype (for example, if *ww* plants have white flowers, but other plants in the population have coloured flowers and are of the *WW* genotype), then any of its progeny with the genotype *Ww* (coloured flowers, in this example) must have been produced by outcrossing, i.e. by fertilizing with a pollen grain carrying the *W* allele. By counting the numbers of the two sorts of progeny (*ww* and *Ww*), we can therefore obtain a quantitative estimate of the frequency of self-fertilization. This method is limited by the fact that suitable recessive visible genetic markers are rarely available.

Another problem is that pollinators will probably notice the different flower colour. Because pollinating insects are sensitive to colour, and have a search image for the kind of flower they are visiting, they may avoid tester plants and this might lead to their having higher selfing rates than other plants that are more heavily visited by pollinators. This happens with colour variants of wild radish (Kay 1978). Furthermore, these mutations might affect survival, which would mean that the numbers of homozygous and heterozygous progeny might reflect not only their frequencies at fertilization (i.e. the selfing rate), but also their mortality after fertilization. It is therefore better to use genetic markers that do not produce visible differences, and do not influence survival. Allozyme variants of enzymes, that can be visualized by electrophoresis, or other kinds of genetic markers, particularly microsatellites (see Section 2.5.1), are therefore used in most modern work, and statistical methods have been developed to make the best possible use of the detailed data that these markers provide (Brown 1979; Ritland & Jain 1981; Schoen & Clegg 1984; Ritland 1990a; Schoen & Lloyd 1992).

2.7.3.4 *Distribution of self-fertilization rates and sex systems in plants*

Some clear patterns have been noted in plant breeding systems, and these can help us understand something about their evolution, a topic we return to in Chapter 9 (see also Holsinger 2000). Apomixis and autogamy tend to occur in populations at the edges of species' ranges. Apomixis is found largely in species whose other populations (or closest relatives) are self-incompatible (Gustaffson 1946; Gustaffson 1947; Charlesworth 1980). Apomicts are also frequently poly-ploids. This makes sense given that polyploids are sometimes sterile as a result of abnormalities in meiosis; however, another possible explanation is that polyploids tend to be larger than diploids and selection in harsh environments may have favoured polyploids in habitats at the edges of species' ranges, where asexual reproduction is also favoured.

Outcrossing mechanisms are much commoner in long-lived plants than in annuals (Baker 1959; Barrett *et al.* 1996). Although some annuals are self-incompatible or dioecious, this is comparatively uncommon, and many weeds are selfers (Baker 1959). On the other hand, few trees and shrubs are highly inbreed-ing, with mangroves being a striking exception (Klekowski & Godfrey 1989). Overall, the distribution of self-fertilization rates includes all possible levels, but they are not all equally frequent (Fig. 2.9). There is a tendency for extreme values to be over-represented, i.e. for the selfing rate distribution to be slightly bimodal, but intermediate selfing rates are common in both plants (e.g. Baker 1959; Lloyd 1979b) and hermaphroditic animals (e.g. Jarne 1995). Self-incompatibility is known from only a minority of flowering plant families. Its distribution is still imperfectly known, but no more than 80 families at the most (out of a total of over 300) seem to have self-incompatible species (Weller *et al.* 1995).

Separation of the sexes (dioecy) is much more widespread than self-incompatibility: it is known in around half of all angiosperm families, though it is rarely common in any family (Table 2.1). This suggests strongly that it has

45

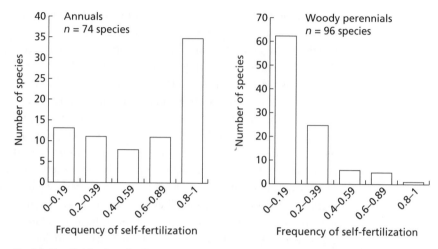

Fig. 2.9 The distribution of selfing rates in samples of annual and perennial plant species (from Barrett *et al.* 1996).

evolved relatively recently, within many different families. Interestingly, dioecy is associated with certain other plant characteristics, being more common in trees and shrubs than in annuals for example. Another association noticed by Darwin (1877) is that many monoecious and dioecious species are wind-pollinated. However, dioecy is much more common in the tropics than in temperate floras (Bawa 1980), despite the fact that in the tropics wind-pollination is rare.

2.8 Consequences of inbreeding, outbreeding and asexual reproduction

Apomictic reproduction and inbreeding have important consequences for populations. Apomixis merely propagates the genotype in which the mutation that caused asexuality arose, whatever it may be, and a single such clone can be of vast size (Grant *et al.* 1992). An apomictic genotype may be heterozygous at some or even many loci, particularly as many apomicts evolved from hybrids (see Gornall 1999; Holsinger 2000). Inbreeding, however, has the additional consequence of leading to homozygous genotypes. In the present section, we will concentrate on the effects of this homozygosity on plant vigour and fertility (inbreeding depression). Inbreeding depression may potentially reduce long-term population survival. Now that phylogenetic relationships between different species can be estimated using the extent of differences in their DNA sequences, evidence is accumulating to suggest that selfing plant taxa tend to have shorter evolutionary life-spans than related outcrossing ones (Schoen *et al.* 1997). This is only one of several consequences of inbreeding for populations. We return to the effects of inbreeding on populations, and how it reduces genetic variability, in Chapter 3 (Section 3.4), after measures of genetic diversity, and some of the major factors that influence it, have been explained.

2.8.1 Population consequences of self-fertilization

Self-fertilization produces progeny than are less heterozygous than their parents. This is easy to see if one considers a single locus in a heterozygote, say A_1/A_2. On selfing, we expect a 1:2:1 segregation ratio of A_1/A_1: A_1/A_2 and A_2/A_2. Thus, the progeny of selfing are half as heterozygous as their parents. If inbreeding continues, half the heterozygosity is lost each generation, on average. The increased frequency of homozygotes can be detected in data on genotype frequencies of individual plants in a population, using genetic markers, and this can help us to infer a population's breeding system (see Section 3.2.1.3), in addition to the methods described in Section 2.7.2.7.

2.8.2 Inbreeding depression

An important consequence of increased homozygosity is an increased chance that a progeny individual will be a homozygote for an allele that is lethal or harmful to health. This is well known in human populations, where close inbreeding leads to increased frequency of health problems, and of genetic diseases. These effects are known as **inbreeding depression**. They are detectable even among offspring of first-cousin marriages, and may be the reason why many societies have restrictive rules about marriages between kin. Inbreeding depression can also be an undesirable consequence of attempts to breed strains with characters chosen by animal and plant breeders. The plant equivalent to this situation is the occurrence of albino seedlings in progeny produced by self-fertilization: the parent plant is green even if it carries an albino allele, because albino alleles are recessive, but selfing such a carrier heterozygote (*Aa*, where *a* is the albino allele) produces homozygotes (*aa*). Mutations occur incessantly and produce alleles that do not function as well as wild-type alleles. Such **deleterious alleles** are often almost without effect in heterozygotes, because the wild-type allele's gene product suffices for most normal functions, even in single dose, i.e. the mutations are recessive or nearly recessive. When homozygotes are produced by inbreeding, however, fitness can be reduced severely (conversely, progeny of outcrosses between inbred populations should have increased fitness).

2.8.2.1 *Detecting and measuring inbreeding depression*

In experiments to detect inbreeding depression, hand pollinations are performed first to produce progeny by selfing and outcrossing. Quality, in terms of fitness-related characteristics such as survival rates and fertility, can then be compared between the selfed and outcrossed progeny. The inbreeding depression resulting from a single generation of selfing is measured by the decrease in fitness values for inbred progeny (w_s), relative to those of the non-inbred controls (w_r):

$$\delta = \frac{w_r - w_s}{w_r} = 1 - \frac{w_s}{w_r}$$

If the survival of self progeny is 30% of that of outcrossed progeny, for instance, the inbreeding depression in survival is $(1-0.3)/1 = 0.7$. Inbreeding can also reduce fertility as well as survival. If progeny from self-fertilization produce 80% of the seed output, compared with progeny from outcrossing, $w_s/w_r = 0.8$, the inbreeding depression in fertility is 0.2. The overall effect of inbreeding on fitness can be determined by assuming that traits expressed at different life history stages are independent of one another, and multiplying the w_s/w_r values together. For our example, we have: $w_s/w_r = 0.3$ for survival and $w_s/w_r = 0.8$ for fertility, thus $w_s/w_r = 0.24$ for overall fitness, giving an inbreeding depression of 0.76, more than if we measured survival alone.

An experiment to estimate the extent of inbreeding depression was carried out in the perennial desert plant, buffalo gourd, *Cucurbita foetidissima*, which is gynodioecious (Kohn 1988). Hermaphrodite plants were hand self-pollinated and the same individuals outcrossed with pollen from different individuals; the progeny quality was compared at the seed stage, and in terms of seedling survival in the harsh conditions of the natural environment. Figure 2.10 shows that selfing yielded fewer seeds than outcrossing (the difference is statistically highly significant), did not affect seed mass, but produced seedlings that survived much less well than those from outcrossing, particularly by their second year (1987), at which stage only about 1% of the seeds produced were still represented by living seedlings.

Inbreeding depression can also be detected indirectly in the field, using genetic markers. If a population is partially selfing, there will be homozygous genotypes among seeds, especially those formed by selfing. Inbreeding depression causing mortality of selfed progeny means that the frequency of homozygous genotypes at marker loci will decrease as the cohort of seeds become seedlings,

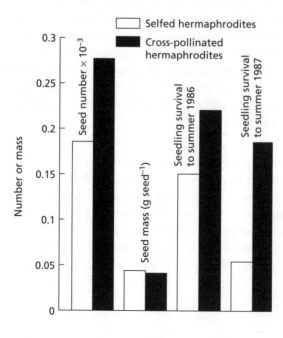

Fig. 2.10 Inbreeding depression in a gynodioecious population of *Cucurbita foetidissima* (from Kohn 1988).

and then young and mature plants. This change in genotype frequencies is the basis for a method of estimating inbreeding depression in field studies, avoiding the labour of the experimental approach described above (Dole & Ritland 1993).

2.8.2.2   *The genetic causes of inbreeding depression*
We saw in Section 2.6 that deleterious mutations occur every generation, probably at an appreciable rate for the whole genome (though at low frequency for any particular genetic locus). Populations must therefore be collections of individuals with various numbers of mutations. Some have zero mutations, while others have one or more. Inbreeding will make some of these mutations homozygous, so that fitness characters—including development speed, survival rate and fertility—will be reduced. Higher numbers will be most common in outcrossed populations, where genotypes are heterozygous and recessive mutations are sheltered from selection, as they do not harm their carriers. In inbreeding populations, the average number of mutations per individual will be reduced by natural selection against homozygotes. So, we can predict that inbreeding depression will be lower in inbreeding than outcrossing populations, and studies in a wide range of plant species have yielded some evidence that this is true (Husband & Schemske 1995). The difference between inbreeders and outcrossers has even been detected between closely related plants, for instance in different *Clarkia tembloriensis* populations with different natural outcrossing rates (Holtsford & Ellstrand 1990). Inbreeding depression will also occur if the population being inbred contains loci whose heterozygotes have higher fitness than either homozygote (**overdominance**). In a random-mating population, this pattern of fitnesses can maintain polymorphisms at the locus, with both alleles present at intermediate frequencies (see Section 3.3.3.3). At the fertilization stage of the life cycle, heterozygotes will therefore often be present at high frequencies, and differential survival of the genotypes will cause heterozygotes to increase in frequency by the adult stage. Inbreeding among the adults causes an increased frequency of homozygotes, with low fitness. It is therefore easy to see that in an outcrossing, highly heterozygous, population, such loci would lead to reduced fitness on inbreeding. There has been prolonged debate over the relative roles of overdominance and mutation in inbreeding depression (reviewed by Wright 1977), and it is likely that both kinds of loci contribute. In highly inbreeding populations, heterozygotes are rare and so their higher fitness cannot easily maintain overdominant polymorphisms (unless the fitnesses of the two homozygotes are almost equal, which seems biologically implausible; see Kimura & Ohta 1971). **Heterosis** (often called hybrid vigour), when very inbred populations or strains of plants are crossed and yield progeny larger than their parents, is therefore probably caused by mutation in the parental populations. Each parental population will carry different mutations, so hybrid vigour results when they are crossed and the (recessive) mutations become heterozygous. Although heterosis and inbreeding depression appear to be opposite aspects of the same phenomenon, they may therefore not have precisely the same causes. Heterosis could, as just explained, largely be a consequence of mutational load, but overdominant alleles maintained

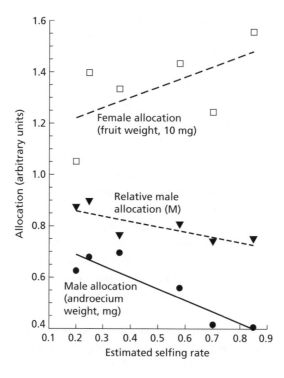

Fig. 2.11 Observed variation in allocation to male functions and in *Gilia achilleifolia* populations with different selfing rates (Schoen 1982b).

within populations may nevertheless contribute significantly to inbreeding depression within outbreeding populations.

2.8.3 Sex allocation

A final important difference between outcrossing and selfing plants is their allocation of reproductive resources to male and female function. Relative allocation to male functions, such as pollen/ovule ratios, has been consistently found to be lower in populations with high selfing rates, compared with largely outbreeding populations. This is found even between different populations of the same species, and an example is shown in Fig. 2.11. We shall return to this topic, and deal with some of the theory that helps us understand sex allocation (Section 9.5.6), but clearly it makes sense for plants that reproduce mainly by selfing within flowers to produce only enough pollen to fertilize the ovules, and this will generally be a much lower amount of pollen than in outcrossers, which have to contend with pollen lost in transfer by pollinators or wind.

2.9 Summary

Inherited variation is the raw material of evolution, but the **phenotype** expressed by an organism always depends upon both its genes and its environment.

Morphological variation in plants and animals can be classified into variability within genotypes (**phenotypic plasticity**) or between genotypes (**genetic variation**), and into **continuous** or **discontinuous variability**. **Heteromorphism** is discontinuous variation within genotypes, whereas the terms **environmental variability**, and **norm of reaction**, are used for continuous variability without genotype differences. Variation between genotypes ranges from discontinuous **major gene** variation to quantitative, or continuous, variability. Quantitatively variable characters are affected by multiple genetic loci, and by the environment, so that the genetic differences cannot be studied by classical genetic methods. Such variability is studied by partitioning the variance into genetic and environmental components. **Broad-sense heritability** is the ratio of the genetic to the phenotypic variance. **Narrow-sense heritability** is the ratio of the additive genetic variance to the phenotypic variance, and measures the amount of heritable variation in a population that is of most importance for responses to natural and artificial selection. When genotypes respond differently to different environments, there is said to be a **genotype–environment interaction**.

The presence of different major gene variants at the same locus at substantial frequencies in a population or species is called **polymorphism**. An important kind of genetic variation is **molecular variability**; such variants can be useful genetic markers that can help in studies of natural populations. Genetic variation ultimately arises from gene **mutation**. Many variants are under selection, and most mutations are detrimental. Some DNA sequence variants, however, do not affect protein sequences, and may therefore be under weak, or no, selection.

The **mating system** of a population describes who mates with whom. An important aspect of plant mating systems is the amount of **inbreeding**, particularly **self-fertilization** in **hermaphrodite** and other **cosexual** plants. This affects the frequency of homozygous genotypes within populations, and the amount of pollen flow between them. The most important features of plant mating systems are: (i) whether individuals reproduce sexually or not; (ii) whether individuals have both male and female sex functions (e.g. sexually monomorphic hermaphrodite populations), or whether some members of the population have just one sex (sexually polymorphic populations, including dioecious populations with male and female plants); and (iii) for hermaphroditic or cosexual populations, the frequency of selfing, and whether there is any outcrossing system, such as **self-incompatibility**, **spatial separation** of male and female reproductive organs, or **temporal separation** of male and female activity within flowers. Self-fertilization is much more common in annual plants than long-lived species; conversely, outcrossing systems such as self-incompatibility are commonest in long-lived plants. Selfing leads to a decreased frequency of heterozygous genotypes, and this may cause **inbreeding depression**, either because of increased numbers of homozygotes for detrimental mutations, or because it causes more homozygotes at polymorphic loci where heterozygotes have the highest fitness. Self-fertilizing species also tend to lose several adaptations that are found in outcrossers, such as large petals and nectar production, and they tend to evolve low pollen output.

2.10 Further reading

Holsinger, K. E. (2000) Reproductive systems and evolution in vascular plants. *Proceedings of the National Academy of Sciences of the USA*, 97, 7037–7042.

Ouborg, N. J., Piquot, Y. & van Groenendael, J. M. (1999) Population genetics, molecular markers and the study of dispersal in plants. *Journal of Ecology*, 87, 551–568.

Proctor, M., Yeo, P. & Lack, A. (1996) *The Natural History of Pollination*. Collins, London.

2.11 Questions

1 What are: major genes, polymorphisms, QTLs, RFLPs, and codominant alleles?

2 What is genotype-by-environment interaction, and how can it be detected empirically?

3 What is heritability and why is it useful in plant breeding? How can one get estimates of the quantities V_G and V_A, and what is the difference between them?

4 Explain the meaning of selection differential and selection response.

5 What are the following: cosexual plant, self-incompatibility, apomixis, inbreeding depression, overdominance?

3 Evolutionary and ecological genetics

3.1 Introduction

Evolution has been defined as a process of change in allele frequencies; **ecological genetics** is the study of genes in natural populations. This is one of the most successful areas in which biology can be treated in quantitative terms, and it has led to a knowledge of the factors that control gene frequencies, including forces such as natural selection that cause them to change. In turn, this means that although we understand a good deal about patterns of genetic diversity (such as which characteristics of species predict how much genetic variability they might be expected to have, and what proportion of the variability will be within populations, vs. between different populations), the question of *why* organisms are so genetically variable is still unclear. Therefore, another theme in this chapter is forces that maintain polymorphism. Nearly half of the enzyme loci studied in plants are polymorphic, having more than a single allelic type (Brown 1979), and the DNA sequences of plant genes also vary (Cummings & Clegg 1998). For example, Raijmann *et al.* (1994) carried out electrophoretic surveys of 10 enzymes in samples from 25 natural populations of the herbaceous perennial gentian *Gentiana pneumonanthe*. Within populations, 3–7 loci (44% overall) had at least two alleles at frequencies above 1%.

In Chapter 2, we took this variability for granted, and outlined ways in which variants are used as 'genetic markers', but we did not discuss in any depth why there should be such variability. Most mutations are deleterious and hence quickly removed by natural selection. Those very rare mutations that are beneficial should spread rapidly in populations, displacing alternative alleles. They will therefore increase genetic diversity only transiently; however, it is implausible that the variants we find so readily are in this transient state, because such a view implies far more ongoing adaptive evolution than seems plausible. Much of the genetic variation found in plant populations (and populations of many other species, including our own) must therefore be persistent. We thus need an explanation that can account for the maintenance of variability.

One possibility is that most variation is selectively neutral, and the diversity and polymorphism observed are the result of slow random changes in the frequencies of alleles (Kimura 1983). This would give an appreciable chance of observing a locus in a transient state between the moment that a new neutral variant arises and the moment when it is lost from the population or has reached a frequency of 100% by the process of **genetic drift**. A very different view is that variability is maintained within populations by natural selection. This controversial issue is still an area of active debate in population and evolutionary genetics.

In this chapter, we will first describe some of the behaviour expected when genetic variants are neutral (Section 3.2.1.1), and then briefly examine deterministic forces that can maintain or eliminate variability.

3.2 Gene and genotype frequencies without selection

3.2.1 Measurement of gene and genotype frequencies in populations

Neutral variants, including allozyme and other genetic markers, were reviewed in Chapter 2. Such genetic variability is quantified in terms of gene and genotype frequencies. To take a simple case, when there are two alleles, say A and a (or F and S for an allozyme locus), we conventionally denote the frequencies of the two alleles by p and q (q, of course, is equal to $1 - p$, because an allele must be one type or the other; if there are more than two alleles, the frequency of the ith allele can be denoted by p_i, where $\sum_i p_i = 1$).

3.2.1.1 *Neutral variants in a large, random mating population*

If two alleles A and a at a single locus are selectively neutral, and individuals mate with each other at random (i.e. simply according to their frequencies in the population, without respect to their genotypes, and without any directed tendency in the mating pattern such as inbreeding), the expected frequencies of genotypes among offspring will depend only on the allele frequencies in the parental generation. If the frequencies of A and a are, respectively, p and q, then the frequencies of the three possible genotypes expected in the next generation are the products of the relevant allele frequencies:

$$
\begin{array}{ll}
AA & p^2 \\
Aa & 2pq \\
aa & q^2
\end{array}
\tag{3.1}
$$

and evidently the total: $p^2 + 2pq + q^2 = 1$. In order to see how these formulae arise, recall that there is a frequency p of the A allele in female gametes, and also in male gametes. The chance that an individual in the progeny generation gets A from both parents is thus p^2 (and so on for the other genotypes; the factor of 2 in the formula for heterozygotes arises because a plant can get A via its maternal parent and a from the pollen donor, or *vice versa*).

These genotype frequency expressions under **random mating** as just defined are called the **Hardy–Weinberg** frequencies (named after the two scientists who derived this result in 1908). Notice that many possible genotype frequencies in the parents can correspond to given values of p and q. For instance, we could have a frequency p of AA plants, and a frequency q of aa, i.e. all homozygotes as in inbreeding populations (which will be discussed next), or there might be many heterozygotes. But, whatever the parents' genotype frequencies, if matings occur randomly, and provided that the population is large enough for the actual frequencies to be similar to those expected, the new generation should have the Hardy–Weinberg genotype frequencies. This is an important result, because it tells us that under these assumptions neither allele nor genotype frequencies have any tendency to change over the generations in a randomly mating population. If deviations from these genotypic ratios are found, we therefore have to think about what might be causing this.

Many plant populations have been surveyed for allozymes, and they often show frequencies that appear to be in Hardy–Weinberg equilibrium. In their study of *Gentiana pneumonanthe* populations, which ranged in size from 1 to 20 000 flowering individuals, Raijmann *et al.* (1994) used the estimated allele frequencies to calculate genotype frequencies expected assuming Hardy–Weinberg equilibrium. Out of 78 tests involving the seven loci found to be polymorphic among the mature plants in these populations, only 7 statistically significant differences were found between observed and expected frequencies. This suggests that the species is largely outbreeding, consistent with the observation that flowers are strongly protandrous so that, although plants are self-compatible, they are unlikely often to self-pollinate. Interestingly, however, 13 of the 39 tests carried out on seedlings deviated significantly, and all showed too few heterozygotes, suggesting that some inbreeding may occur but that inbred seedlings infrequently survive to maturity. It would not be surprising for such a highly outbreeding population to show such inbreeding depression.

3.2.1.2 *Measurement of genetic diversity*

There are several ways to measure genetic diversity, such as allozyme diversity, using genotype frequencies in populations. One measure, discussed in Section 2.5.1, is the fraction of loci at which polymorphism is found (often denoted by P). Clearly, this depends on the number of individuals studied, because a small number studied can at most include only a few different genotypes, while a more extensive survey should reveal more of the variability that exists. In a very small sample, only the most frequent type may be seen. Furthermore, we need a rule to decide whether a locus is to be classified as polymorphic or not. The criterion used is often that the commonest allele is present at a frequency of less than 99%, which is arbitrary (sometimes 95% is used), and thus somewhat unsatisfactory. For the *G. pneumonanthe* populations discussed above, based on samples of 6–71 plants per population, from 3 to 6 of the 16 loci studied were polymorphic; the mean number of polymorphic loci was 29%, averaging over the 25 populations studied (Raijmann *et al.* 1994). Alternatively one can count the numbers of alleles found at each locus surveyed, and calculate the mean value, including all loci, whether polymorphic or not. This measure is often called $\bar{A}$. This is also strongly sample size dependent.

A much better measure is the chance that two randomly chosen alleles will be different. This **gene diversity**, H_e, at a locus can easily be calculated from the allele frequencies as:

$$H_e = 1 - \sum_i p_i^2, \tag{3.2}$$

where p_i is the frequency of the ith allele type. Here p_i^2 is the chance that the two alleles picked will both be of that type; this is summed up over all allele types present, so that one minus this sum is the chance that two alleles in the sample are *not* the same. As one might expect, this diversity measure is equal to zero if the frequency of one allele at the locus equals 1 (i.e. if there is no variability at the locus in question, a situation that is referred to by saying that the allele is fixed at the locus). It is usually averaged over all the loci studied (including non-polymorphic loci), and can then be used in quantitative comparisons and tests of differences between different populations, or averaged over a set of populations. For the *G. pneumonanthe* populations discussed above, the mean H_e, based on all 16 loci studied, was 0.12 (Raijmann *et al.* 1994).

It is very important to understand that H_e is simply a measure of a locus's variability, and that it can be applied to populations with any breeding system, not just to random mating systems, as it depends only on allele frequencies and not genotype frequencies. Confusion has been caused because H_e is sometimes re-ferred to as 'heterozygosity', because it is the same as the expected frequency of heterozygotes in a random-mating population. The actual frequency of heterozy-gotes in a population (H_o, to stand for observed heterozygosity) may be very different, as we shall see in the next section, although in *G. pneumonanthe* they were similar (this is just another way of stating, as described in the previous section, that these populations all conformed to the Hardy–Weinberg frequencies, for almost all loci and populations).

Diversity measures are useful for many purposes. For instance, one study quantified the effects of postglacial recolonization on allozyme variation in 19 North American populations of the herbaceous perennial *Asclepias exaltata* (Broyles 1998). In these northern populations, 46% of the loci were polymorphic, the mean number of alleles per polymorphic locus was 1.84, and $H_e = 0.133$. These values are significantly lower than for populations studied in the Pleistocene refugium in the southern Appalachian mountains, suggesting that a loss of vari-ability occurred as small populations were founded during the northwards range extension after the last Ice Age.

3.2.1.3 *Inbreeding populations*

In many situations, Hardy–Weinberg genotype frequencies are not expected. We have discussed breeding systems, including inbreeding in Chapter 2. Many plants, such as the annual weed *Arabidopsis thaliana*, do not mate randomly, but are highly self-fertile and can reproduce by inbreeding. Some frequency of self-fertilization occurs in many cosexual plants, and other inbreeding (e.g. matings between sibling plants) must also occur if seed dispersal is limited. An important, relatively simple, approximation to the breeding system of a self-compatible population is the **mixed mating model**. All individuals are assumed to reproduce with the same self-fertilization rate. A selfing rate of S means that a proportion S of the ovules will be self-fertilized, while the rest $(1 - S)$ will receive pollen from unrelated plants. The **outcrossing rate** (t) of such a population is $t = 1 - S$.

How does self-fertilization affect genotype frequencies at marker loci? Complete self-fertilization leads to a halving of the frequency of heterozygotes in each generation, but does not, of course, change the frequencies of the alleles at a neutral locus. All homozygous plants reproduce their own homozygous genotype when they self-fertilize, but only half the progeny of each Aa plant are heterozygous, and the rest are either AA or aa. Eventually, the population will therefore become completely homozygous. There may still be genetic variability, if both alleles remain present, but there will be no heterozygotes. Thus the population will not have the Hardy–Weinberg frequencies.

In the British *A. thaliana* populations whose allozyme diversity was described in Section 2.5.1, the diversity measure within populations ranged from 0.031 for the least variable locus, to 0.275 for the most variable one (Abbott & Gomes 1988). Averaged over all loci (10 of which were not polymorphic in the species), the allozyme gene diversity within *A. thaliana* populations was about 0.02, considerably lower than the value for *G. pneumonanthe*, or other outcrossing plants (Hamrick & Godt 1990). The actual frequency of heterozygotes, H_o, was much lower in *A. thaliana* (only 0.00024), i.e. almost every plant sampled was homozygous, as expected because this species is highly self-fertilizing.

Many plants have mixed mating systems, and are much less selfing than *A. thaliana*. In cases where the selfing rate, S, is more moderate, the frequency of heterozygotes in a population will not become zero. There is a simple relationship between the selfing rate and the expected homozygosity, assuming that the population has had the same selfing rate for many generations and reached an equilibrium. The genotype frequencies expected in such populations can be expressed using simple modifications of the Hardy–Weinberg expressions, involving, in addition to the allele frequencies, one further quantity, the **inbreeding coefficient**, F (see Brown 1979). F is defined as the probability that an individual's two alleles are **identical by descent** (i.e. descended from a common ancestor; for further explanation, see Hartl and Clark 1997). For instance, a seedling produced by self-fertilization of its parent plant has a 50% chance of being a homozygote for one of its parent's alleles. The equations for the expected genotype frequencies under inbreeding, in terms of the inbreeding coefficient, F, are:

AA $p^2 + Fpq$

Aa $2pq(1 - F)$

aa $q^2 + Fpq$ $\hspace{3cm}$ (3.3)

You can check that the equations above sum to 1, just as for the random mating formulae. A population reproducing by complete self-fertilization will have $F = 1$.

With a more moderate rate of selfing, the inbreeding coefficient will equilibrate at an intermediate value (Fig. 3.1). With selfing rate S in the simple 'mixed mating model' described above, a population is expected to reach an equilibrium $F = S/(2 - S)$ (Brown 1979). F predicts the extent to which heterozygotes are deficient. $F = 1$ implies that there are no heterozygotes, as is almost true in *A. thaliana*.

These genotype frequency formulae, like the Hardy–Weinberg formulae, assume that the allelic differences have no selective effects, i.e. they are neutral. The equations describe the genotype frequencies expected for neutral variants, such as alleles at marker loci, in a large population. Data on the frequency of heterozygotes in a population can therefore be used to estimate the rate of self-fertilization, S, if we are prepared to assume that the population has reached this equilibrium state, and if we have some kind of marker that we can score (see Section 2.5.1). There has thus been a great deal of work estimating genotype frequencies in plant populations, usually for allozyme genetic markers (which it is assumed are reasonably close to neutral in their effects on fitness), and this has enormously improved our knowledge of mating systems in the field (see Chapter 2). In the herbaceous plant, *Mimulus platycalyx*, for instance, F based on 7 polymorphic loci was about 0.75 for seeds produced in 1988, and about 0.6 in the two following years, yielding estimates of the selfing rates (S) in these years of 0.86 and 0.75, results which are quite similar to estimates from sets of progeny using the method described in Section 2.7.3.3 (Dole & Ritland 1993).

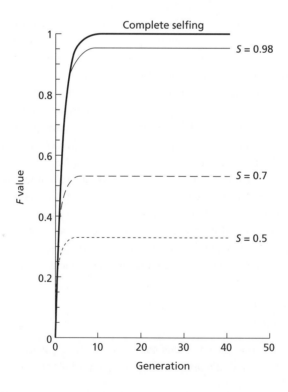

Fig. 3.1 Equilibrium inbreeding coefficients in populations with different self-fertilization rates (S) (see Falconer & Mackay 1996).

Genotype frequencies for genetic markers are, of course, estimated at particular stages of life, and differences can be informative. For example, such data can provide evidence of the occurrence of inbreeding depression (see Chapter 2). In many plant populations, older age classes have higher frequencies of heterozygotes than younger stages, suggesting that homozygotes produced by inbreeding have lower survival (Cheliak *et al.* 1985; Ritland 1990b; Tonsor *et al.* 1993). In a tropical tree species, *Cecropia obtusifolia*, for instance, the mean H_o for 8 polymorphic loci increased from below 0.3 for seeds to 0.38 in mature individuals (Alvarez-Buylla & Garay 1994). In the work on *M. platycalyx* just described, the inbreeding coefficients of the adult plants were lower than those of the seeds, in both years studied, and Dole and Ritland (1993) estimated that the survival of progeny produced by selfing was only 15% of that of outcrossed seeds in 1989, and 36% in 1990. Changes of this kind in H_o mean that a population reproducing by partial inbreeding (with a low heterozygote frequency in seeds) might show no significant deviations from Hardy–Weinberg equilibrium by the adult stage. In order to discover the true breeding system, it is thus best to have data on seeds or seedlings, before any differential mortality of inbred plants.

We will return to genetic diversity and examine patterns of variability within and between populations of outcrossing and inbreeding plants, in Section 3.4; however, before doing so, we must first discuss genetic drift.

3.2.2 Genetic drift

Chance factors affect the frequencies of alleles in natural populations of all organisms, because the transmission of alleles from generation to generation is prone to sampling that causes changes in frequencies in a random direction. If there is no selection (i.e. alleles are neutral, having no phenotypic effects; see Section 2.5.3), the process of reproduction is equivalent to taking a random sample each generation, in which each allele has an equal chance of being picked. It is quite unlikely that precisely the same frequency of each allele will be present in a new generation as in its parents. Thus, allele frequencies in real populations fluctuate randomly over the generations. This is **genetic drift**. When the size of a population is very large, these chance fluctuations may be negligible for alleles that are present in many individuals, i.e. at high frequency. In such populations, each new generation is expected to have an allele frequency very similar to that of its parents, and the possibility of a large difference in allele frequencies can safely be ignored.

There are two contexts in which random processes have evolutionary importance. First, even in large populations, new alleles that arise by mutation are always rare initially. Thus, the survival of new mutations is a matter of chance, and a mutation will often fail to be transmitted to the next generation. The most likely fate of a newly arisen neutral mutation is to become extinct within a few generations, purely by chance; even a selectively advantageous mutation can easily fail to become established in a population (see Section 3.5.2.1).

Secondly, it is also possible for the frequency of a new mutation to increase over time by chance, and even a neutral variant may occasionally reach a high enough frequency that chance loss from the population is very unlikely, i.e. the variant will remain polymorphic in the population for a long time (Fig. 3.2). Such 'drift' of neutral variants in small populations has two important consequences. One is that small populations tend to become genetically uniform over time, because after sufficient generations every allele will be a descendent of just one of the alleles that were present in the ancestral population. In addition, because this process occurs independently in different isolated populations, such populations will tend to diverge in their allele frequencies over time. We will return to this in the next section.

Consider first a single population that contains several different alleles at a locus, for instance, different DNA sequence types. If, by chance, an allele becomes over-represented in successive generations, the population has 'drifted' to a higher frequency of that type, and other alleles must have drifted to lower frequency. Sooner or later, one allele will reach a frequency of 100%. This is called 'fixation' for the new allele or, equivalently, all other alleles have been lost as a result of genetic drift, and this locus is no longer polymorphic in the population; the genetic drift process has then caused uniformity of genotypes within the population (Fig. 3.2). Only if new alleles enter the population by mutation or migration from elsewhere will diversity exist.

If the allelic differences are neutral, each allele is as likely as any other to be the lucky one that is fixed. For a new allele that has arisen by mutation, i.e. a new neutral variant present as a single copy, the chance that after enough time the new allele will be the only type remaining in the population is simply its frequency in the starting population, i.e. $1/N$, where N is the number of gametes in the population ($N/2$ is the number of reproductive individuals). Thus, an allele initially present in 1 out of a total population size of 10 gametes has a 10% chance of some day becoming fixed, but one in a population of 10 000 has a chance of just 10^{-4}, because it is much more likely that an allele of another kind is the one destined for fixation. Of course, if the new allele is disadvantageous, it is much less likely to become fixed, though if the disadvantage is slight, and the population very small, fixation is not impossible. Thus, both the population size and the selective effects of variants are important in determining the effects of genetic drift.

Genetic drift has been detected in several plant populations. An example is the loss of floral morphs in tristylous species. In one such species, *Lythrum salicaria* (Fig. 2.8), one of the three flower morphs is lacking from some populations. This occurs frequently in Canadian populations, where the plant is an alien invasive weed of wetlands, but is rare in native populations of the species in Europe: 23% of 102 Ontario populations had lost short- or mid-styled plants, and one population was monomorphic long-styled, whereas only 5% out of the same number of French populations examined were not trimorphic, and large populations in France invariably had all three morphs (Eckert *et al.* 1996). Most populations lacking a morph were small (3–50 plants). Rare flower morphs of heteromorphic species have a frequency-dependent advantage (see Section 3.5.6.1), so selection

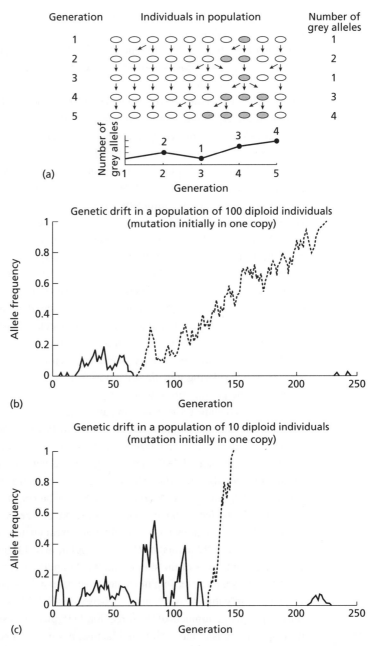

Fig. 3.2 Genetic drift. (a) The transmission of alleles from one generation to the next. A variant that arises in generation 1 as a single mutation is present in four copies in generation 5, but the number fluctuates as a result of random differences in transmission every generation. (b, c) Examples of computer simulations of finite populations of two different sizes, showing the allele frequency fluctuations of new neutral variants that arise. Note that most new variants quickly go extinct soon after they arise (solid lines show repeated new variants arising by mutation), but the occasional one can increase by chance and become fixed in the population (dashed lines). This process is very slow if the population is not very small.

should prevent morph loss from populations, but clearly drift overwhelms selection in these small populations.

The mathematical theory of genetic drift of allele frequencies in finite populations is complex, and this is only an outline of the simplest and most fundamental concepts. A further important point is that the amount of time it will take for a new mutant allele to become fixed also depends on the population size. As one would expect intuitively, if the population is large it takes a very long time for a common allele to be lost and replaced by a new neutral variant, purely by random sampling. The relevant population size that determines the effects and speed of genetic drift is not simply the number one can count in a population, because evidently plants that do not contribute to the next generation are irrelevant to the process; it is the number and reproductive output of breeding individuals which is relevant. Genetic drift can be described in terms of an **effective population size** (N_e), which accounts correctly for factors that affect the genetic drift process. The effective population size concept was introduced by Sewall Wright (1931). In the context of genetic drift of neutral variants, as here, it is defined as the size of an 'ideal' panmictic population which fluctuates in gene frequencies from generation to generation to the same degree as observed in an actual population (see Falconer & Mackay 1996). An ideal population is one in which the new zygotes are formed randomly each generation by sampling from the parent generation, just as we assumed in deriving the Hardy–Weinberg genotype frequencies (see section 3.2.1.1). In considering genetic drift, however, there is one important difference: we assume the population to be finite.

Values of N_e can be estimated from data on ecological variables such as temporal variability in number of plants and variance in flower production (see Caballero 1994). Estimates of effective sizes of natural populations can also be calculated using data on marker allele frequency fluctuations between successive generations of the same populations. The values estimated are often many times smaller than actual numbers of plants, N (Table 3.1). For instance, in 10 populations of the tristylous plant *Eichhornia paniculata*, ranging in size census from 50 to 5000 individuals, N_e estimates based on allozyme frequencies ranged from 3.4 (for a population of $N = 37$ plants) to 71 (for a population of 1310 plants); on average, N_e values were only 10% of the simple counts of population sizes, N. These values are also slightly smaller than values calculated taking into account five ecological variables (the value of N, changes between years in N, variance in numbers of flowers per plant, numbers of the three floral morphs, and estimates of selfing rates in the populations). Thus, additional factors, not detected ecologically, must have affected genetic drift in these populations (Husband & Barrett 1992).

3.2.3 Population subdivision and identity by descent

In real plant populations, even in outcrossing species, random mating is unlikely because in sessile organisms matings tend to occur between neighbours. This produces genetic subdivision or **genetic structure**. Here, we shall consider how

Table 3.1 The ratio of effective population size N_e, and population N for some herbs and trees.

Species	N_e/N	Source
Herbs		
Papaver dubium	0.07	Crawford (1984)
Eichhornia paniculata	0.11	Husband & Barrett (1992)
Various annuals	0.15–0.68	Heywood (1986)
Trees		
Pinus sylvestris	0.12	Burczyk (1996)
Abies balsamea	0.75–0.81	Dodd & Silvertown (2000)
Astrocaryum mexicanum	0.18–0.43	Eguiarte *et al.* (1993)

subdivision of a population into discrete patches or **demes** can cause genetic divergence between them through the random sampling process of genetic drift. Population genetic structure can also arise in spatially continuous populations through **isolation by distance** and we shall discuss this in Section 3.3.5.

A population's genetic structure can be quantified in terms of the probabilities that alleles come from a common ancestor (this is identity-by-descent, already defined in Section 3.2.1.3 for the two alleles within individuals). Here, we need to think of this measure on different spatial scales. The probability of identity-by-descent for two alleles from a single individual is increased by inbreeding. We can also measure this probability for two alleles from different members of the same population, or from two different demes. If migration between the demes is low, alleles within a deme are more likely to be identical by descent than alleles sampled at random from the species as a whole. Thus, we can think of population structure as analogous to inbreeding; even if plants do not self-fertilize, they nevertheless mate with relatives (of distant degree) if one compares the situation with random mating throughout the species range.

3.2.3.1 *Genetic drift in isolated demes*

Because local demes often include small numbers of plants, genetic drift can have great importance. It is easy to see that, in a set of isolated demes all initially with the same frequency of a neutral allele at a locus, the allele frequencies will tend to 'drift' apart over time. If the frequency of an allele is low, it will be lost in some subpopulations, while in others it may happen to rise in frequency. If the process of genetic drift is continued for sufficient generations, eventually each deme will become fixed for one allele or another. After this point in time, there will be no genetic variation in any deme (unless mutation occurs), but the set of populations as a whole could remain polymorphic.

3.3 **Gene flow**

Gene flow controls the extent to which populations evolve independently of one another, and develop genetic structure in gene frequencies. A method for

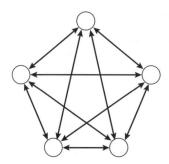

Fig. 3.3 The island model. Among its assumptions are that demes (circles) are all the same size and that each exchanges individuals with all the others at the same rate (arrows).

quantifying this structure is introduced in Section 3.3.1 and is based upon a simple model used to study isolation: the '**island model**'. In this model, migration is equally likely between any pair of demes, i.e. subdivision is incorporated into our thinking, but geographical separation is ignored (Fig. 3.3). In this, and several other kinds of model, very low rates of gene flow are sufficient to prevent allele frequency differentiation (Wright 1948). It is therefore not surprising that many plants and mobile animals have very similar allele frequencies in different populations. For instance, in a study of allozyme diversity in the wind-pollinated outcrossing species *Plantago lanceolata*, the *Got-1* locus had three alleles (*F*, *I* and *S*). The first two alleles were common, with *I* alleles at frequencies ranging from 0.648 to 0.747 in six populations, while the seventh population had a lower frequency (0.418). Allele *S* was absent in one population, and rare in all others (maximum frequency 0.079; van-Dijk *et al.* 1988).

Studies of pollen movement are important in modern agriculture. For instance, we might want to know whether genes from genetically modified crops such as *Brassica* spp. will 'escape' into and contaminate wild plants of the same or related species. If the crop is near a population that can acquire genes from the crop by pollen flow, herbicide resistance genes could spread to the weed, which might then become hard to eradicate (Ellstrand *et al.* 1999). For many plant populations, isolation is not complete. Migration occurs by pollen flow as well as by seed dispersal (Sections 3.3.2 and 3.3.3), and both kinds of migration can re-introduce genetic variation into a deme. Migration occurs not only between populations, but also at different distances within continuous plant populations, with more distant individuals being less likely to exchange genes than closer ones (isolation by distance; Section 3.3.5).

3.3.1 Fixation indices based on genetic markers

The excess homozygosity caused by genetic structure or by inbreeding can be detected using marker loci. We can compare the observed frequency of heterozygotes with the frequency expected at Hardy–Weinberg equilibrium, and use the values to estimate the value of *F* in Eqns 3.3:

$$1 - F = \frac{\text{actual heterozygote frequency}}{\text{expected heterozygote frequency}} = \frac{H_o}{H_e}$$

which is equivalent to:

$$F = 1 - \frac{H_o}{H_e} = \frac{H_e - H_o}{H_e} \qquad (3.4)$$

The F-value thus calculated is often called F_{IS}; this notation is to remind one that individuals are being compared with a sample of genotypes from the same locality and treated as a single subpopulation. It is sometimes called the correlation of genes within individuals.

Genetic differentiation between populations resulting from any ecological cause can be quantified using marker loci, by measuring the proportion of the genetic diversity that is present within and between populations. The gene diversity measure H_e was defined above (Eqn 3.2) for a population such as an individual deme, and is therefore often written as H_S (where the S stands for subpopulation) The same calculation can also be performed for a sample from a set of demes, giving a total population value (H_T). This may be higher than the subpopulation value, H_S, if diversity within demes is low, owing to founder effects or isolation and genetic drift. A useful measure of the between-population proportion of genetic diversity (the extent to which individuals in subpopulations are more similar than in the total set of populations) is thus:

$$F_{ST} = \frac{H_T - H_S}{H_T} \qquad (3.5)$$

For example, the *Arabidopsis thaliana* diversity study based on allozymes de-scribed above yielded an H_T value of 0.148 for the species (Abbott & Gomes 1988), more than seven times higher than the average diversity within populations; averaged over the polymorphic loci, F_{ST} was 0.62 (compared with 0.19 for *Gentiana pneumonanthe*; see Raijmann *et al.* 1994).

These types of quantities (F_{IS} and F_{ST}) are called **fixation indices**, or F-statistics. They are very frequently estimated in plant populations, using readily available genetic markers, such as allozymes and microsatellites, and provide a convenient way to summarize the structure of a set of populations (Nei 1987). Because F_{ST} (and the closely similar G_{ST}) measures the proportion of genetic variability that exists between populations, out of the total variability (i.e. it is scaled by the total variability), it gives us a useful way to compare population structure of different plants, even between studies using different genetic markers.

F-statistics calculated from genetic markers can also tell us something about the behaviour of plants in the populations. F_{IS} tends to be high in inbreeding populations; for example, in *Polygala vulgaris*, which has a mixed mating system, Lack and Kay (1988) found values ranging from 0.57 to 0.94. In the highly inbreeding *A. thaliana*, it was 0.99 (Abbott & Gomes 1988). By contrast, in the outcrosser, *Tachigali versicolor*, Loveless *et al.*'s (1998) values of H_o and H_e can be used in Eqn 3.4 to obtain F_{IS} values for the six different populations studied ranging from 0.08 to 0.19; in the outcrossing *G. pneumonanthe* the mean was 0.10 (Raijmann *et al.* 1994).

F_{ST} increases with increasing isolation between populations. Its expected value can be derived, assuming simple models of sets of subpopulations that exchange genes by migration. It can therefore be used to estimate migration rates, *provided* that the model is a reasonable approximation to the true mode of gene flow in the set of populations. For the island model mentioned in Section 3.3, and assuming a large number of demes, the relationship is:

$$F_{ST} = \frac{1}{1 + 4N_e m} \tag{3.6}$$

assuming a constant migration rate, *m* (Slatkin 1981).

In the *G. pneumonanthe* populations described above, the F_{ST} values consistently suggest considerable genetic differentiation between populations, averaging 0.19, which yields an estimated $N_e m$ value of about 1. One must bear in mind, however, that this estimate is based on the specific assumptions of the island model, and on assuming that populations have reached an equilibrium state. A problem is that this approach estimates $N_e m$ and not the actual proportion of migrants, *m*, or their actual number. Thus, migration rate estimates do not necessarily tell us the actual amount of migration (Whitlock & McCauley 1999). However, the *G. pneumonanthe* data are from a set of discrete subpopulations, so the island model may be a reasonable approximation to the true migration pattern. Measurement of actual seed and pollen dispersal is discussed in Sections 3.3.2 and 3.3.3.

Allele frequency differences between newly founded populations, because of founder effects, are another cause of high F_{ST} values, but these will tend to be reduced as pollen flow homogenizes allele frequencies. Such a pattern was detected in a demographic study of white campion, *Silene latifolia*, populations (McCauley *et al.* 1995). A similar kind of effect may explain the low population structure of trees, compared with herbs, even in populations that are known to be colonizing new territory. Trees have a long prereproductive stage, which allows time for repeated colonization of the same locality, and their populations should thus show fewer bottleneck effects (Austerlitz *et al.* 2000).

3.3.2   Seed dispersal

As outlined in Chapter 1, seeds are dispersed in several different ways, but even in wind-dispersed plants most seeds do not move far from their maternal parent (e.g. Fig. 1.10) (Levin & Kerster 1974). Occasionally, however, seeds may move great distances and the upper tail of the dispersal curve may be very long. The tail is arguably the most important part of the dispersal curve because the seeds that reach this far may potentially found new populations (Portnoy & Willson 1993). This is how distant volcanic oceanic islands such as Hawaii, which emerged from beneath the ocean and started off devoid of vegetation, were first colonized. Long-distance

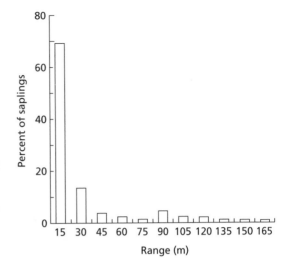

Fig. 3.4 Distribution of dispersal distances for a random sample of 94 saplings of bur oak *Quercus macrocarpa* inferred by analysis of microsatellite markers (data from Dow & Ashley 1996).

dispersal events can be detected using genetic markers: Fig. 3.4 shows an example of a study in bur oak *Quercus macrocarpa*, using microsatellite markers.

Interestingly, all plant species that have cleistogamous flowers (see Section 2.7.3) have two modes of dispersal. Seeds of cleistogamous flowers are poorly dispersed, while chasmogamous flowers disperse their seeds more widely (Lord 1981). This limited cleistogamous dispersal is taken to its extreme in peanut grass *Amphicarpum purshii* and other 'amphicarpic' species, whose seeds are produced from subterranean flowers (Cheplick 1987). Thus, progeny genetically similar to their parents remain close to their parents, perhaps benefitting from local adaptation (see below), while outcrossed progeny are dispersed (Schoen & Lloyd 1984).

3.3.3 Pollen movement

In plants it is even harder than in animals to determine the fathers of progeny, because pollen movement depends upon the movements of insect pollinators and the vagaries of wind dispersal, and matings cannot be observed directly. Pollen movement can be estimated using dyes of similar particle size to pollen, or substances that can be used to label pollen (Dudash & Fenster 1997). In species with pollinia, pollen movements in small populations can even be followed by physically labelling pollinia, giving unsurpassed details of pollinator movements. Using this method Nilsson *et al.* (1992) found that most pollination of the orchid *Aerangis ellisii* by hawkmoths occurred between plants no more than 5 m apart. Like seed movement, pollen dispersal curves for both insect and wind pollinated species tend to be highly skewed, with most movement being relatively local, but with occasional very long distance movement (Fig. 3.5).

An important aspect of dispersal via either seeds or pollen (see below) is establishment. Physical movement is not the same as migration and does not necessarily cause successful establishment in the population where the pollen

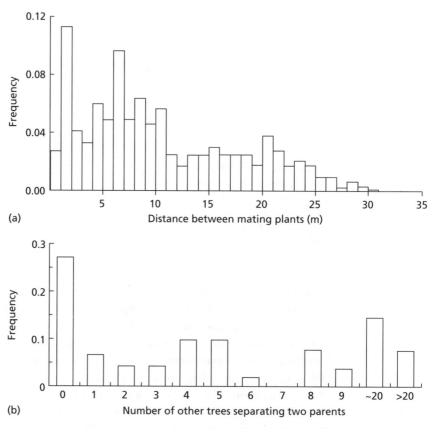

(a)

Distance between mating plants (m)

(b)

Number of other trees separating two parents

Fig. 3.5 Frequency distribution of distances between seed and pollen parents derived from parentage analysis using molecular markers in two insect pollinated species. (a) the dioecious lily *Chamaelirium luteum* (Meagher & Thompson 1987) and (b) the tree *Magnolia obovata* (Isagi *et al.* 2000).

arrives. In order to detect and estimate gene flow, genetic markers are needed. One approach uses paternity analysis to distinguish among possible male parents and discover which one most probably fathered each particular offspring. The maternal genotypes at the marker loci used must be known, so that the paternal allelic contributions to each progeny can be deduced. For each seed or seedling, the likelihood of each potential pollen donor being the true male parent can then be calculated, based on the marker genotypes. The likelihood for each potential paternal plant is compared with the likelihoods for randomly chosen individuals, and a paternal parent estimated for a given progeny plant (Meagher 1986). Marker loci that are highly polymorphic, with many different alleles, help such estimates, because potential male parents may then be identified uniquely by their genotypes. Microsatellite loci (see Section 2.5.2.3) are therefore being used increasingly in the study of pollen dispersal. For example, in bur oak (see Section 3.3.2), the four microsatellite loci used had 13–21 alleles, providing enough variation to exclude all local trees as potential fathers of saplings in more than half the individuals sampled, implying high levels of long-distance pollination (Dow & Ashley 1996, 1998). Seed dispersal, in contrast, was quite local (Fig. 3.4). A similar study of the oaks *Q. robur* and *Q. petraea* in France found that over 60% of seedlings sampled in a 240-m square plot were fertilized by pollen that had travelled more than 100 m (Streiff *et al.* 1999).

Paternity studies work particularly well if all the seeds in a fruit are fertilized by the same paternal plant, because in this situation, given sufficient marker loci, a great deal of genotype information about the true father can be obtained. Plants whose pollen is bundled into pollinia, as in orchids and Asclepiads, are thus good candidates for paternity studies. A study in a population of *Asclepias exaltata* at Mountain Lake, Virginia, using allozyme variants, showed that the distribution of pollen dispersal distances was similar to that of interplant distances, whereas interplant pollinator flights tended to be over much shorter distances (Fig. 3.6) (Broyles & Wyatt 1991). Clearly, many short-distance pollinator flights did not result in seedlings. This may be the consequence of the mechanism of pollinium transport by the butterfly pollinators, which ensures that, after a pollinium has been removed, many flowers are visited before a pollinium is inserted into another flower.

3.3.3.1 *Pollen movement and patterns of mating*

The behaviour of pollinators determines the distance pollen is carried, and hence the pattern of mating. In self-compatible species, transfer of pollen between

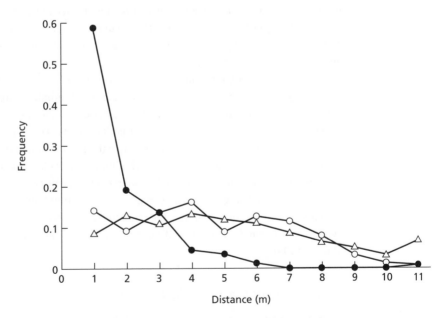

Fig. 3.6 Distributions of interplant pollinator flight distances (●), realized pollen dispersal (O), and interplant distances (Δ) in the Mountain Lake population of *Asclepias exaltata* in 1986 (from Broyles & Wyatt, 1991).

different flowers on the same plant is known as **geitonogamous selfing** (see Schoen & Lloyd 1992). The outcome of pollination is determined by an interaction between pollinator behaviour, individual plant traits and population properties. For example, Crawford (1984) found higher selfing rates in plants of the herb *Malva moschata* with large numbers of flowers, presumably because bees spend more time on such plants than on ones with few flowers. In the wind-pollinated plant *Plantago lanceolata*, gene flow and outcrossing rates were, not surprisingly, higher along transects orientated in some directions than others, corresponding to the direction of the prevailing wind (Tonsor 1990).

A related consequence of the fact that pollen movement is finite is that, in outcrossing species, plants growing in low-density situations may fail to receive enough pollen to set as much seed as they could potentially mature (van Treuren *et al.* 1994). By hand-pollinating flowers we can test whether or not pollen supply limits seed set: greater seed production compared with control plants would suggest **pollen limitation** of seed set. Such experiments show that pollen limitation sometimes occurs in nature (Burd 1994) and it may lead to selection for self-fertilization as a means of **reproductive assurance**. This can explain patterns of breakdown of outbreeding systems, such as the breakdown of dioecy to monoecy at the edges of the range of the buffalograss *Buchloe dactyloides* (Huff & Wu 1992).

3.3.4 Relative contributions of seed and pollen movement

There are important differences between pollen and seed dispersal in hermaphrodite plants. Only seeds can found new populations, and migration of genes via pollen flow can only occur into populations that already exist. Seeds carry maternally inherited genes, such as those in the mitochondrial and chloroplast genomes of many angiosperm species (Milligan 1992), which are not transmitted through pollen in these species (in gymnosperms, chloroplasts are usually paternally, and mitochondria maternally, transmitted). For nuclear genes, seeds carry twice the number of alleles than haploid pollen grains. These differences are the basis for methods that can be used to estimate the ratio of pollen flow to seed flow, based on the island migration model. This ratio is calculated as the ratio of F_{ST} (Section 3.3.1) estimated using nuclear markers (transmitted by seeds and pollen) to that estimated using cytoplasmic markers (Ennos 1994). In *Q. petraea*, a highly outcrossing, wind-pollinated tree, the estimated pollen-to-seed flow ratio was 196, while for a population of the selfing grass *Hordeum spontaneum* it was only 4 (Ennos 1994).

3.3.5 Isolation by distance in continuous populations

Even in a continuous large population, the probability that two individuals will mate tends to decrease with the distance between them. Because of this, and the limited dispersal of offspring from their parents, neighbours tend to be related (i.e.

the probability of two alleles from nearby plants being descended from the same parent is greater than under random mating). Population geneticists have studied simple models of populations in order to understand this kind of process quantitatively, and to establish the important parameters of such processes, so that they can be estimated (Slatkin 1993). A population's genetic structure can be quantified in terms of the probability that two alleles drawn at random come from a common ancestor (the identity by descent defined in Section 3.2.1.3, where we considered only physically subdivided populations). This can be used to measure the population's **neighbourhood** size by the size of an unstructured population that would produce the same probability of identity by descent. For plant populations, if seeds and pollen are assumed to move equally in all directions in a circle around the parent plants, with a variance σ^2 of the dispersal distance between parents and offspring, the genetic neighbourhood area is (Crawford 1984):

$$A = 4\pi\sigma^2. \tag{3.7}$$

If the dispersal distances are normally distributed, this defines a circle within which there is an 86.5% probability of finding the parents of the focal individual. The neighbourhood area can be used to calculate a corresponding effective population size, N_e (see Section 3.2.2), by multiplying A by the density of reproductive plants in the population:

$$N_e = NA. \tag{3.8}$$

Although this N_e value is calculated for a continuous population, and there are no isolated demes, it can be used in a similar way to the effective sizes for patchy populations to help understand population structure, in this case reflecting the extent to which genetic differentiation will depend on physical distance.

3.3.5.1 *Pollen movement and neighbourhood area*

Consistent with the effect of plant size on selfing, high population density also tends to reduce neighbourhood area in bee-pollinated species, because flight distances are shorter in denser populations (Levin & Kerster 1974; Fenster 1991). Pollinators that routinely carry pollen over long distances will thus tend to produce large values of N_e and A, and low population differentiation, while restricted pollinator movements will have the opposite effect.

Several studies have attempted to estimate directly the relative contributions to gene flow of the different elements of pollen and seed dispersal, including pollen siring ability and fitness differences of dispersed seeds (Beattie & Culver 1979; Gliddon & Saleem 1985; Eguiarte *et al.* 1993). In a natural population of the annual prairie legume *Chamaecrista fasciculata*, Fenster (1991) found the outcrossing rate to be high ($t = 0.8$), but gene dispersal was so limited that the average A corresponded to a circle of radius 2.4 m containing only about 100 individuals. Seed dispersal contributed remarkably little to gene flow (Fig. 3.7) (Fenster 1991). The seedling survival and fruit production of progeny plants

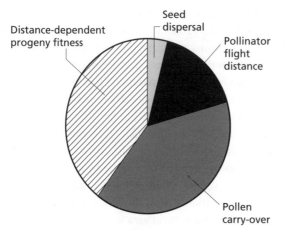

Fig. 3.7 The relative contributions of seed dispersal, pollinator flight distance, pollen carry-over and distance-dependent progeny fitness to gene flow in *Chamaecrista fasciculata* (from Fenster 1991).

depended upon how far apart the parent plants were growing from one another. This suggests that plants growing close together may often be related, so that their progeny may suffer some degree of inbreeding depression. Direct estimates of inbreeding depression showed that it occured in the population: progeny of selfing had half the fitness of progeny with two different parents from the same neighbourhood, and seeds with parents more than one neighbourhood diameter apart had the highest fitness of all. When these relative fitness values were used to weight the contribution of pollen dispersal to overall gene flow, the total estimated pollen dispersal distance was nearly doubled (Fig. 3.7).

3.3.6 Founder effects, extinction and colonization in patchy populations

Realistic models of gene flow for plant species with many demes should also take into account the possibility that demes might go extinct, even though the species persists in other localities. Extinction and recolonization are regular processes in some plant species (Chapter 7). Recolonization can occur only if seeds reach the empty sites, but extant demes may be subject to gene flow via pollen transfer as well as further seed migration. All of these migration processes increase the genetic similarity between different populations.

Random events may also lead to changes in allele frequencies when populations are impermanent. New populations of plants are usually founded by just a few seeds. This has two important implications. One is that the founders carry a small sample of the alleles in the source population; the allele frequency in the founders is thus likely to differ slightly from that in the source. The changes in allele frequencies caused by sampling at the foundation of new demes are known as **founder effects**. The same small-sample effect will occur if a population becomes very small but then recovers in size. This is called a population size **bottleneck**. Bottlenecks occur both at the foundation of new demes, and after events that reduce population size, e.g. fires.

The second important effect of colonization events, or reduced population size from some other cause that forces the population through a bottleneck, is that genetic diversity is decreased, in addition to allele frequencies being changed. Rare alleles are less likely to be included in a small sample than are more common ones, so they are particularly likely to be lost in a bottleneck event, and many examples are known of this happening. A study of 51 British populations of the self-pollinating herb *Polygala vulgaris* detected these effects (Lack & Kay 1988). High genetic variation was detected in an allozyme study of populations in long-established grassland habitats ($H_e = 0.166$) but several roadside and sand dune populations, each probably founded by just one or a very few individuals, had no polymorphisms ($H_e = 0.083$ for roadsides and 0.109 for sand dune populations, both statistically significantly lower than the grassland value). Bottlenecks may be particularly important in rare species, and it is important in the context of conservation biology to understand their effects. Low levels of genetic diversity in many plant populations may be attributable to this cause (Barrett & Kohn 1991).

The expected loss of genetic diversity when a population passes through a bottleneck is, however, surprisingly small (Nei *et al.* 1975). This is because the alleles that are lost when a population goes through a bottleneck are most likely to be those that are present at low frequency to start with; alleles at intermediate frequencies are very likely to persist. In the *Asclepias exaltata* study described above, 19 uncommon alleles previously found in southern Appalachia, USA, were absent from the population samples from the northern region, which probably experienced repeated bottlenecks as they migrated northwards (Broyles 1998). If a population expands again quite rapidly in the first generations after a bottleneck, genetic diversity may be little affected, even when rare alleles have been lost. The reason for this is that rare alleles make a negligible contribution to the measure of gene diversity H_e, as you can see by thinking about its definition (Eqn 3.2), which shows that alleles with low p_i values can almost be neglected.

A final implication is that plants founding new demes may often have few neighbouring individuals. If the species is self-incompatible, colonists may have no compatible neighbours. If it is dioecious, it may have neighbours only of the same sex as itself. In either situation, the plant will be unable to reproduce. It is thus expected that successful colonizing populations should often be self-fertile (Baker 1955, Pannell & Barrett 1998); this is called 'Baker's rule'.

3.3.7 The Wahlund effect

Genetic structure within a population leads to an overall deficiency in the frequency of heterozygotes. Imagine a set of subpopulations, each with Hardy–Weinberg genotype frequencies based on their local allele frequencies, but with different allele frequencies in different subpopulations. A sample of genotypes taken indiscriminately from the whole population will have more homozygotes than the Hardy–Weinberg genotype frequency formulae predict. This is because each subpopulation, with its own characteristic allele frequency, will have more homozygotes for its commonest allele than predicted from the average frequencies of the alleles over the whole set of subpopulations (in demes where only one allele is present at the locus, for instance, homozygotes will be the only genotype). This is termed the Wahlund effect. In other words, genetic structure is another departure from the assumption of panmixia used to derive the Hardy–Weinberg genotype frequencies.

Inbreeding is thus not the only possible cause of excess homozygote frequencies. Wahlund effects are the most likely explanation for excess homozygosity in outcrossing populations, such as those of the semelparous, neotropical tree *Tachigali versicolor*, studied by Loveless *et al.* (1998). For an excess of homozygotes to be caused by a Wahlund effect, differences in allele frequencies must exist between populations. These differences may be spatial or, as in the case of the semelparous populations of *T. versicolor*, exist between cohorts growing in the same locality (Loveless *et al.* 1998). This makes it possible to test whether an observed case can be explained in this way (Brown 1979). If no spatial or other differences can be detected, inbreeding resulting from mating of related plants, e.g. selfing, must be causing the departure from Hardy–Weinberg genotype frequencies. Of course, inbreeding populations may be structured spatially, in which case both of these factors contribute to excess homozygosity. It is then less easy to tell what is happening, and evidence about the mating system may be necessary to clarify the situation.

3.4 Patterns of genetic diversity in plant populations

In a set of data from 468 plant populations, Hamrick and Godt (1990) found that 34% of allozyme loci were polymorphic. There are, however, pronounced differences between different types of populations in both the magnitude of diversity and in its distribution within and between populations. These can be understood in terms of differences in effective population sizes and degrees of isolation. First consider causes of low genetic variation within populations. Endemic plant species typically have low levels of diversity (Hamrick & Godt 1990, 1997; Gitzendanner & Soltis 2000). We saw an example in Section 3.2.1.2 in which northern populations of *Asclepias exaltata* lacked alleles present in southern populations, probably because of reduced population sizes during colonization bottlenecks (Broyles 1998). However, some widespread plants, such as the colonizing species *Iris lacustris*, are genetically depauperate (Simonich & Morgan 1994). Such cases are often associated with asexual reproduction (as in *I. lacustris*) or with inbreeding. Like asexual populations (reviewed in Holsinger 2000), highly self-fertilizing populations, like those of *Arabidopsis thaliana*, typically have lower levels of polymorphism than outcrossing ones (Fig. 3.8; Table 3.2) (Charlesworth *et al.* 1997; Hamrick & Godt 1990, 1997). We explained above how random processes cause genetic differentiation between isolated populations or demes, and loss of variability within demes. Because inbreeding populations exchange genes infrequently with other demes, this process is faster than in outcrossers. In

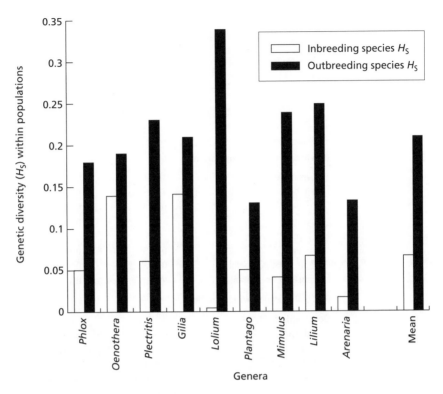

Fig. 3.8 Allozyme diversity in related inbreeding and outcrossing species pairs belonging to the same genus.

addition, inbreeding reduces effective population sizes (Charlesworth *et al.* 1997). Finally, because inbreeding populations are highly homozygous, different genotypes rarely come together and recombine, and consequently genetic diversity will be reduced by the occurrence of natural selection changing allele frequencies at loci involved in adaptation to local environments, or responding to ecological or climatic changes (see Section 3.5.2.1). Within inbreeding populations we therefore expect low variability. Diversity of the entire species may, however, be high. For instance, if local adaptation is important, we expect different populations to have different alleles at selected loci (see below) and populations may be isolated enough to maintain differences at marker loci also.

In addition to low within-population diversity, highly self-fertilizing populations often show high F_{ST} values, compared with outcrossing populations (Hamrick & Godt 1990). This is easy to understand: if H_S is very small, F_{ST} must be close to 1 (unless H_T is also small; see Eqn 3.5). This may not, of course, be the only reason for strong population structure in inbreeding species. It is also likely that such populations, with their isolation from others of the same species, may evolve stronger adaptation to local ecological conditions (see below), compared with outcrossing species, and therefore genetic differences could be greater between than within populations.

Table 3.2 (a) Mean proportion of genetic variation at allozyme loci (equivalent to F_{ST}; see Section 3.3.1) found among populations within species possessing different combinations of breeding system and dispersal mode and life form. (b) Genetic diversity (H_T) (see Section 3.2.1) within species broken down by breeding system and geographic range. (Data from Hamrick & Godt 1997.)

	Breeding system		
Traits	Selfing	Mixed-mating	Outcrossing
(a) Variation among populations			
Dispersal mode			
Wind	–	0.175[cd]	0.101[d]
Gravity	0.533[a]	0.248[c]	0.189[c]
Ingested	–	0.269[c]	0.223[c]
Life form			
Annual	0.553[a]	0.343[b]	0.191[cd]
Short-lived perennial	0.442[b]	0.238[c]	0.218[cd]
Long-lived perennial	–	0.145[de]	0.094[e]
(b) Total variation within species			
Geographic range			
Endemic	0.034[a]	0.100[ef]	0.142[cdef]
Regional	0.121[def]	0.164[abcd]	0.171[abc]
Widespread	0.165[abcd]	0.206[a]	0.183[ab]

Values within (a) or (b) that have different superscript letters are significantly different, $P < 0.05$.

3.5 Natural selection

Natural selection was proposed by Darwin in his book *The Origin of Species* as an explanation for adaptation (Darwin 1859), and this is still thought to be the chief mechanism of adaptive evolutionary change. An important form of evolution in plant populations is adaptation to local conditions. Knowledge of how often this occurs would provide valuable information about how often genetic variants are affected by selection, as opposed to being neutral and transiently present in populations as a result of genetic drift (see above).

Plants are ideal organisms in which to study adaptation to local environmental conditions, because they are sessile and thus we know what environment they are exposed to. Although there is abundant evidence for local adaptation in plants, as we shall see, it is important to remember that not all phenotypic differences between populations growing in different environments are adaptive, and that not all adaptive differences reflect genetic differences. As we explained in detail in Section 2.3, interpopulation differences are frequently phenotypic responses to the environment. For instance, plants growing at high altitude are often short in stature. To demonstrate that such a difference is genetic, we must test the phenotypes under a uniform environment, by transplanting individuals (or growing their seeds) from the wild into **common gardens**, as was done in the classic work on *Achillea lanulosa* from different altitudes (Clausen *et al.* 1948) described in Chapter 2. If natural selection has led to genetic differences between populations, genotypes should do best in their native environments, and we shall see below how such differences can be measured. Before doing so, we will briefly describe how selection can act, and how it affects gene frequencies.

3.5.1 Types of selection

Natural selection can alter the frequency distribution of phenotypes in three basic ways, depending upon whether selection favours one tail of the distribution, intermediate phenotypes, or both tails of the distribution (Fig. 3.9).

3.5.2 Directional selection

Perhaps the simplest type of selection, most closely related to the selection that Darwin had in mind, is directional selection, when selection favours one extreme

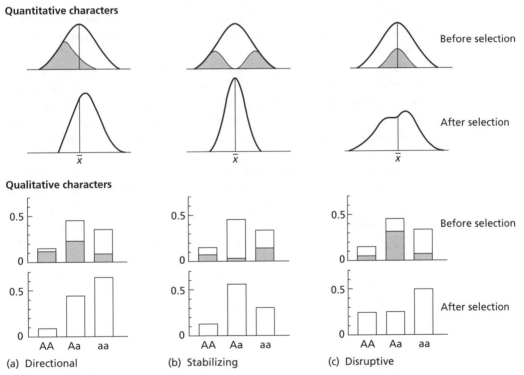

Fig. 3.9 Three modes of selection operating on quantitative and qualitative characters: (a) directional selection; (b) stabilizing selection; (c) disruptive selection. Open areas are genotype frequencies, filled areas are those genotypes which are selected against (from Endler 1986).

of the phenotype distribution. Directional selection operates when a new advantageous allele arises in a population. For instance, a plant population living in an environment in which pollinators are rare may be selected to evolve a higher rate of self-pollination than before, and a small-flowered mutant type that is highly self-pollinating would have higher seed production than the initial phenotype; there would then be directional selection for small flowers. An example is the small-flowered inbreeding species of *Mimulus* described in Chapter 2.7.3.2. In such a situation, a gene for small flowers would be predicted to increase in frequency, until ultimately the entire population would be made up of homozygotes for this allele.

3.5.2.1 *Directional selection on alleles at a single locus*

An example of how selection can alter allele frequencies is provided by an experiment on cyanogenic and acyanogenic genotypes of *Trifolium repens* (see Chapter 2.2). Ennos (1981) sowed seeds in a field, and scored allele frequencies among the survivors after six months. Although most seeds sown possessed the *Ac* allele and produced a cyanogenic glucoside, only plants with the *Li* allele at a second locus possessed the enzyme that releases cyanide when leaves are damaged. The *Li/Li* and *Li/li* genotypes are both cyanogenic, but *li/li* seedlings are not and have poorer survival. Releasing cyanide apparently increased plants' survival (probably because it protects them against herbivores), leading to selection against *li/li* genotypes.

This process of selective replacement of an allele at one locus by another can be described by a simple extension of the model used in the derivation of the Hardy–Weinberg genotype frequencies, with the addition of selection. Table 3.3 gives the genotype frequencies and their fitnesses (which, for present purposes, can be taken as the number of progeny produced, taking into account the survival rates of the genotypes).

From such a table, it is straightforward to calculate the frequency of the two alleles in the progeny generation from those in the parent generation, given the fitnesses (this, of course, assumes nonoverlapping generations, as in annual plants). In the table, and in Eqn 3.9 just below, $\bar{w}$ denotes the mean fitness of the progeny. It simplifies the calculations to work in terms of **selection**

Table 3.3 Genotype frequencies before and after selection for one locus with two alleles.

Genotype	A_1A_1	A_1A_2	A_2A_2	Total
Frequency	p^2	$2pq$	q^2	1
Fitness	w_{11}	w_{12}	w_{22}	
Contribution to next generation	p^2w_{11}	$2pqw_{12}$	q^2w_{22}	$\bar{w} = p^2w_{11} + 2pqw_{12} + q^2w_{22}$
Genotype frequencies after selection	$p^2w_{11}/\bar{w}$	$2pqw_{12}/\bar{w}$	$q^2w_{22}/\bar{w}$	1

coefficients (*s*), defined as one minus the fitness values of the genotypes. This allows us to work simply in terms of the relative advantages and disadvantages of different genotypes. Suppose, for instance, that A_2 is a recessive advantageous allele, giving A_2A_2 an advantage compared with the A_1A_1 and A_1A_2 genotypes. Such a situation is a simple case of directional selection. We can assign A_1A_1 and A_1A_2 fitnesses equal to 1. In this case, the selection coefficient for the A_2A_2 homozygotes is $s = 1 - w_{22}$. Note that in this case, A_2 is advantageous ($w_{22} > 1$), so that *s* is a negative number. Selection against a deleterious mutation is also directional selection; if A_2 is disadvantageous, w_{22} would be less than 1, and *s* would be positive. In the *T. repens* example just described, the selection coefficient estimated in the experiment for the *li/li* genotype was $s = 0.3$ (Ennos 1981).

The allele frequency change from one generation to the next depends on the selection coefficients (i.e. on the fitness differences) and on the allele frequencies. The expected change (Δp) is:

$$\Delta p = pq \frac{(w_{11} - w_{12})p + (w_{12} - w_{22})q}{\bar{w}}. \tag{3.9}$$

The expected frequencies of the new allele in a very large random mating population are expected to change over time as in Fig. 3.10. As one would expect, weak directional selection implies slow evolutionary change, but note the important result that the selection coefficient is not the only factor determining the speed of evolutionary change. Another important factor is the dominance of the alleles.

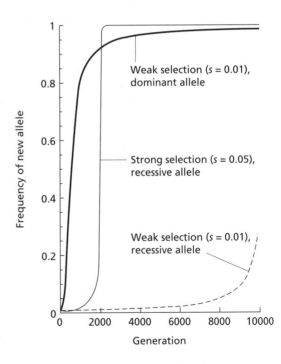

Fig. 3.10 Allele frequencies during the spread of an advantageous allele in a very large random mating population. Only the magnitudes of the selection coefficients are shown, as they are all negative, since we are dealing with advantageous alleles.

When a new advantageous recessive allele is rare, it is very slow to increase in a random mating population. This is because the change in allele frequency is driven by the advantage in homozygotes (i.e. by the second term in the numerator of Eqn 3.9, since the first term is zero for a recessive allele). For a rare allele, homozygotes will be extremely rare, as is seen readily by considering the genotype frequencies after a generation of random mating. Only when the allele has reached an intermediate frequency will homozygotes become common and the change in allele frequency become appreciable in a generation. A dominant allele gives an advantage to heterozygotes as well as homozygotes, so its initial increase is faster.

If the population is not random-mating, however, but partially inbreeding, as is true of many self-compatible plants (see Chapter 2), there is less difference between the speeds of increase of dominant and recessive advantageous alleles, because inbreeding generates homozygotes (Charlesworth 1992). A difference between evolution in outcrossing and inbreeding populations is thus that adaptations in the former are expected generally to be the result of alleles that are either dominant or partially dominant, whereas inbreeders are not subject to any such restriction.

Another important difference between outcrossing and inbreeding populations is that selection in inbreeders or asexual populations affects other genes as well as the gene that is the target of selection (i.e. the one that itself affects the phenotype of the plants and changes in frequency). When recombination is absent, as in asexual populations, or rare, as in highly inbreeding ones (see Section 3.4), the spread of an advantageous allele means that its genotype will increase in the population, causing alleles present in this genotype at all its loci to increase as well. In such populations, selection operates between different multilocus genotypes, rather than acting on individual genes as it does in outcrossing populations. After an advantageous gene has spread through a population, there clearly will be no variability at the locus that was selected, and, unless recombination happened, the same will be true at other loci. This is called a **selective sweep** and may be a further reason for the low genetic diversity observed in inbreeders (see Section 3.4).

A final important point is that not every advantageous gene will follow the curves shown in Fig. 3.10. While a gene is rare, it is not guaranteed to increase in frequency, even if it is advantageous, because the number of progeny that receive it is a matter of chance, because not every gene in a parent generation is transmitted to the progeny generation. It is quite possible that a newly arisen advantageous mutant allele may never rise to an intermediate or high frequency, but instead is simply lost from the population. Taking into account variance in offspring number, the chance for such an allele to rise to high frequency and ultimately be fixed in a population turns out to be low, about twice the allele's selection coefficient (Hartl & Clark 1997). Thus, only 10% of alleles that give a 5% fitness advantage are expected to succeed, in an evolutionary sense.

3.5.2.2 *Directional selection on quantitative characters*

Selection against the homozygous *li/li* noncyanogenic genotype in Ennos' experiment with *T. repens* is an example of directional selection on a qualitative character. Directional **phenotypic selection** often occurs on quantitatively variable traits. When survival or reproductive output correlates positively or negatively with the trait value, individuals are favoured at one end of the distribution of phenotypes, causing the character distribution among the breeding set of individuals to have a higher (or lower) value than the population as a whole (Fig. 3.9a). For example, recall the study of corolla flare in the Rocky Mountain populations of *Polemonium viscosum* (Section 2.4.4), which showed both heritable variation and selection by pollinators (Galen 1997).

Transplant experiments offer one way to detect and measure selection (Bell 1997). Selection can be detected by measures of survival and/or fertility of different phenotypes, and fitnesses and selection coefficients can be estimated for the phenotypes studied, in a similar fashion to the way fitnesses of genotypes at a single locus were defined in the simple single-locus case above. In the experiment on *Achillea lanulosa* (Clausen *et al.* 1948), plants from one locality (Mather) grown at a different locality (Timberline) had survival rates only 0.57 of those of native Timberline plants; grown at Mather, plants from Timberline had survival rates 0.26 those of Mather natives (Fig. 2.3a). Selection against non-local genotypes can be surprisingly strong even over distances separating subpopulations of less than 10 m (Table 3.4). Such short distances might be expected

Table 3.4 Mean selection coefficients (*s*) against non-local genotypes in some reciprocal transplant experiments between sites. See Chapter 1 for definitions of measure of fitness.

Species	Mean *s*	Measure of fitness	Habitat	Source
Annuals				
Bromus tectorum	≈ 0	λ	Steppe & forest	Rice & Mack (1991)
Impatiens pallida	0.35	l_x	Forest	Schemske (1984)
Phlox drummondii	0.43	λ	Grassland	Schmidt & Levin (1985)
Salicornia europaea	0.68	λ	Salt marsh	Davy & Smith (1988)
Perennials				
Anthoxanthum odoratum	0	m_x	Grassland	Platenkamp (1990)
Dryas octopetala	0.47	l_x	Tundra	McGraw & Antonovics (1983)
Delphinium nelsonii	0.23	λ	Meadow	Waser & Price (1985)
Polemonium viscosum	0.83	l_x	Alpine tundra	Galen *et al.* (1991)
Ranunculus repens	0.43	Ramet production	Grassland/ woodland	Lovett Doust (1981b)

to give little opportunity for local adaptive specialization because of gene flow from nearby demes. Many transplant studies have involved populations that straddle the boundaries of sites polluted with heavy metals (Antonovics 1968; Bradshaw & McNeilly 1981), but many other ecological variables can cause local selective differences (Linhart & Grant 1996), including the physical environment, such as soil type (Nevo *et al.* 1988) and salinity (Antlfinger 1981), and biological factors such as herbivory (Schemske 1984: see Chapter 10.2).

A convincing way to test whether variation in particular physical factors is responsible for local genetic differences is to alter the environment experimentally and monitor the resultant genetic changes. Significant adaptation to recently imposed environmental treatments can occur in under a decade, even in outcrossing perennials exposed to gene flow from other types of environment nearby (e.g. Snaydon & Davies 1982). In the outcrossing, perennial grass *Festuca ovina* Prentice *et al.* (1995) found associations between allele frequencies at four allozyme loci and habitat variables in populations sampled on the Swedish island of Öland in the Baltic Sea. In particular, one of the alleles at a locus known as *Pgi-2* was negatively correlated with soil moisture and was especially common at dry sites. Experimental plots at a range of sites were watered for nine years in the summer months, when drought often occurs. Prentice *et al.* (2000) scored allele frequencies at the end of the experiment and found a reduced frequency of the relevant allele of *Pgi-2* in watered plots compared to controls, suggesting selection on loci linked to this marker. Another approach that can enable one to estimate the strength of selection acting on a particular trait is from the regression of the trait's values on components of fitness such as survival or seed production. A significant relationship between, say, leaf area and fertility shows that there is both phenotypic variation between individuals and fitness differences between phenotypes, so that evolutionary change might be expected to occur. As explained above, however, it is important to understand that, for phenotypic selection to be translated into natural selection, the variation must be heritable (see Chapter 2). For quantitative traits that are correlated with each other, multiple regression methods have been developed to identify which trait or set of traits is selectively most important (Lande & Arnold 1983; Mitchell-Olds & Shaw 1987; Wade & Kalisz 1990).

3.5.2.3 *Clines*

In discussing selection, we have so far considered only a single large population, and have ignored gene flow between subpopulations. When some migration occurs between populations, allele frequencies may show a gradient over a set of populations, with a high frequency of one allele in populations in one region, grading down to a low frequency in another. Such **clines** are commonly found in plants. These allele frequency differences often seem to be the result of selection, i.e. a consequence of the occurrence of locally adapted genotypes, following soil or other habitat gradients or patchiness. For instance, *P. viscosum* (see Section 2.4.4), is variable for flower scent: sweet-smelling flowers are visited by bumblebees, and skunky-smelling ones by flies. The frequency of the bumblebee-pollinated morph increases with altitude, correlated with decreasing abundance of flies (Galen *et al.* 1987). This small-scale cline in scent type is therefore probably caused by natural selection.

A larger scale example, also very likely to have been caused by natural selection, is seen in many trees in the Northern hemisphere which have clines related to day length. An example is photoperiodic control of bud set and branch growth in Norway spruce, *Picea abies* (Clapham *et al.* 1998). Such clines in the response of tree survival and growth to climate are of great practical importance in forestry and many studies have transplanted trees between different source populations, called **provenances**, sometimes extending to reciprocal transplants across the entire natural range of the species (Campbell 1974; Matheson & Raymond 1986; Morgenstern & Mullin 1990). The results of these experiments, which are little known outside forestry science, tell us a great deal about clinal variation and adaptation to climate; information which should be useful in predicting the consequences of global warming for the health and distribution of different tree species (Mátyás 1994; Rehfeldt *et al.* 1999). Provenance trials demonstrate, for example, that genotypes transferred southwards, or to lower elevations, may survive better than genotypes native to the area of the receiving plantation (Fig. 3.11), but that their growth may be slower than that of natives (e.g. Morgenstern & Mullin 1990; Mátyás & Yeatman 1992). The relative performance of trees of different provenances typically changes as they age, so that the genotypes doing best at five years of age may often perform poorly when the same trees are re-evaluated at 20 or 30 years of age. This means that local adaptation in some trees can only be properly evaluated using lifetime studies, rather than survival measured over only a few years.

In addition to clines in phenotypic characters, gradients are also often found in the frequencies of genetic marker alleles, such as allozymes, that are thought to be neutral or under weak selection. In the example of northern migration of *Asclepias*

Fig. 3.11 Increase in cumulative mortality rate over five years for provenances of Scots pine, *Pinus sylvestris*, transferred across different latitudes and altitudes to 17 plantations located in the north of Sweden. Source populations are at zero and distances indicate transfers in km north or in m above this point. Negative distances indicate transfers south or below the source populations (Campbell 1974).

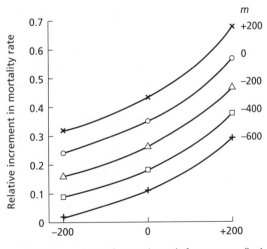

exaltata populations since the last ice age (Section 3.2.1.2), seven common alleles were found to exhibit clinal changes in allele frequency (Broyles 1998). Some of these differences may have developed by genetic drift in isolated populations (Section 3.2.3.1). The term cline is also used when there is a gradient in diversity levels between populations. The north–south difference in variability between *A. exaltata* populations is an example; allozyme diversity decreased linearly from south to north (Broyles 1998).

Evidence that a cline is the result of selection (rather than genetic drift during a period of population isolation, followed by gradual mixing) comes from situations in which similar clines are repeated under similar ecological conditions. For instance, the frequency of hairy leaved *Spergula arvensis* in Continental Europe, and in Britain, increases with latitude, suggesting that it is better adapted to colder climates than the non-hairy morph. This is supported by the finding that hairy plants increase in frequency with altitude, both in Derbyshire and in Shropshire, as well as by some direct experimental evidence for better germination at low temperatures (New 1958). A cline affecting seed traits in this species is described in Section 5.5.3 and Fig. 5.18.

Both allozyme and phenotypic differences can exist in the same cline, and this was found in *P. viscosum* (Galen *et al.* 1991), but sometimes a cline is found in a phenotype that has a clear relationship with a difference in environmental circumstances, such as the coldness of the winters, without an accompanying cline for marker loci. This discordance between a character under natural selection, and loci that are probably neutral in their fitness effects, suggests that gene flow between the different populations is extensive and thus that a persistent phenotypic difference is probably maintained by natural selection, even in the face of gene flow.

3.5.3 Deleterious mutations and mutation-selection balance

An important form of directional selection is selection against mutations that lower fitness. Even though selection tends to lower the frequencies of deleterious alleles every generation, populations always have some low frequency of mutations, at various loci, because selection cannot reduce their frequencies to zero, especially in the extreme case of recessive mutations (see Eqn 3.9 for allele frequency change). Such alleles are effectively hidden from the action of natural selection, and their presence in populations is revealed only by inbreeding (for instance, when one inbreeds plants in an inbreeding depression experiment see Chapter 2). Mutant alleles will therefore reach equilibrium at a low frequency, at which the losses resulting from selection equal the input from mutation (**mutation–selection balance**).

Selection against deleterious mutations is called **purifying selection**. One would expect purifying selection to operate most strongly on genes crucial to fitness. This is borne out by the extreme conservation of the coding sequences of many genes. For example, guanylate kinase is an essential enzyme for nucleotide metabolism, which is expressed in all plant tissues with highest activity in roots. The amino acid sequence of a functional kinase domain of the plant protein shares 46–52% identity with guanylate kinases from yeast, *Escherichia coli*, human, mouse and the nematode worm, *Caenorhabditis elegans*. It is even possible to express this gene from a plant (*Arabidopsis thaliana*) in yeast cells, and rescue cells carrying a nonfunctional yeast version of the gene, i.e. the plant gene can replace the yeast gene function (Kumar *et al.* 2000).

In comparisons between species, this conservation of amino acid sequences is detectable. It is usual for sites in DNA sequences where base changes will alter the amino acid to change much more slowly than 'silent sites' (such as third positions of codons) where changes leave the amino acid unaffected. For example, comparing *A. thaliana* alcohol dehydrogenase with that of its close relative *A. lyrata*, about 2% of amino acid replacement sites were different, compared with about 19% of non-replacement, or silent, sites (Miyashita *et al.* 1998). Comparing nuclear genes of maize and rice, amino acid replacements are estimated to have occurred at 7% of the possible sites, while 57% of silent sites are estimated to have changed (Li 1997); the higher divergence between these species is presumably because they are distantly related, although divergence also depends on the rate of sequence mutations in the genes compared. Chloroplast and mitochondrial genes diverge more slowly than nuclear genes, for example (Gaut 1998; Li 1997). Similarly, within species, variants most often occur at non-replacement sites or in introns, and amino acid replacement variants are rarer, although they do exist (recall the allozyme variants described in Section 2.5.1). Conversely, when selection does not operate, or is weak, sequence change will occur because of mutation. The chloroplast genomes of tobacco, rice and a bryophyte all possess the same set of genes for photosynthetic functions, but none of 22 of these genes tested was present in beechdrops *Epifagus virginiana*, a heterotrophic plant parasitic on beech trees. In this nonphotosynthetic plant, selection on the photosynthetic genes has relaxed sufficiently to have allowed them to lose function, or to be lost entirely (Pamphilis & Palmer 1990).

3.5.4   Stabilizing and disruptive selection

Stabilizing selection is selection that favours individuals near the mean in a population, while those in the tails of the frequency distribution have lower fitness (Fig. 3.9b). Many examples of this type of selection have been studied in animals, including selection for intermediate birth weight in humans (Endler 1986). Much of the natural selection that operates in nature is probably selection against individuals that deviate from the mean, and this may explain stability of phenotypes over long evolutionary time periods. Stabilizing selection on phenotypes can be detected using regression of fitness related characters, such as fecundity, on phenotypic measures (Section 3.5.2.2). If a quadratic regression fits the data significantly better than a linear one, this suggests stabilizing selection (Lande & Arnold 1983). Using this approach, Widén (1991) detected stabilizing selection for flowering time in the grassland perennial *Senecio integrifolius*. The precise nature of phenotypic selection varied from year to year in this plant, and in one season it was entirely directional, favouring the earliest plants to flower (Fig. 3.12).

3.5.5 Balancing selection: heterozygote advantage

Selection that has the effect of preserving genetic variation is particularly interesting, given that, as we have seen, natural populations of plants and animals are genetically diverse. Forms of selection having this property are called **balancing selection**. With balancing selection, an allele that is perturbed to a lower frequency for any reason will tend to increase, but if it goes to a higher value its frequency tends to decrease; there is thus a **stable equilibrium** allele frequency under this form of selection.

One type of balancing selection occurs when heterozygotes have higher fitness than either homozygote. This is known as **heterozygote advantage** (or overdominance). The fitnesses of the three genotypes in Table 3.3 might be denoted as: $1 - s$, 1 and $1 - t$, where $w_{11} = (1 - s)/w_{12}$ and $w_{22} = (1 - t)/w_{12}$; overdominance occurs when both selection coefficients are positive numbers. Because both homozygotes have lower fitness than the heterozygote, this type of selection helps maintain genetic variation at the locus; this may account for some of the polymorphisms found in populations of many plant and animal species (especially if they outcross, so that heterozygotes are common). However, as direct evidence for this form of selection has proved very difficult to find, only a handful of examples are known (see, e.g., Lewontin 1974; Hartl & Clark 1997) and its importance remains uncertain. Heterozygote advantage may be responsible for the persistence of a mutant gene found in oats in Finland which, in the homozygous condition, causes gigantism and such a long delay in flowering that no seeds are set, while heterozygotes can have seed yields per plant that are 20–30% greater than the wild-type (Ahokas 1996).

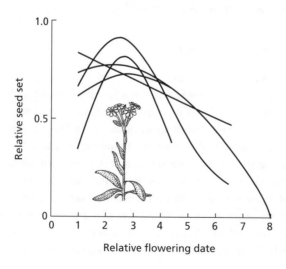

Fig. 3.12 Relationship between seed set (a component of fitness) and flowering time in a population of *Senecio integrifolius* (pictured) in Sweden. Each line represents a different year. Straight lines indicate directional selection, and curved lines indicate stabilizing selection (after Widén 1991).

Polymorphisms in haploid plants, such as the high levels of genetic variability in mosses such as *Plagiomnium ciliare* (Wyatt *et al.* 1989), cannot be attributed to overdominance, because there are no heterozygotes in haploids. Polymorphisms in highly inbreeding populations are also unlikely to be the consequence of overdominance. If the two selection coefficients are not the same (i.e. $s \neq t$), the equilibrium frequencies of the A_1 and A_2 alleles become increasingly asymmetrical with increased levels of inbreeding, until at some level of inbreeding one of them is lost and there is no longer a polymorphism.

3.5.6 Frequency-dependent selection

Frequency-dependent selection can be understood in terms of what happens in plant populations that are polymorphic for flower colour. Pollinators often have a 'search image' for the kind of flower that has been rewarding to visit; for instance, bees quickly learn the species currently yielding the best nectar or pollen. They can learn the appearance of the best plants to visit, and, as many pollinators have colour vision, they can learn the flower colours (Kay 1978). Individual plants with the most common flower colour will be encountered most frequently and bees will learn this colour and mostly be faithful to it. Rare variants, such as white-flowered plants, will not be encountered so often and will therefore be under-visited, relative to their frequency in flower populations, receiving occasional visits largely from newly arrived, or forgetful, bees. Rare morphs will thus tend to be more selfing than average, as was found for white flowered morning glory *Ipomoea purpurea* (Brown & Clegg 1984; Schoen & Clegg 1985). In other words, high frequency increases the chance of mating with others, so that fertility depends on frequency (Kay 1978). Such **positive frequency-dependent selection** means that rare flower colour morphs have a disadvantage that cannot be detected just by growing and measuring survival, or even fertility in experimental situations, but can be observed only in natural conditions, with pollinators present and behaving normally. This form of frequency-dependent selection tends to keep populations uniform.

3.5.6.1 *Balancing selection: negative frequency dependence in mating system polymorphisms*

Frequency-dependent selection acting in the opposite direction, to the benefit of rare genotypes, is important in maintaining some kinds of genetic variability in populations. A classical example of negative frequency-dependent selection, again involving fertility differences, is the selection on self-incompatibility (SI) alleles (see Chapter 2). The fecundity of an SI type depends upon the frequency of compatible morphs. When a morph is rare it has many potential mates, but when it is common many other individuals share incompatibility alleles with it and so it has lower fertility. The advantage this confers on rare SI morphs explains the large number of alleles at the incompatibility loci in natural populations of self-incompatible plants. A new allele that arises by mutation will be at an advantage and be favoured by selection, and an allele that happens to become rare is unlikely to be lost from a population, because its rarity gives it a higher fitness than other alleles, so it tends to increase again. In this kind of system there is thus a stable equilibrium with many alleles at equal frequency. Self-incompatible plant populations fit this model well. In the poppy *Papaver rhoeas*, for example, 45 alleles were found in a study of three populations, 15 of them in all three populations (Lawrence *et al.* 1993).

3.5.7 Selection that varies in space or time

It is intuitively appealing to think that genetic diversity should readily be maintained in natural populations because environments are heterogeneous, both spatially and temporally. We saw in Widén's study of flowering time in *S. integrifolius* that selection may vary in its strength, form or direction from one year to the next (Fig. 3.12), and selection in natural populations must often vary in time. Local adaptation in relation to flooding and competition within vernal pool populations of *Veronica peregrina* was described in Chapter 1. Even when the reasons for differences in the direction of selection are not as obvious as this, they can sometimes be detected. In a set of common garden experiments using the self-fertilizing species *Impatiens capensis*, microgeographic genetic differences in several life history traits were detected. Some characters showed spatial structuring, while others did not, suggesting that selection caused the structuring, rather than some historical process that would affect all characters similarly (Argyres & Schmitt 1991).

One might think that this environmental variation could preserve genetic variation for the traits in question, and there are indeed circumstances under which this is true, but they are usually quite limited (Hartl & Clark 1997, Hedrick 1995). Here, we will briefly discuss a few of the possibilities.

3.5.7.1 *Frequency dependence caused by competition*

When different genotypes have different ecological requirements, and the environment is patchy, frequency-dependent selection may result. For example, suppose that there are some dry and some wet patches in an environment, and different genotypes survive and reproduce best in one or the other type of patch. If one genotype is very frequent, say the dry-environment adapted form, many such plants will be present in the dry patches, and will suffer high competition with one another, which will probably reduce their fertility and/or survival, compared with the rarer wet-environment adapted form. Overall, therefore, when either form is the rarer it will tend to increase, i.e. there is negative frequency dependence.

Examples of this mechanism operating in natural populations are hard to find. Stratton (1994, 1995) found genotype–environment interactions affecting fitness in the apomictic annual *Erigeron annus* sown into an old field in New Jersey, USA, that operated at a very fine spatial scale. Sites 10 cm apart were found to be sufficiently different to reverse the relative fitness of different genotypes sown there. However, although this kind of fine-scale specialization by genotypes is needed to generate frequency-dependent selection, further studies have revealed no evidence of this (Bennington & Stratton 1998), and simulation models have shown that the observed spatial and temporal variation in genotype fitness would reduce rather than maintain genetic variation in the population (Stratton & Bennington 1998).

3.5.7.2   *Host–pathogen systems*
Resistance of plants to many pathogens is determined genetically, and is sometimes specific for particular genetically determined strains of the pathogen, which, in turn, are specialized for particular host strains (Pryor 1987). Thus, one host plant genotype is resistant to one pathogen genotype, but susceptible to others, while a different plant genotype is susceptible to infection by the pathogen strain to which another plant is resistant. These differences (in both pathogens and host plants) are often controlled by different genotypes at single genes. We will discuss the evidence for such **gene-for-gene** interactions below, but first let us consider how such systems will behave. When a plant genotype is common, it provides a pool of susceptible plants for pathogen genotypes that are specialized for it, thus conferring high fitness on these pathogen genotypes. Pathogen genotypes that are virulent on this plant genotype therefore increase in the pathogen population at the expense of other genotypes whose hosts are less common. This lowers the survival of the common host genotype, but not that of other genotypes (on which this particular pathogen genotype is nonvirulent, owing to the specialization). As the susceptible plant genotype becomes rarer as a result of pathogen attack, other, initially rare, resistant host genotypes increase in the plant population. This implies that the set of pathogen genotypes that are virulent on the initially common plant genotype will lose their advantage over other pathogen genotypes. Once a pathogen type that can infect the newly common plant genotypes arises, by mutation or immigration into the pathogen population, the process repeats itself. Both host genotype survival and pathogen genotype fitnesses are therefore constantly changing (Roy & Kirchner 2000).

It has been suggested that cyclical selection by pathogens may create long-term cycles in gene frequencies in populations, and that this can play an important role in the evolution and maintenance of sexual reproduction (see Chapter 9). If this kind of process actually happens, it may be detectable in clonal species, where genotypes at marker loci could enable one to detect the predicted genotype frequency fluctuations, or at least to test whether clones that are at high abundance in populations tend to be more heavily or frequently infected with pathogens than rare clones, and this has sometimes been found, for example in *Chondrilla juncea* populations in Turkey (Chaboudez & Burdon 1995). However, the predictions are not entirely clear (for discussion and references; see Dybdahl & Lively 1998), because the time lag in the cycles means that common clones should sometimes have low infection rates, because the pathogen has not yet caught up.

This model depends strongly on the assumption of specificity (Parker 1994). Gene-for-gene relationships have been demonstrated or postulated in at least 27 different host/pathogen systems affecting crop plants (Michelmore & Meyers 1998), but their importance in wild systems is so far much less clear (Burdon 1987; Pryor 1987; Burdon & Jarosz 1988). Even in crops, many examples are known of host genes that confer resistance to multiple pathogens (e.g. Century *et al.* 1995). It is thus not clear whether the assumptions of the model described above hold for plants or whether pathogens can often attack a wide range of host genotypes (Parker 1994). This may simply reflect the relative lack of knowledge about disease in wild plants.

3.6 Summary

Many plant populations contain large amounts of genetic variation, with quantitative differences between different types of plants. Many **polymorphisms** are probably **selectively neutral**, but some are likely to be maintained by **natural selection**. In the simplest possible case, when there is no selection, and we are dealing with a large, randomly mating population, the **Hardy–Weinberg** frequencies of genotypes are expected. Deviations from Hardy–Weinberg equilibrium frequencies may be caused by **nonrandom mating, subdivision of the population**, or **natural selection**. **Nonrandom mating** occurs when plants mate with other individuals that are more likely to carry the same alleles as themselves. **Inbreeding** is an important kind of nonrandom mating in many plants. It includes self-fertilization, and also mating with relatives that happens because restricted seed dispersal causes the progeny of each maternal individual to be more likely to mate with one another than with nonrelatives (biparental inbreeding, including sib-mating, and inbreeding as a result of restricted neighbourhood size).

Gene flow within and between plant populations is often restricted. **Population subdivision** can result from physical subdivision (i.e. there are distinct subpopulations or **demes**), caused by local absence of suitable habitat, or by chance local extinction of plants, or can effectively occur when mating occurs chiefly between neighbouring individuals. Gene flow, or migration, occurs by **seed and pollen dispersal**. Most pollen and seeds do not travel far from their source, but occasional dispersal events may be over long distances. Seeds can colonize new or empty sites, but pollen dispersal can occur only where there is a pre-existing plant population. Seeds bring maternally transmitted genomes, such as the mitochondrial genomes of many angiosperms, but pollen brings only nuclear genes unless organelles (such as the chloroplast genomes of many conifers) are transmitted paternally. Seeds may fail to disperse to a site successfully because the site is unsuitable for the species, or because their genotypes are unsuccessful in competition with conspecific plants already present there. Pollen dispersal may fail for the second reason, or because the seeds generated are inbred and survive poorly.

When a population is founded by a limited number of individuals (a population size **bottleneck**), its allele frequencies may exhibit a **founder effect**, i.e. they may by chance differ from the frequencies in the source populations; typically, rare alleles are unlikely to be present in a small sample, so they are often absent in a new colony (though they may arrive later by pollen flow).

Finite population size also leads to chance (stochastic) variations in allele frequencies each generation, particularly for variants that are neutral in their effects on fitness, and in small populations. The effect of population size on the speed of such **genetic drift** depends on the **effective population size** (N_e), or number of breeding individuals. Genetic drift leads to loss of genetic variation, so small populations tend to have less diversity than large ones. When gene flow is small, subdivision allows allele frequency differences to develop as a result of genetic drift (genetic structure). The extent of such differences for neutral variants can be predicted in terms of an overall effective population size that takes into account the pattern and rate of migration between demes.

If subpopulations or demes with different allele frequencies are sampled as a single unit, the genotype frequencies overall may show an excess of homozygotes, i.e. depart from Hardy–Weinberg equilibrium, even when mating is random within each deme. This is the **Wahlund effect**. Inbreeding also causes excess homozygotes, but this will cause excess homozygotes within demes.

Natural selection can produce adaptive evolutionary changes in gene frequencies. **Selection coefficients** can be estimated from transplant experiments, and **phenotypic selection** on traits can be detected from the regression of trait value on measures of characters related to fitness. Three types of selection are distinguished: **directional selection** favours one tail of the frequency distribution of phenotypes, **stabilizing selection** favours the mode against the tails of the distribution, and **disruptive selection** favours both tails of the distribution against the mode. Selection may also vary over time, either because fitnesses depend on changing environmental conditions, or because fitnesses are **frequency-dependent** and genotype frequencies change over time.

3.7 Further reading

Bell, G. (1997) *Selection.* Chapman & Hall, London.

Caballero, A. (1994) Developments in the prediction of effective population size. *Heredity*, 73, 657–679.

Futuyma, D. J. (1999) *Evolutionary Biology*, 3rd edn. Sinauer Associates, Sunderland, MA.

Hartl, D. L. & Clark, A. G. (1997) *Principles of Population Genetics*, 3rd edn. Sinauer Associates, Sunderland, MA.

3.8 Questions

1 A survey of variation at 7 protein and enzyme loci of 192 individuals from a plant population showed that 5 loci had a predominant allele with a frequency of at least 99%. At one of the other 2 loci, glucose-6–phosphate dehydrogenase, the numbers of plants of the different genotypes were as follows:

Locus	Genotype	Number of individuals
G6PD	S/S	25
	F/S	43
	F/F	124

Calculate:
(a) the proportion of polymorphic loci
(b) the allele frequencies at the G6PD locus
(c) the expected genotype frequencies at the G6PD locus, assuming Hardy–Weinberg equilibrium.
Examine the data to ask whether the G6PD data deviate from the Hardy–Weinberg genotype frequencies.
(d) Calculate the gene diversity, H_e, per individual for this locus.

2 What are: neutral variants, amino acid polymorphism, allozyme polymorphism, identity-by-descent?

3 Explain how genetic drift causes allele frequency differences between isolated populations.

4 Describe how plants disperse and colonize newly available habitat, and how this may affect the breeding system of a plant species.

5 Explain the Wahlund effect.

4 Intraspecific interactions

4.1 Introduction

Anyone who has grown a bed of vegetables, or has negelected one and turned around to find it filled with weeds doesn't need to be told that plants will fill any available growing space. This observation is the basis of most plant population models. The underlying assumption is that the primary resource limiting plant growth is space. Even in arid environments, where there appears to be plenty of bare ground between plants, roots spread out well beyond the area of the canopy and plants encounter each other below ground. A plant must occupy space to capture the essentials of life: light above ground and water and nutrients beneath. Space appears in population and genetic models in two guises: as an explicit statement of the area over which a plant is able to influence or be influenced by others (the ecological and the genetic neighbourhoods), and as density—a simple count of the number of plants per unit area. In most of this chapter, we will use density as an average measure of space occupancy, but we shall also consider under what circumstances this is an over-simplification and when we need to measure the influence of plants upon one another in terms of individual neighbourhoods.

The simplest populations in which to study interactions between plants are monocultures, where differences between individuals are reduced to those within the phenotypic range of one species. As we shall see, this still leaves plenty of scope for differences between individuals, because plants are so genetically variable and phenotypically plastic. In order to reduce still further the complexities that this creates, experiments are often carried out on monocultures of even-aged plants. Because plantation forestry and much of agriculture and horticulture operate with even-aged monocultures (so long as weeds can be kept out!), such populations are of practical as well as of theoretical interest. In fact, as we shall see, plant population biologists have learnt a great deal from experiments with crops.

4.2 Yield and density

As plants grow they occupy more and more space, and sooner or later the gaps between plants are filled and they begin to interfere with each other's access to resources. The total yield (weight of plant material per unit area) then approaches a ceiling whose level is determined by the species of plant and its growing conditions but which, above a threshold density, is *independent of plant density*. When the final yield of a crop is plotted against the density of plants sown we usually get a curve like that shown in Fig. 4.1(a). The characteristic way in which this curve

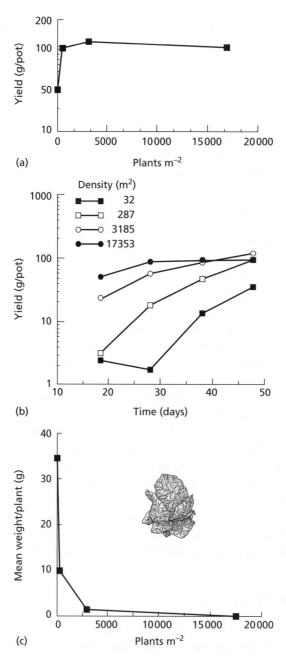

Fig. 4.1 The growth of a cos lettuce crop (pictured) expressed as (a) final yield from pots sown at different densities (b) yield at different dates of harvest for different sowing densities and (c) mean weight per plant at final harvest for different densities (data from Scaife & Jones 1976).

flattens off is referred to as the **law of constant final yield.** Remember that all plants need time to grow, so the full description of how yield changes with density should have a time dimension as well. If the crops planted at each density are followed through time, the relationship between total yield and time follows a similar pattern to the yield/density curve (Fig. 4.1b). The higher the initial density of plants, the sooner a constant final yield is reached.

The asymptotic relationship between total yield Y and density N (Fig. 4.1a) can be described by a simple equation of the form:

$$Y = w_m N(1 + aN)^{-1}, \tag{4.1}$$

where w_m is the maximum potential biomass per plant and a is the area necessary to achieve w_m. At high densities Y is proportional to $w_m N(aN)^{-1}$ and density cancels out of the relationship to give final yield a constant value of $Y \propto w_m a^{-1}$. If final yield is constant it follows that mean yield *per plant* (w) must be inversely related to density (Fig. 4.1c). In fact, $w = Y/N$ so by dividing both sides of Eqn 4.1 by N we get an expression for the relationship between final yield per plant and density (Watkinson 1980):

$$w = w_m(1 + aN)^{-1}. \tag{4.2}$$

The curve of yield per plant corresponding to Eqn 4.2 has the hyperbolic shape shown in Fig. 4.1(c).

The relationship between yield and density has been studied for many different crops, and while most conform to the law of constant final yield when the weight of the whole plant is measured (Fig. 4.2a,b,d), the yields of plant parts such as maize grain (Fig. 4.2c) or parsnip roots graded by size (Fig. 4.2d) may *decrease* at high density. Notice that the precise shape of the yield/density curve may vary with growing conditions (maize in Fig. 4.2c), and that the ceiling on the yield/density curve may vary from year to year (potatoes in Fig. 4.2b). Variation in the ceiling value of yield is the result of variation in the maximum possible plant size w_m, but to allow for other *shapes* of yield/density curve we need to modify Eqns 4.1 and 4.2. This can be simply done by substituting the exponent -1 in the equations by a variable $-b$.

$$Y = w_m N(1 + aN)^{-b} \tag{4.3}$$

$$w = w_m(1 + aN)^{-b}. \tag{4.4}$$

Equation 4.4 describes how mean plant weight (w) will vary between plots or populations at different densities at *a particular moment in time* (e.g. Fig. 4.1a). The relationship of w to density is called the **competition–density effect** or C–D effect (Firbank & Watkinson 1990).

The parameters w_m, a, and b all change during the growth of a crop (Watkinson 1984). When $b = 1$, these equations produce the curves appropriate for constant final yield (Fig. 4.1a,d). Biologically, b may be interpreted as the rate at which the

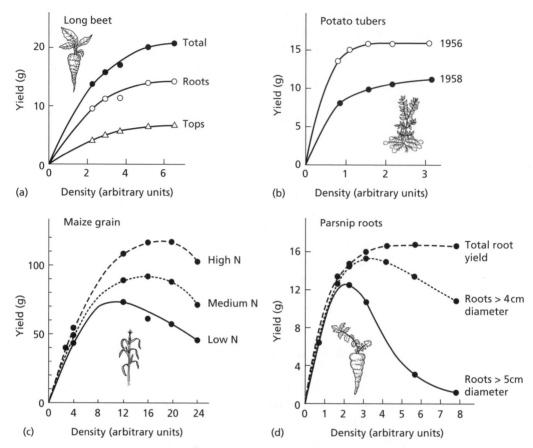

Fig. 4.2 Yield/density relationships in four crops (from Willey & Heath 1969, after various authors).

effects of competition change with density (Vandermeer 1984). Figure 4.3 illustrates the effect of varying the value of b between 0.3 and 2, on total yield and on yield per plant. When $b = 1$, w is linearly related to the reciprocal of density ($w = Y/N$), but at lower and higher values of b this relationship is nonlinear. Compared with the $b = 1$ scenario, at values of b greater than unity yield per plant diminishes more as density rises, and at values less than unity yield per plant diminishes less as density rises. We will see in Chapter 5 that these nonlinear responses to density can have interesting effects upon population dynamics.

4.3 Self-thinning

The relationships between plant density and yield described by Eqns. 4.1–4.4, treat plant density N as a quantity that varies between populations, all recorded at the same time. This is appropriate for even-aged stands of plants that will be harvested or measured together, but it does not describe fully how w will vary *through time*

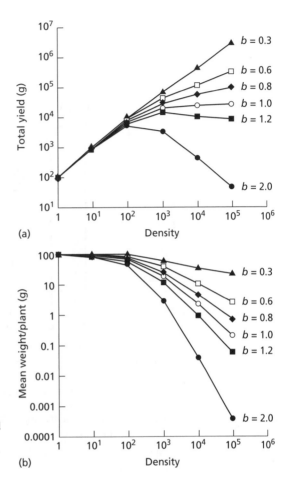

Fig. 4.3 Effect of varying the value of b in Eqns 4.3 and 4.4 upon (a) total yield and (b) yield per plant, when other variables in the equation are held constant.

in a population where N may change as a result of mortality. This may be unimportant when dealing with annual crops deliberately sown at fairly low density, by say a farmer who wants to avoid plant mortality and wastage of seed. But what of natural populations where there is no artificial limit on the starting density, or perennial populations such as those of trees where the individuals start off very small and end up giants? If Eqn 4.4 applied literally to *any* density, then an infinite density would produce infinitesimally small plants. Obviously this cannot happen, and at some point as density rises or crowding intensifies with plant growth, plants die, reducing N with time. This decrease in N due to crowding is described as **self-thinning**. Self-thinning mortality is **density-dependent**. An example can be seen in Fig. 4.4, which follows the course of a number of populations through time.

Note that as time progresses, w rises and then, at a certain point, N begins to fall. The fall in N happens soonest, and at lower values of w, in the densest plots. Although often confused with it, self-thinning is quite distinct from the C–D effect. Self-thinning describes how density and yield change *in the same plot followed*

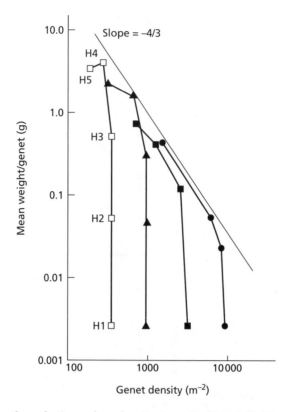

Fig. 4.4 Self-thinning in four populations of *Lolium perenne* planted at different densities. H1–H5 are replicates harvested at five successive intervals (from Kays & Harper 1974).

through time when density-dependent mortality is occurring. It sets an upper boundary on the C–D effect, which describes the relationship at *one moment in time* between mean plant weight and density *across a range of plots.* Both can be seen in Fig. 4.4.

As mortality and growth proceed, the trajectory of populations on the self-thinning graph approaches a line of constant slope. There is an empirical relationship between mean plant weight w and surviving density N that is described by the equation:

$$w = cN^{-k}. \tag{4.5}$$

To see the meanings of the terms in Eqn 4.5, we plot log w against log N (Fig. 4.5):

$$\log w = \log c - k \log N. \tag{4.6}$$

It is apparent from this equation that $-k$ is the slope of the self-thinning line in Fig. 4.5, and that log c is its intercept with the vertical axis. The term log c in Eqn 4.6 is a constant that has an empirical range of values across plants of all types between 3.5 and 5 (White 1985). Note that this is really quite wide, and that the range 3.5–5 on a log scale is nearly 10^5 when expressed on a linear scale. It is important to realize that mortality as a result of self-thinning often occurs *in the*

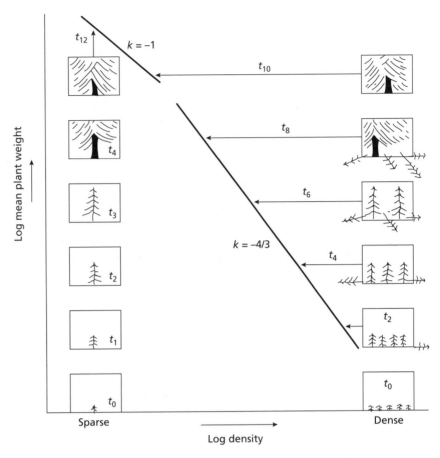

Fig. 4.5 The progress of a sparse and a dense tree population through time, illustrating the main features of the self-thinning process (modified from Silvertown 1987a).

approach to the thinning line, before it is actually reached (Osawa & Sugita 1989). The thinning line is a *boundary* and its slope (k) has a value of approximately 4/3 in many populations (Weller 1989; Lonsdale 1990; Franco & Kelly 1998). Once the line described by Eqn 4.6 is reached, the population should follow it (e.g. Fig. 4.4). While following a line of slope $k = 4/3$, N decreases but w and total yield Y ($= Nw$) increase. Eventually the population hits a yield ceiling, and then the value of k changes to 1. From then on, Y is constant and any increase in mean plant size w is accompanied by a corresponding decrease in plant density N, so that Nw does not change.

In addition to the features of the self-thinning process already mentioned, Fig. 4.5 shows that dense populations reach the thinning line before sparse ones, and that it is possible for very sparse populations to reach maximum final yield (where $k = 1$) without any density-dependent mortality whatsoever. Below the self-thinning line the relationship between yield and density is governed by the C–D effect discussed in Section 4.1. In clonal plants, genets can control rates of shoot production and the spatial distribution and density of their ramets; thus, as one might expect, densities are usually kept below the point that would cause mortality. When clonal plants do occasionally reach self-thinning densities, their behaviour appears to be similar to that of nonclonal plants (Hutchings 1979; Pitelka 1984; White 1985).

Yoda *et al.* (1963) orginally proposed that the exponent in Eqn 4.5 had a constant value of $-3/2$ and they suggested a geometric explanation for this, based upon the ratio of plant volume (the cube of linear dimensions, or l^3) to the area occupied (the square of linear dimensions, or l^2). This became known as the '$-3/2$ power law', but subsequent research has shown that the more correct value of k is $-4/3$. A line with this slope also governs how average plant mass varies with maximum density between species as different as redwoods *Sequioa* and duckweed *Lemna* (Fig. 4.6). An interesting question is why a power law with an exponent of $-4/3$ apparently governs all maximum size/density relationships in plants. In fact, Enquist *et al.* (1998) have shown that the $-4/3$ power law in plants is no different from its equivalent in animals when both are expressed as relationships with mean body mass as the independent variable. Relationships that scale with body mass are described as **allometric**. Re-expressed allometrically the proportionality $w \propto N^{-4/3}$ becomes $N \propto w^{-3/4}$ which is identical to the way maximum population density scales with body mass in many animals (Damuth 1981, 1998). The explanation for this remarkably general relationship appears to be that the $-3/4$ exponent arises from fundamental constraints on how resource use and metabolic rate scale with body size in all organisms (West *et al.* 1997; Enquist *et al.* 1998; Franco & Kelly 1998). Thus, self-thinning turns out to be just the botanical manifestation of a more general rule in ecology.

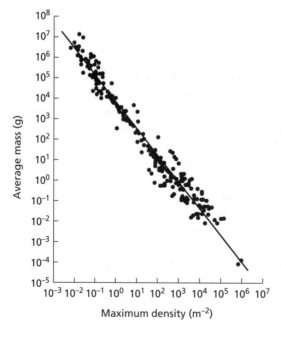

Fig. 4.6 Relationship between mean plant mass and maximum population density in 251 populations of plants, ranging from *Sequoia* to *Lemna*. The slope of the regression line is -1.34, which is not significantly different from $-4/3$ (from Enquist *et al.* 1998).

4.4 Size variation

Crowded populations typically begin life with a more-or-less normal distribution of plant size, which quickly skews to an L-shaped distribution, with many small individuals and a few large ones (Fig. 4.7). When self-thinning occurs it is the smallest plants that are the first casualties. This of course contributes to the increase in mean plant weight (w) during self-thinning. In even-aged stands of trees where self-thinning continues over a long course of time, mortality may totally remove the smallest individuals, producing a more symmetrical size distribution once more (Fig. 4.8).

Why does size variability change in this way as plants grow? There are two simple alternative explanations. The first assumes no intraspecific competition and is based only on the observation that plant growth is exponential during its early phase. If initial size and, hence, relative growth rate is normally distributed within a population, a rapid divergence can occur between plants with different starting sizes, turning an originally normal size distribution into a log normal, or L-shaped one. This occurs because size increases growth rate, which increases size, which increases growth rate.

An alternative explanation is that large plants suppress the growth of small ones. Of course, it is possible that the truth lies somewhere between these two alternatives, but there is a simple observation which can tell us if one mechanism is more important than the other. If the change in plant size distribution with time is a consequence of large plants suppressing small ones, it should not occur when plants are grown at low density. If, on the other hand, the change is the result merely of exponential growth then it should occur in populations of all densities (Turner & Rabinowitz 1983). In order to apply this test to the two hypotheses we

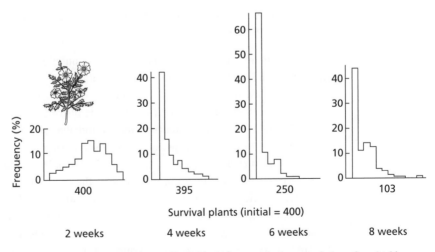

Fig. 4.7 The changing distribution of individual plant weight in a population of marigolds (*Tagetes patula*, pictured) with time. Frequencies are shown in 12 equal intervals in the range of weights present at each date (from Ford 1975).

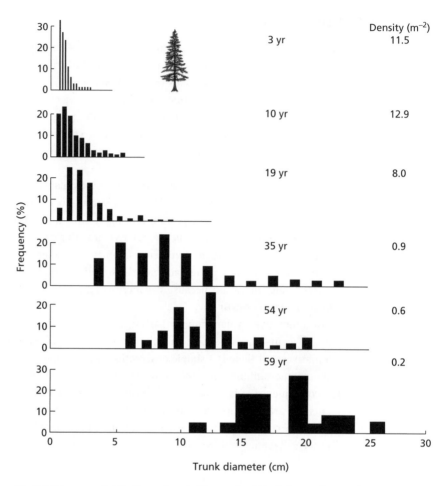

Fig. 4.8 Frequency distributions of trunk diameter for *Abies balsamea* (pictured) stands divided into 12 equal intervals (from Mohler *et al.* 1978).

need a measure of the size inequality in a plant population. The simplest effective one is the coefficient of variation (CV) of plant size, which is the standard deviation divided by the mean. (This is highly correlated with another measure of inequality called the Gini Coefficient that is also often used; see Weiner & Solbrig 1984.) Weiner and Thomas (1986) found that size inequality increased with density in 14 out of 16 experiments they surveyed. In the two tree species in their sample, *Abies balsamea* (Fig. 4.8) and *Pinus ponderosa*, size inequality fell again once self-thinning took place because of the death of small individuals.

Clearly, size inequality in plant monocultures is strongly influenced by density, and we can therefore conclude that intraspecific competition is important. A number of studies have thrown further light on the mechanism of this process by looking at the effects of neighbours on each other (Schwinning 1996; Schwinning & Weiner 1998). The effects of competitors on each other depend upon their relative size, but it is important to know precisely how competitive effects vary with size. When the effect of neighbours on each other is proportional to their relative sizes, competition is said to be **relative size symmetric**. When this kind of symmetric competition occurs, a plant weighing 100 g should reduce the growth of a 50–g neighbour by twice as much as the effect of the 50–g plant on the 100 g one. When the effect of neighbours on each other is disproportionate to their relative sizes, competition is said to be **relative size asymmetric**, or just 'asymmetric' for short (Weiner 1990).

The extreme case of asymmetric competition is where competition is completely **one-sided** and the growth of a small plant is reduced by the presence of a larger neighbour, but the larger plant is unaffected. Cannell *et al.* (1984) found an example of one-sided asymmetry in plantations of sitka spruce *Picea sitchensis* and lodgepole pine *Pinus contorta*. Tall trees suppressed the height growth of shorter neighbours, but were themselves unaffected. However, height increment is an insensitive measure of growth for suppressed plants in dense stands because they become etiolated in response to shade. Using growth in stem volume rather than height as a measure of neighbour effects, Brand and Magnussen (1988) found competition in plantations of red pine *Pinus resinosa* growing in southern Ontario, Canada, to be two-sided, but still asymmetric.

In what circumstances is plant competition asymmetric? Asymmetric competition is most likely to occur when light is the resource limiting plant growth, simply because light is directional and taller plants shade shorter plants far more than *vice versa*. Wayne and Bazzaz (1997) estimated the light received by individual seedlings of yellow birch *Betula alleghaniensis* growing in competition with one another at high density and found that, after only 6 weeks growth, light interception per unit biomass increased steeply with seedling size, indicating the presence of strong size-asymmetric competition for light (Fig. 4.9).

In a study with cultivated carrots, Benjamin (1984) experimentally lowered the canopy of some of the plants in a population by pinning their normally erect leaves to the ground, while leaving other plants in the population with erect leaves. The root and shoot growth of pinned and unpinned plants in mixtures of the two treatments was compared with growth in populations with all plants pinned and with no plants pinned. Neither root nor shoot size at harvest was significantly affected by treatment when all plants in a treatment were treated the same way (pinned or unpinned), but in mixtures pinned plants performed significantly less well than unpinned ones, producing roots nearly 60% smaller than unpinned competitors. Asymmetry in competition for light was therefore crucial in determining which plants grew large and which remained small. A plant that, by being an exception, proves the rule that asymmetric shoot competition produces size inequality is the small annual salt marsh herb *Salicornia europaea*. This plant has no leaves, but only a succulent stem and branches. When growing in dense stands of $10\,000\,\text{m}^{-2}$ or more it branches very little. By weeding out plants, Ellison (1987a) thinned field populations to a range of densities at the beginning of the growing season and compared their size inequality at three harvests until the season's end. Size inequality was not related to density at any harvest, and

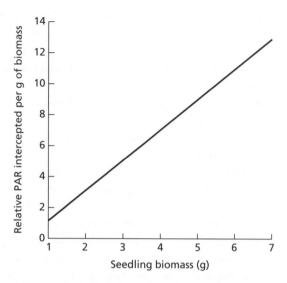

Fig. 4.9 Light (PAR, Photosynthetically Active Radiation) captured per gram of biomass by yellow birch seedlings growing at a density of 265 m^{-2} in relation to size after 6 weeks growth, showing strong relative size asymmetry (data of Wayne & Bazzaz 1997).

there was no density-dependent mortality. *Salicornia*'s peculiar growth habit apparently minimizes the type of competition between shoots that generates asymmetric relationships between neighbours.

By contrast with competition for light, competition for soil resources is usually size-symmetric (Casper & Jackson 1997; Weiner *et al.* 1997). Weiner (1986) grew the climbing herb *Ipomoea tricolor* in experimental monocultures to compare the effects of root and shoot competition. Thanks to this species' twining habit he was able to separate the effects on size inequality of competition between roots and competition between shoots. The roots of two plants could share a pot, while their canopies could be separated or, in another treatment, the canopies of plants growing in separate pots could be made to overlap by growing them up a shared stake. Root competition was much more severe than shoot competition. Regardless of shoot competition, plants grown in pots where they shared root space were smaller than plants without root competition. Nevertheless, shoot competition and not root competition was the deciding factor in size inequality. There was significantly greater inequality among plants competing for light than among those competing for soil resources. This is compelling evidence that above-ground competition is asymmetric but that below-ground competition may not be.

It has been suggested that underground connections between plants might ameliorate the effect of root competition on small plants by transferring nutrients from larger ones. This does occur in some clonal species (Section 10.5.2) and it has been observed to reduce size inequality through time between connected ramets of the common reed *Phragmites australis* (Ekstam 1995). In nonclonal plants nutrient transfer can occur either directly via root grafts, or indirectly via mycorrhizal connections. Root grafts are common in forest trees, and intact neighbours can support the growth of stumps for several years after felling (Graham & Boorman 1966). The vast majority of vascular plants have mycorrhizal associations with fungi, and a large group of these–the endomycorrhizal fungi–tend to be nonspecific in their host associations. The network of mycelia in the soil has been shown to channel mineral nutrients between plants of different size and different species (see Section 8.3.3). Nonetheless, rather than equalizing intraspecific competition between neighbours, as one might expect, mycorrhizas appear to increase inequality of size and/or fecundity when mycorrhizal populations are compared with nonmycorrhizal controls (e.g. Shumway & Koide 1995; Moora & Zobel 1998; Facelli *et al.* 1999). This is presumably because the benefit of mycorrhizal infection to larger plants causes increased shading of smaller ones that outweighs any nutritional benefits small plants may receive from mycorrhizal infection.

4.5 The influence of neighbours

The relationships described so far in this chapter have been presented in terms of population density, which averages out local interactions. However, to understand better how plants interact in crowded populations we now focus on individuals and look at how they perform in relation to their local neighbourhood.

4.5.1 The ecological neighbourhood and intraspecific competition

By analogy with the concept of the genetic neighbourhood (Section 3.3.5), a plant's **ecological neighbourhood** is the local area around it which contains all the conspecifics that significantly influence its size, and consequently its fecundity and survival (Antonovics & Levin 1980). Density relationships, such as the hyperbolic curve that shows how mean individual plant weight varies with density in lettuces (Fig. 4.4c) as described by Eqn 4.4, reflect local interactions averaged over all the ecological neighbourhoods in a population. So, for example, in experimental populations of *Arabidopsis thaliana* Silander and Pacala (1985) found a hyperbolic decline in the number of seeds produced per plant and the number of neighbours within a 5-cm radius (Fig. 4.10) which could be fitted to an equation essentially similar to Eqn 4.4:

$$s = s_m(1 + cn)^{-1}, \tag{4.7}$$

where s is mean seeds per plant, s_m is the maximum number of seeds produced by a plant with no neighbours, c is a constant and n is the number of neighbours. Enlarging the radius beyond 5 cm or reducing it below 3 cm resulted in a poorer fit of Eqn 4.7. The 5-cm radius defined the size of the ecological neighbourhood in this population of *A. thaliana*. In this particular experiment, weighting n by the distance between neighbours and the focal plant did not improve the fit of Eqn 4.7. Other studies of neighbourhood competition have variously weighted n

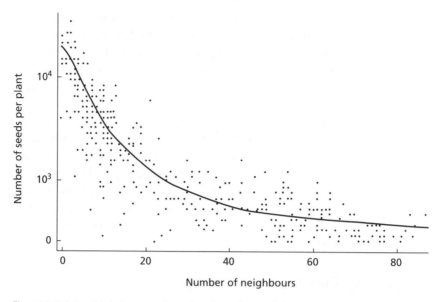

Fig. 4.10 Relationship between seeds produced per plant and the number of neighbours within a 5-cm radius in an experimental population of *Arabidopsis thaliana* (from Silander & Pacala 1985).

by the distance, mass, relative height and angular dispersion of neighbours (Bonan 1993).

Another way to quantify the influence of neighbours upon each other is to construct a map of the population and then to divide the space into polygonal 'tiles' around each individual (Fig. 4.11). In an experiment in which seeds of *Lapsana communis* were sown randomly Mithen *et al.* (1984) found that polygon area explained 60% of the variance in plant biomass, but the relationship between polygon area and plant size is more complicated than this would suggest. Franco and Harper (1988) planted seedlings of the annual herb *Kochia scoparia* in density gradients, but instead of the direct relationship between local density and plant size that one might expect, they found that plant size varied in a wave fashion both along the gradients and across them. The largest plants were on the edge of the plot, these suppressed their immediate neighbours, and *their* neighbours were consequently that much bigger. This effect rippled through the population, so that the size of a plant was always negatively correlated with that of its immediate neighbours and positively correlated with the size of neighbours one plant away. Such effects result from asymmetry in competition between neighbours and are accentuated by a density gradient, but they can occur along rows of plants sown at equal spacing too.

Waller (1981) mapped ramets of four *Viola* species in natural populations in New England, USA, and Canada, and found that the size of individuals was negatively correlated with the size and number of neighbours in only 2 out of 11 populations. This suggests that neighbours may have little influence on each

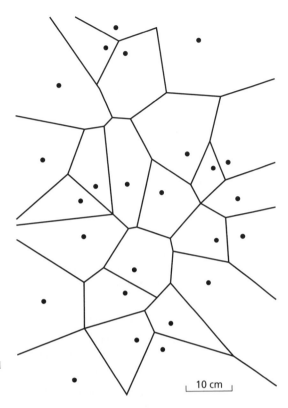

Fig. 4.11 A map of a plant population. Polygons were constructed around each individual by connecting all adjacent plants by a chord, bisecting each chord with a perpendicular line, and extending the perpendiculars till they met (from Mithen *et al.* 1984).

10 cm

other; however, that conclusion would be premature because asymmetric competition can cause small initial differences between competitiors to become magnified, so introducing nonlinearity into their interrelationships. A small difference in the time of seedling emergence can be hugely magnified by asymmetric competition, if seedlings are crowded. Also studying a species of violet in New England, Cook (1980) marked two cohorts of seedlings of *Viola blanda* that emerged 15 days apart in the same plots. The older plants remained consistently larger than the younger ones over the following three years and, during an episode of severe mortality at the end of this period, older plants suffered significantly less mortality than plants only 15 days younger. Mortality depended on size rather than age. The fateful event that made some plants large and others small was the late emergence of one cohort in the midst of another three years prior to the experiment.

The effect of **priority of emergence** on size and subsequent fate is an extremely common phenomenon that has been observed in numerous populations (Miller 1987). For example, emergence delays of only a day or two affected the survival of the small sand dune annual *Androsace septentrionalis*, and 10 days made a difference to *Tragopogon heterospermus* (Fig. 4.12) (Symonides 1977). Suggestive though these patterns are, that older seedlings are suppressing younger ones, an experiment is needed to rule out the possibility that there is some other cause, such as a deterioration in the weather, that confers a disadvantage on

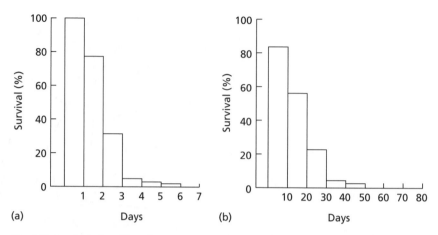

Fig. 4.12 Survival of seedlings from successive cohorts of (a) *Androsace septentrionalis* emerging at daily intervals and (b) *Tragopogon heterospermus* emerging at 10–day intervals in natural populations on dunes in Poland (from Symonides 1977).

late-emerging seedlings. Weaver and Cavers (1979) carried out such an experiment in which they attempted to separate the effects of emergence *order* (who was first) from emergence *date* on the survival of successive cohorts of the dock *Rumex crispus*. Mortality was greatest in later cohorts, and emergence order was more important than emergence date in determining the contribution of each cohort to the final population.

The effect of emergence order on plant fate is largely a result of the influence this has on plant size. This is greatly accentuated by cumulative differences in growth initiated by asymmetric competition between large plants and small ones. Seedlings that emerge later are suppressed by competition from those that emerge earlier, which have a greatly disproportionate advantage because the competition for light is strongly asymmetric.

Taken at face value, then, it would seem that the effects of neighbours should be detectable in the field, even where emergence order determines the size of individual plants, because it is size that really matters. However, things are not as simple as this. Although emergence order may *cause* size differences, asymmetry in competition for light makes the interaction between emergence order and size a strongly nonlinear one.

Waller's (1981) finding that the size of neighbours was only rarely negatively correlated is not unusual (Wilson & Gurevitch 1995). In experiments with the grass *Dactylis glomerata* sown in a random arrangement Ross and Harper (1972) found that emergence date accounted for 95% of the variance in final harvest weight of individuals, and that there was no consistent relationship whatsoever between the size of an individual and the size and number of its neighbours. They suspected that the effects of neighbours were obscured by the effects of emergence time.

Using a computer simulation of neighbourhoods in which competition from neighbours and emergence time were *known* to be the only variables affecting individual plant size, Firbank and Watkinson (1987) found that the variance in plant size explained by these two factors *decreased* as the asymmetry of competitive interactions and, in particular, the variance in emergence time *increased*. This confirms Ross and Harper's (1972) suspicion that the effect of neighbours on each other is obscured by a nonlinear interaction between emergence order and competition. Also using theoretical models of monocultures, Hara and Wyszomirski (1994) found that increasing competitive asymmetry decreased the influence of initial spatial patterns upon the ultimate size-structure of a population and weakened the fit of neighbourhood models such as Eqn 4.7.

Although emergence time is undoubtedly important, competition may also magnify other causes of initial difference in plant size. In experiments with velvetleaf *Abutilon theophrasti* sown in heterogeneous and homogeneous seed beds at three densities, Hartgerink and Bazzaz (1984) found that aspects of soil heterogeneity explained up to 76% of the variance in final seed output of plants, but they found no effect of neighbours, although there must have been competition between them. The effect of soil heterogeneity in this experiment was found to be greater than the effect of neighbourhood competition on plant size in experiments with the same species by Pacala and Silander (1987). In general, the goodness-of-fit of neighbourhood models varies enormously between studies, accounting for between almost 0% and over 90% of the variance between neighbourhoods in plant performance (Bonan 1993; Hara & Wyszomirski 1994). Hara and Wyszomirski (1994) suggest that this is the consequence of variation between studies in the degree of competitive asymmetry.

The neighbourhood approach to competition has thus proved disappointing in reliably predicting the performance of individual plants in monocultures. For example, Pacala and Silander (1990) found that neighbourhood models of *A. theophrasti* were no more accurate predictors of this annual's population dynamics than were models based upon density. However, as we shall see in Chapter 8, the story is quite different for the neighbourhood approach to *interspecific* competiton.

4.5.2 Mortality and spatial pattern

We already know, from looking at size–frequency distributions, that density-dependent mortality isn't random but instead it affects small plants in preference to large ones. Is this mortality spatially random, or are some small plants more susceptible than others? The spatial pattern of shrubs, such as the creosote bush *Larrea tridentata* growing in deserts, has been studied by many authors interested in intraspecific competition (e.g. Woodell *et al.* 1969; Phillips & MacMahon 1981; Wright 1982). The canopies of adjacent shrubs rarely meet, although their extensive root systems often do, so competition between neighbours, when it occurs, must be for below-ground resources—most probably water. The argument goes that the intensity of competition between individuals should be greatest between the closest neighbours. If this is intense enough, then individuals with close neighbours will have higher mortality than individuals with more distant ones, and over time the distribution of individuals in the population will acquire a more regular spatial arrangement.

The crucial word in the last sentence is *more*. A once-off analysis of pattern can show regularity, but unless you know from what previous pattern this developed, any inference about the process that produced it is severely weakened; a failure to find a regular pattern does not mean that competition is absent. Ideally, spatial pattern should be measured before and after an episode of mortality. The distributions of young and old, or live and dead plants in desert populations have been compared by a small number of authors who have usually found patterns that suggest competitive interactions. For example, Wright (1982) found that living plants of the perennial *Eriogonum inflatum* were spaced more regularly than in a sample that included dead individuals. Most studies have not made such a comparison. In a review of the extensive literature on plant competition and spacing in deserts, Fowler (1986) found that only 7 out of 33 species studied exhibited regular distributions, and in 5 of these 7, some populations were clumped or random. By contrast, it is noteworthy that many studies have found positive correlations between plant size and distance to nearest neighbour (Fowler 1986).

As one might expect, self-thinning, which is driven by above-ground competition for light, alters the pattern of a population, tending to produce a more regular spatial distribution with time (Cannell *et al.* 1984; Mithen *et al.* 1984; Bonan 1988). Because this equalizes the space around surviving plants, the correlation between plant size and neighbourhood effects must become weaker, making it difficult or impossible to detect the influence of neighbours in retrospect if only living plants can be mapped. Kenkel (1988) compared the spatial distributions of living and dead trees in a natural stand of Jack pine *Pinus banksiana* at a site in Canada. The population was even-aged and had been recruited after a fire some 65 years before the study. Dead stumps of *P. banksiana* are slow to decay, so although there was no remaining clue as to the location of the first and smallest trees to die, there were still 916 stumps and standing dead trees to 459 living ones. The spatial distribution of the whole population, living and dead, was random but the distribution of survivors was highly regular and the distribution of dead trees was clumped. Surviving trees exerted a detectable influence on the fate of neighbours within a 3.5–m radius, which on average included six other trees. A polygonal tile (Fig. 4.11) has an average of six sides, suggesting that trees were competing only with immediately adjacent neighbours. Similar results have been obtained for a population of *P. banksiana* followed through time (Kenkel *et al.* 1997).

Another comparison of spatial patterns at two stages of the life cycle was made by Sterner *et al.* (1986) for four species of trees, two of them palms, in a tropical evergreen forest at La Selva, Costa Rica. By extreme contrast with the *P. banksiana* study site, La Selva is a species-rich forest and populations of the species studied contained individuals of all ages. In three of the four species, adults were distributed more regularly than juveniles, suggesting that mortality was spatially nonrandom. To test for this, a random pattern of juvenile mortality was applied to the map of living plants, and the adult spatial distribution was then compared with this simulation. The comparison showed that the spatial distribution of adults was more regular than that of randomly thinned juveniles in the same three species. Although all of the species in this study were abundant, every tree must have had neighbours of other species, greatly complicating the possible neighbourhood interactions. The fact that an analysis of pattern that ignored other neighbour species found evidence in 75% of cases that mortality was spatially nonrandom is strong evidence for the role of intraspecific competition, be it real or apparent (Connell 1990). We will return to the implications of this for tropical forest diversity in Section 8.4.

4.6 Size, density and fitness

The fitnesses of individual plants are strongly related to their size. Small plants not only bear the brunt of mortality in crowded populations but also, if they survive, produce fewer seeds than large plants. If the relationship between plant size and seed production is linear, the L-shaped size distribution that develops in a dense population will create an L-shaped distribution of fecundity. Samson and Werk (1986) examined the relationship between plant size and the biomass of reproductive parts in 30 species, most of them annuals, and found that the relationship varied with species. However, in perennials fecundity per unit plant weight often increases with plant size, with large plants producing more seeds per unit biomass than small ones (Fig. 4.13). When this is the case, the hierarchy of plant fecundity will be even stronger and more unequal than the hierarchy of plant size.

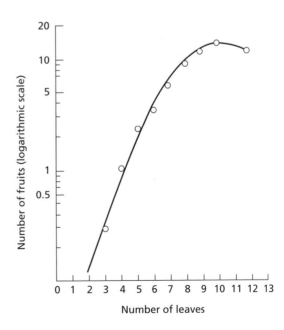

Fig. 4.13 Relationship between plant size and fruit production for *Viola fimbriatula* (from Solbrig *et al.* 1988).

Hierarchies of plant fecundity created by crowding reduce the effective population size N_e because a small fraction of the population makes a disproportionately large contribution to future generations. This has important genetic consequences. Levin and Wilson (1978, Wilson & Levin 1986) used simulation models to compare the effects of L-shaped and Poisson fecundity distributions on a number of genetic processes. They found that a fecundity hierarchy resulted in faster response to directional selection, faster loss of rare alleles from the population and a greater propensity for genetic drift.

Heywood (1986) estimated the ratio N_e/N for 34 species of annuals, using estimates of the variance in seed crop size to take account of the unequal genetic contributions of different individuals, and compared these N_e/N values with those expected from a Poisson distribution of contributions to the seed pool. In most cases, allowing for the observed variance in seed crop size reduced N_e/N by more than half and substantially increased the potential for genetic drift. However, N_e/N was probably still overestimated because variance is a poor measure of the inequality in crop size when this is not normally distributed. Dodd and Silvertown (2000) calculated the effect of variance in lifetime fecundity on N_e/N in the tree *Abies balsamea* and found this to be much less than in Heywood's annuals. This appeared to be because self-thinning and the presence of a size threshold for reproduction prevented small individuals from reaching the reproductive size class (Fig. 4.8). Above a threshold size, fecundity was linearly related to size (diameter at breast height = 1.4 m); thus, the failure of small trees to reproduce resulted in a distribution of fecundity among reproductively mature trees that was not greatly skewed. Annuals tend not to have a size threshold for reproduction and thus show a greater skew, as found by Heywood (1986). Variance in male reproductive success has so far been ignored in such calculations, although it may well be higher than for female reproductive success because of the possibility of an additional component of variability resulting from differences in pollination success (Chapter 9).

A plant's response to density is affected by its shape, size and pattern of growth. If these characters are heritable, density-dependent processes may be selective (Bazzaz *et al.* 1982). Davy and Smith (1985) transplanted *Salicornia europaea* between subpopulations in the upper and lower levels of a salt marsh in Norfolk, UK, and found that, among several genetic differences between the two levels, plants showed different responses of fecundity to density. Transplants from either level to the other had lower fitness than natives replanted into home ground, and the degree of disadvantage varied with density. Families of *Salvia lyrata* from different parents growing in a meadow in North Carolina were planted back into the home turf at a range of densities by Shaw (1987), who found evidence for differential growth and fecundity in response to density. These experiments suggest that intraspecific competition may be a source of natural selection. However, the evidence is weak that plant size, which is a major determinant of individual success in crowded populations, has high heritability; if it is not heritable, then selective differences will not lead to evolutionary change.

Gottlieb (1977) compared the allozyme genotypes of large and small individuals in a dense, natural population of the annual *Stephanomeria exigua* and found no evidence that size correlated with genotype. Antlfinger *et al.* (1985) estimated the heritability of characters associated with plant size and shape in *Viola sororia* and found heritabilities between 0.09 and 0.39. Because these measures were made in a greenhouse environment they tell us little about the heritability of plant size in the field where it matters. A field study of the annual *Erigeron annus* found that the broad-sense heritability of seedling size was less than 0.1 (Stratton 1992). Although this is a low value, it could lead to large changes if selection were continued in a given direction over several generations.

Another important effect of density is that dense plant stands are particularly susceptible to fungal disease. In experimental mixtures of susceptible and resistant genotypes of skeleton weeds *Chondrilla juncea*, Burdon *et al.* (1984) found that the presence or absence of the rust *Puccinnia chondrillana* determined which genotype was at the top of the size hierarchy, and which at the bottom (Fig. 4.14). Interestingly, in disease-free mixtures the resistant genotype was inferior to the susceptible one, demonstrating that there was a '**cost of resistance**' in terms of size (Harlan 1976; Bergelson & Purrington 1996). A cost of resistance is a lowered relative fitness of resistant genotypes in populations no longer exposed to disease. It would result in selection against them unless disease reappears, when selection will change to favouring resistant plants. This is an example of natural selection that changes in direction over time.

4.7 Population regulation: density dependence

We have seen that crowding affects individual plants by limiting their growth, that this determines mean plant size and size *variation*, and that in turn size may affect individual mortality and fecundity. We will now see that density also has an important influence on population dynamics.

As we saw in Chapter 1, the simplest statement one can make about how numbers in a population change between one occasion and the next is that the

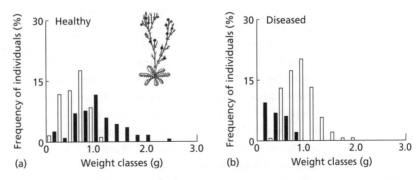

Fig. 4.14 The frequency distribution of plant size for resistant (open columns) and susceptible (filled columns) genotypes of *Chondrilla juncea* (pictured) in (a) healthy and (b) diseased stands (from Burdon *et al.* 1984).

number at time $t+1$ is equal to the number at time t plus the number born over the time interval (plus any immigrants), minus the number that died (minus any emigrants). Using B for births, D for deaths, I for immigrants and E for emigrants, this gives us the equation presented at the begining of Chapter 1:

$$N_{t+1} = N_t + \text{B} - \text{D} + \text{I} - \text{E}.$$

If the time-step is one year, the ratio N_{t+1}/N_t calculated from the values in Eqn 1.1 is the **annual rate of increase** (λ) (Chapter 1). When the population is stable, neither increasing nor decreasing in size, by definition $N_{t+1} = N_t$ and therefore $\lambda = 1$. When this is true it follows that $\text{B} - \text{D} + \text{I} - \text{E} = 0$, and of course $\text{B} + \text{I} = \text{D} + \text{E}$. In fact, the number of births and deaths in a population tends to be related to N_t, so it is better to express these parameters in terms of *rates*, or in other words numbers per head of population per time interval. We will use the lower case letters b, d, e, i to signify *rates* of birth, death, immigration and emigration, respectively. They are defined as follows:

$$b = \frac{\text{B}}{N_t} \qquad d = \frac{\text{D}}{N_t} \qquad e = \frac{\text{E}}{N_t} \qquad i = \frac{\text{I}}{N_t}$$

Using these definitions Eqn 4.1 can be rewritten:

$$N_{t+1} = N_t(1 + b - d + i - e) \tag{4.8}$$

Divide both sides of Eqn 4.8 by N_t and, remembering $\lambda = N_{t+1}/N_t$, you will find that:

$$\lambda = 1 + b - d + i - e. \tag{4.9}$$

When $b + i > d + e$ the population will increase at a rate λ each generation. Now, consider what will happen if λ has a fixed value. First, if $\lambda > 1$ the population will increase indefinitely at an exponential rate; this is obviously impossible because such a population must eventually run out of space. Secondly, if $\lambda < 1$ the population will decline rapidly to extinction. Thirdly, if $\lambda = 1$ exactly the population will, according to this equation, remain stable in size. However, for the latter case, random variation in b and d make it impossible for λ to remain forever at unity. To stay at this value $b + i$ must *always exactly* match $d + e$, but this is very unlikely in reality. A severe attack by some herbivore or disease will raise d or decrease b which, without a compensatory change by the other parameters, will eventually lead to extinction. Therefore, we must conclude that *no fixed* value of λ, be it more than one, less than one *or* exactly one can adequately account for how real populations behave in the long term. On the other hand, for most populations the *average* value of λ measured over long periods should be unity, unless the population is heading for extinction. Given the hazards of life, how do populations hang on with an average $\lambda = 1$?

There are two ways in which populations can be cushioned from the consequences of catastrophes that alter b and d. Either (i) by the immigration of plants from areas not affected by the catastrophe; or (ii) by changes in fecundity or mortality that compensate for the losses caused. Compensatory changes operate in a **density-dependent** manner, altering the value of λ so that overall density is **regulated** within upper and lower limits.

We have already seen many examples of how the components of λ, that is b and d, change with density. For changes in b and d to *regulate* a population they *must* be density dependent. Density-dependent mortality increases the *per capita* death rate as density rises and density-dependent fecundity decreases the *per capita* birth rate as density rises, with the net result that a wide range of starting densities is reduced to a narrower range of final population size (Fig. 4.15).

Notice that the experimental populations of *Bromus tectorum* in Fig. 4.15(b) obey the law of constant final yield: all populations produced about the same total number of seeds, regardless of the sowing density. Constant final yield and population regulation are simply different facets of a single process: the response of plants to density. There is an important distinction between the two, however, because final yield is expressed in terms of plant size (or weight), whereas the effects of population regulation depend not upon biomass, but upon seed production. As long as seed production remains a constant fraction of total biomass this distinction is unimportant, but in some plants the fraction of biomass allocated to

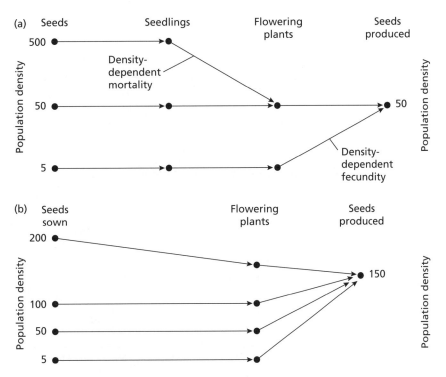

Fig. 4.15 Population regulation by density dependent mortality and fecundity in (a) idealized plant populations (b) experimental populations of an annual grass *Bromus tectorum* (data from Palmblad 1968, figure from Silvertown 1987).

reproduction alters with plant size (Samson & Werk 1986). Intraspecific competition is not the only source of density dependence, and *any* factor that operates in a density-dependent manner can regulate a population. Infection by fungal diseases, such as damping-off caused by *Pythium* spp. in seedlings, is commonly density-dependent (Fig. 4.18c) (Burdon & Chilvers 1975, 1982). Packer and Clay (2000) found that seedlings of black cherry *Prunus serotina* growing in woods in Indiana were killed by *Pythium* spp. in the soil when growing at high density beneath adults, but survived better at low density or far from adults.

The effectiveness of an alteration in birth or death rates in regulating a population depends upon how strongly these rates react to density. This can be determined from a graph of *b* or *d* against density. Large ranges of density are normally required to show an effect, so double-log axes are used. If changes in vital rates are density-independent, the graph will be a straight line of zero slope (Fig. 4.16a). The intercept of this line on the *y*-axis tells you what the average value of the density-independent rate is. If changes in *b* or *d* are density-dependent, the graph will have a positive slope for mortality, or a negative slope for fecundity. The strength of the regulatory effect is determined by the absolute value of the slope (the steepness) of the line (Fig. 4.16b–d). It is of course possible for birth or death rates to be inversely density-dependent also (Fig. 4.16e), for

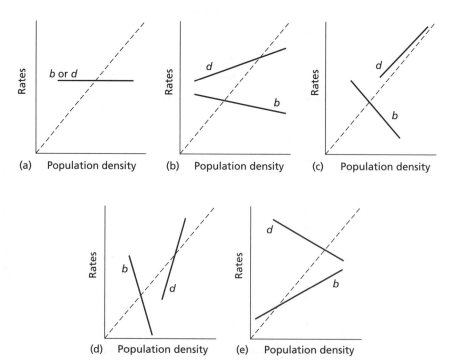

Fig. 4.16 Birth and death rates that are (a) density-independent, (b) density-dependent and undercompensate, (c) density-dependent and exactly compensate, (d) density-dependent and overcompensate, and (e) inversely density-dependent.

example if seed set depends upon cross-pollination and this is ineffective at low density.

Only in the biologically highly unlikely case when density dependence *exactly compensates* for changes in density is a population perfectly regulated and will it exhibit constant population size between one generation and the next. A population where density dependence is not exactly compensating may fluctuate from one generation to the next, but this does not mean that the population is not regulated.

A population is at **equilibrium** when the death rate and birth rate are equal. The equilibrium density of a population can therefore be determined from the intersection of the death rate and birth rate curves in Fig. 4.17. As this figure shows, a change in one parameter by a fixed amount, for example the density-independent birth rate, can lead to widely different population densities depending upon the degree of regulation (the slope of the curve) by the other (e.g. the death rate).

We have argued for the existence of population regulation from first principles, on the grounds that unregulated populations would go extinct. In the field there are essentially two ways that population regulation can be detected. The first is by looking for changes in b and d that correlate with natural fluctuations in population density through time or differences in density between plots (Fig. 4.18). Because regulated populations are cushioned against changes in density, observation alone is not a very effective way of measuring density dependence because the range of observable densities will be small. It is paradoxical but true that density dependence is most difficult to detect in those populations that are regulated most strongly.

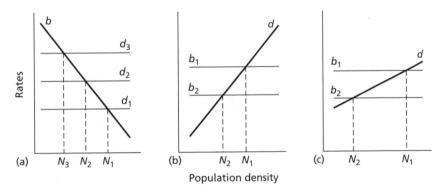

Rates

(a) N_3 N_2 N_1 (b) N_2 N_1 (c) N_2 N_1

Population density

Fig. 4.17 A variety of ways in which birth and death rates interact to determine equilibrium population size N. (a) When the birth rate is density-dependent, a density-independent change in death rates (d_1 to d_3) will shift N. (b) A density-independent change in birth rates (b_1 to b_2) when the death rate is density-dependent will also change N, but (c) if the strength of density dependence (shown by the slope of d) is reduced, the same density-independent change B_1 to B_2 will have a bigger effect upon N (after Watkinson 1986).

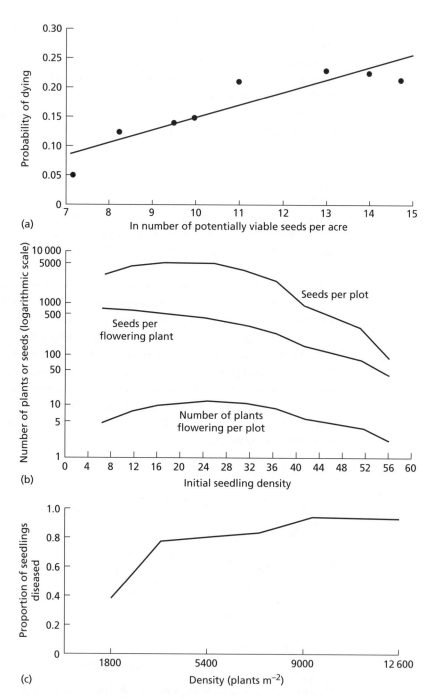

Fig. 4.18 Density-dependent responses: (a) seedling mortality in a population of sugar maple *Acer saccharum* (Hett 1971); (b) fecundity in the annual *Erophila verna* (data from Symonides 1983); (c) the proportion of *Lepidium sativum* seedlings infected by the damping-off fungus *Pythium irregulare* seven days after sowing and experimental innoculation with the disease (from Burdon & Chilvers 1975).

The second, and better, method of detecting population regulation in the field is to alter density experimentally, and to measure the response of N, b and d to the change. A good example that shows how density-dependent and density-independent factors combine to regulate a population, and how they may create different equilibria at different sites, is provided in the experimental study by Keddy (1981, 1982) of sea rocket *Cakile edentula*, an annual species growing on a sand dune in Nova Scotia. There were three sites: on the seaward side of the dune at the top of the beach, near the dune crest, and on the landward side of the dune (Fig. 4.19). A range of densities was sown at each site, and survival and fecundity were determined for each of them.

Plants were large at the seaward end where there was no competing vegetation and where there was decaying eelgrass that supplied nitrogen. Here, increasing density resulted in a plastic adjustment of plant size and fecundity without causing mortality. Plants were small at the landward end and many of them were too small to reproduce. At this site, increasing density did not alter fecundity but did increase mortality. We shall look at further examples of population regulation in the general context of plant population dynamics in the next chapter.

4.8 Summary

Most population models are based on the assumption that **space** is the primary resource limiting plant growth. At the level of the individual, a plant interacts with others in its **neighbourhood,** but **density** (*N*, plants per unit area) usually gives a good guide to the average behaviour of plants in a crowded population. At high density the total yield from a crop tends to approach a constant value. This is known as the **law of constant final yield** (Eqn 4.1), and from it we can develop an expression (Eqn 4.4) to describe how mean plant weight (*w*) varies between plots growing at different densities, called the **competition–density effect** (or C–D effect). Intraspecific competition at very high densities leads to **density-dependent mortality** or **self-thinning.** As plants in very crowded stands grow, mean plant weight rises and mean density falls. This relationship takes the form $w = cN^{-k}$ (Eqn 4.5), where *c* and *k* are constants and the value of *k* is often in the region of 4/3. When density is expressed as an **allometric** function of mass it scales with the power $-3/4$, a value found in animals as well as plants.

Competition in crowded stands initially increases the **inequality** of size among members of the population, producing a **size-heirarchy.** When self-thinning occurs, the smallest plants are usually the first to die, and this may eventually reduce size variation. Size heirarchies tend to develop when the effect of neighbours on each other is disproportionate to their relative size, and competition is **relative size asymmetric.** Competition for light is usually asymmetric, but competition for soil resources is more likely to be **relative size symmetric.** Asymmetric

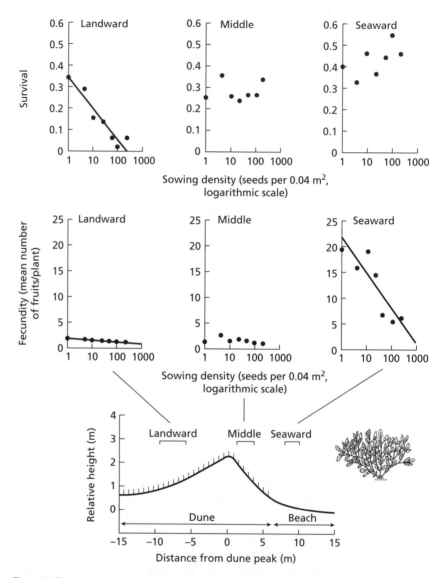

Fig. 4.19 The relationship of fecundity and mortality to density in experimental plots sown with *Cakile edentula* (pictured) at three sites on a sand dune. Statistically significant density-dependent relationships are shown by a regression line. There is no significant difference in levels of density-independent mortality at the three sites but there is a significant difference in the level of density-independent fecundity. The size and fecundity of solitary plants are far greater on the beach than elsewhere (from Keddy 1981).

competition magnifies slight initial differences between crowded seedlings, such as caused by **priority of emergence**, into much larger differences in plant size. This can weaken the correlation that might otherwise be expected between the size of neighbours, but nevertheless intraspecific competition can produce nonrandom patterns among neighbours in the field.

The **fitness** of a plant is strongly related to its size, but because seed output may be nonlinearly related to total biomass, heirarchies of seed output and fitness can be even more marked than inequalities in size itself. These fitness heirarchies reduce N_e because a few plants make a quite disproportionate contribution to the total seed production of a population. This effect may be more marked in annuals than in perennials and trees that have a size-threshold for reproduction.

Population size is **regulated** by **density-dependent** processes that affect mortality and fecundity. Density dependence often arises from intraspecific competition, but other causes such as **disease** may also operate in this manner. The absolute abundance of a population is set by the interaction of **density-independent** factors with density-dependent ones. Population regulation is most readily detected in the field by the experimental manipulation of density.

4.9 Further reading

Schwinning, S. & Weiner, J. (1998) Mechanisms determining the degree of size asymmetry in competition among plants. *Oecologia*, 113, 447–455.

Weiner, J. (1990) Asymmetric competition in plant populations. *Trends in Ecology* and *Evolution*, 5, 360–364.

4.10 Questions

1 Why do neighbourhood models of intraspecific competition tend to be less successful in practice than one might expect them to be from theory?

2 Program a spreadsheet (or write a simple computer program) to plot how total yield varies with density according to Eqn 4.3. Experiment with different values of the parameters, varying them systematically one at a time, to see how they influence the shape of the yield/density curve.

3 Discuss the ecological causes of relative size symmetry/asymmetry in intraspecific competition among plants. What are its ecological and evolutionary consequences?

4 Why is natural variation in population densities a poor method for detecting density dependence?

5 Using Fig. 4.16 as a guide interpret the results of Keddy's (1981) experiment with *Cakile edentula* shown in Fig. 4.19.

5 Population dynamics

5.1 Introduction

Plant populations are dynamic. You may think that this is the least we can tell you in a chapter with this title, but the fact is not obvious. As we have seen in the last chapter, plants occupy space to its limit and so plant communities often appear full. If you live in the temperate zone or humid tropics, a cursory glance at a lawn or pasture that you last saw a year ago will probably suggest that little has changed. Nothing could be further from the truth. Grass populations have high turnover rates, and it is likely that hardly a single tiller of last year's population is still alive. All will have been replaced. This will also be true of more than half the buttercups (*Ranunculus* spp.), and other broad-leaved herbs. Because these populations are regulated, their total numbers are relatively constant, despite the flux of individuals through them (Fig. 5.1).

Marking or mapping individual tillers or rosettes in permanent plots will tell you what the turnover rates are for ramets, and how net population sizes change. What it will not tell you so easily for most species is how many genets have been lost from the population, because it is usually difficult to tell which ramets belong to which genet. If, as is sometimes the case, seedlings are difficult to distinguish from small ramets, genet birth rates may also be difficult to estimate. The genetic variation in a plant population resides in genets, so we must know genet birth and death rates to discover whether these rates affect genotypes differentially. If they do, we have evidence of natural selection.

Unravelling the intricacies of plant population dynamics will take us three chapters. We will start by constructing a simple model of population dynamics using the demographic parameters that you met in Chapter 1 and the yield–density equation introduced in the last chapter. This will be applied to some examples of the simplest kind of population—annuals with no bank of dormant seeds. Next, we shall look at the ecology of seeds in the soil and then, in Chapter 6, we shall see how the timing of germination and the pattern of recruitment generate age structure in populations. That chapter examines populations whose dynamics are influenced by age or size structure, and Chapter 7 looks at dynamics at the regional scale, including metapopulations.

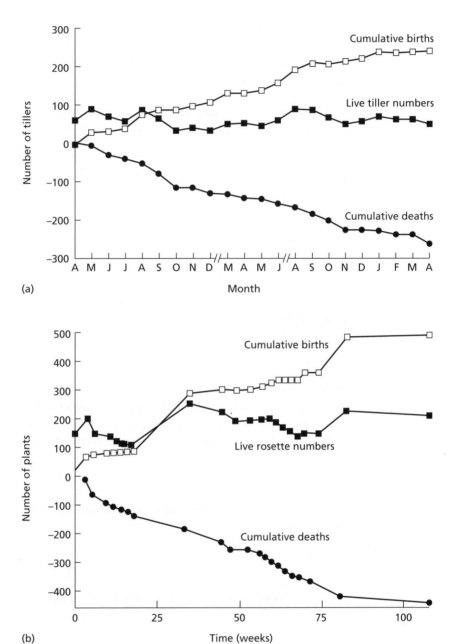

Fig. 5.1 Population flux in two grassland species: (a) *Lolium perenne* (J. Bullock, B. Clear Hill & J. Silvertown, unpubl. data) and (b) *Ranunculus repens* (Sarukhán & Harper 1973).

5.2 Demographic parameters

In Section 4.7 we took the simple model of population dynamics introduced in Chapter 1 (Eqn 1.1) and modified it to take account of the fact that births, deaths and migration are best measured as rates per head of population per unit time. This gave us demographic parameters for the birth rate (b), the death rate (d), and immigration (i) and emigration (e) rates, and equation (4.8):

$$N_{t+1} = N_t(1 + b - d + i - e).$$

Technically, this kind of equation is known as a **difference equation**. Population dynamics becomes more complicated than Eqn 4.8, but this equation is fundamental to the population models used in this book, so you should remind yourself of it if ever things look too complicated. In this section we will see that much can be learned about a population from the basic demographic parameters in Eqn 4.8.

To calculate N after x years, for a population that starts with N_t individuals, we use the equation that describes exponential increase.

$$N_{t+x} = N_t e^{rx}, \tag{5.1}$$

where e is the base of natural logarithms (not to be confused with 'e' for the rate of emigration) and r is the **intrinsic rate of natural increase** of the population. In truth there is nothing very 'intrinsic' about r, because it is affected by all the ecological influences that alter b, d, i and e. If we let $x = 1$, divide both sides of Eqn 5.1 by N_t and express the result in terms of natural logarithms, we get:

$$\ln \lambda = r. \tag{5.2}$$

Sometimes it is more useful to measure the change in N_t over the length of a generation rather than over one year. A **generation** is defined as the average time τ (pronounced 'tau') between a mother giving birth and her daughter doing the same. This is the same as the average age of mothers, which can be estimated by:

$$\tau = \frac{\sum x l_x m_x}{\sum l_x m_x}. \tag{5.3}$$

The rate of population increase over a generation is called the **net reproductive rate** R_0 and is:

$$R_0 = \frac{N_{t+\tau}}{N_t} \tag{5.4}$$

Because R_0 is measured over τ years, the *annual* rate of increase is given by R_0 root τ:

$$\lambda = R_0^{1/\tau}. \tag{5.5}$$

R_0 may also be calculated from a life table (Chapter 1) using the formula:

$$R_0 = \sum l_x m_x. \tag{5.6}$$

No population can increase exponentially for long, for even quite low values of r, or values of λ significantly greater than 1, lead to very rapid population growth which must eventually be checked by lack of suitable habitat, if nothing else. In plant populations we are most likely to see exponential population growth when a species is a new arrival in an uncolonized habitat, exemplified by the situation at the end of the Pleistocene when trees were able to reinvade higher latitudes (Fig. 5.2a), or when a weed is allowed to proliferate without control in a crop (Fig. 5.2b).

With the definitions of λ and R_0 in our hands, we are now equipped with the basic tools to begin an investigation of the dynamics of some real populations. The simplest are annuals with no seed bank.

5.3 Annuals with no seed bank

Annual plants have a generation time of one year, so $R_0 = \lambda$. The dynamics of such a simple population can be described with the equation $N_{t+1} = R_0 N_t$, but if R_0 has a fixed value greater than 1 then this relationship only applies at low density when there is sufficient space for the population to increase exponentially. To be realistic, as density rises the value of R_0 must fall. We can quite easily construct an equation to describe this, based upon what we have already learned about yield–density relationships in Chapter 4. There we used equation 4.3 to describe the density-dependent adjustment of yield:

$$Y = w_m N (1 + aN)^{-b}.$$

Seeds are a component of plant yield. The number per plant is a function of plant size and the total number is a function of plant density. Therefore, an equation of

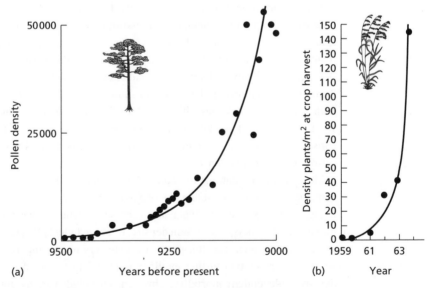

Fig. 5.2 The exponential increase of (a) *Pinus sylvestris* between 9500 and 9000 BP, when this tree was invading Hockham Mere, Norfolk, England. The abundance of *P. sylvestris* is plotted as the density of pollen grains in peat samples, which it is assumed is correlated with the historical size of the tree population. (b) *Avena fatua* infesting a barley crop at Boxworth experimental farm, Cambridge, England. (a) From Bennett (1983); (b) data from Selman (1970).

similar form to Eqn 4.3 can describe the total seed output of a population (Watkinson 1980). First, remember that $Y = wN$, so Eqn 4.3 can be written:

$$wN = w_m N(1 + aN)^{-b}. \tag{5.7}$$

Next, we translate the w terms representing plant size in the equation into seed outputs. Let us say that a plant of maximum size w_m produces s_m seeds, and that the seed output of the average plant w is s. Replacing the weight terms in Eqn 5.7 with their equivalents in seed production we get:

$$sN = s_m N(1 + aN)^{-b}. \tag{5.8}$$

The left-hand side of this equation is the seed output of a population at density N. If we assume for simplicity that there is no seed mortality and all seeds germinate, then this also represents the density of the population next year: $sN = N_{t+1}$. By placing a subscript t on N in Eqn 5.8 to make it clear which generation we are referring to, we now have an expression for the density-dependent dynamics of our population:

$$N_{t+1} = s_m N_t (1 + aN_t)^{-b}. \tag{5.9}$$

In Chapter 4 we remarked that the yield–density response for whole plants and for their parts may be different. Seeds are only a part of total plant yield, so we should not expect the parameters a and b in Eqn 5.9 to have the same values as they do in Eqn 4.3, even for the same populations. The parameters for yield and for seed production must be determined separately. Parameters for the seed production of four species of annuals, including populations of *Cakile edentula* at the three sites studied by Keddy (Chapter 4) are given in Table 5.1. Figure 4.3 illustrated the relationship between the value of the parameter b in the density–yield Eqn 4.3 and the shape of the curve relating total yield to density. Because Eqn 5.9 has the same form as this relationship, Fig. 4.3 also shows how the shape of the curve for the graph of N_{t+1} vs. N_t varies with the value of b. Refer to Fig. 4.3 now, and you will see that when $b = 1$, seed production behaves like the law of constant final yield: a constant seed output is achieved, regardless of density. This is **exactly compensating density dependence.**

Most of the populations in Table 5.1 have values of $b < 1$ and exhibit **undercompensation,** but it is notable that two populations have values of $b > 1$. Referring to Fig. 4.3 you will see that when $b > 1$ the relationship between N_{t+1} and N_t has a hump, and that, beyond a certain density, total seed yield (N_{t+1}) actually drops. We will see in a moment that this produces some very interesting dynamics. However, we need to make two further modifications to Eqn 5.9. First, we must allow for **density-independent mortality.** This can be included by assuming that only a fraction of the total s_m seeds that an uncrowded plant can produce will survive density-independent mortality; thus, we will replace s_m by s. Secondly, Eqn 5.9 only accounts for one kind of density-dependent response – the adjustment of fecundity with density – but, as we know, at very high density another response occurs too –

Table 5.1 Estimates for the parameters of Eqns 5.9 and 5.10 for four species of annuals. Calculated by Watkinson & Davy (1985) from the sources indicated below.

Species	s_m	a	b	m
Cakile edentula[1]				
Seaward site	20.8	1.17×10^{-3}	0.43	–
Middle site	1.97	7.86×10^{-3}	0.28	–
Landward site	2.21	1	0.14	–
Rhinanthus serotinus[2]	19	3.59×10^{-3}	1.11	–
Salicornia europaea[3]				
Low marsh	1100	7.57×10^{-4}	1.63	8×10^{-5}
High marsh	33.8	9.23×10^{-4}	0.36	–
Vulpia fasciculata[4]	3.05	1.93×10^{-4}	0.90	10×10^{-6}

Sources of data: [1]Keddy (1981), [2]ter Borg (1979), [3]Jefferies *et al.* (1981), [4]Watkinson & Harper (1978)

density-dependent mortality. This can be incorporated by dividing sN_t by an additional term msN_t, where m represents the reciprocal of the maximum value that density can achieve, limited by self-thinning (Watkinson 1980). Adding these extra terms to Eqn 5.9 gives:

$$N_{t+1} = \frac{sN_t}{(1 + aN)^b + msN_t} \tag{5.10}$$

Only two of the populations in Table 5.1 achieved densities sufficient to cause density-dependent mortality, but the values of m calculated for these populations will be useful as a means of investigating the effect of self-thinning upon population dynamics.

5.4 Density–dependent dynamics

In this section we introduce a simple graphical method that can be used to predict how a population whose dynamics are described by Eqn 5.10, or any function relating N_{t+1} to N_t, will behave (Fig. 5.3). N_t is represented on the horizontal axis of the graph, and N_{t+1} on its vertical axis. The diagonal line shown in Fig. 5.3(a), passes through all points where $N_{t+1} = N_t$. Any population falling on this line will by definition be at equilibrium at that point. The curved line in Fig. 5.3(a) shows the relationship between N_{t+1} and N_t for the grass *Vulpia fasciculata*, calculated by using the values for s_m, a, b and m given in Table 5.1 into Eqn 5.10. This curve, which we shall call the **recruitment curve**, can be used to predict how the population will behave through time, because the slope of the curve, which is $(N_{t+1})/N_t$, gives us the value of λ for any value of N_t. Box 5.1 describes how to use a method called **cob-webbing** to predict the behaviour of a population from its recruitment curve. You should read this now, before proceeding.

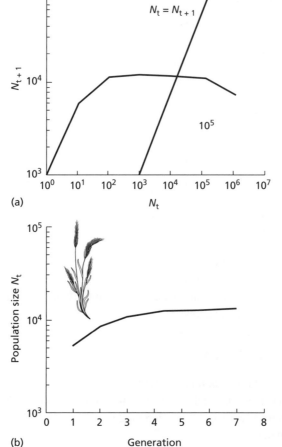

Fig. 5.3 Density-dependent population dynamics of *Vulpia fasciculata* (pictured): (a) The recruitment curve obtained by applying the values in Table 5.1 to Eqn 5.10. (b) Dynamics of a population starting with $N = 5000$ and no density-independent mortality, predicted from the recruitment curve.

Analysis of recruitment curves by cob-webbing (see Box 5.1) shows that the behaviour of a population is affected radically by the slope of the recruitment curve at the point where it is intersected by the line $N_{t+1} = N_t$. Unlike the simple examples given in Box 5.1, real recruitment curves vary in slope with density. If the recruitment curve is intersected where its slope lies between 0 and $+1$ (i.e. between the horizontal and the diagonal), as is the case for example in *V. fasciculata*, deviations in population size caused by some mortality factor such as herbivory will be followed by a smooth return to equilibrium called **monotonic damping** (Fig. 5.3). An intersection where the slope lies between 0 and -1 leads to **stable oscillations**, as is seen in *Erophila verna* (Fig. 5.6a). When the slope is less than -1, unstable oscillations occur which are described as **chaos**. Chaotic oscillations are extremely difficult to tell apart from random fluctuations caused by density-independent factors, yet this kind of analysis shows that they are entirely the result of deterministic density dependence. The dynamic behaviour of a population can be predicted from

Box 5.1: Cob-webbing the recruitment curve

For simplicity in this example we will use a recruitment curve that is a straight line, representing a population with a slope of 0.5. The equation for this recruitment curve is $N_{t+1} = C + 0.5N_t$, where C is the intercept of the curve on the vertical axis and has a value of 2.5. We will start the population off with $N_t = 9$ individuals.

Step 1: *Find N_{t+1} by drawing a vertical line from the starting value of N_t until it meets the recruitment curve. A horizontal line from this point on the recruitment curve to the vertical axis gives us the value of N_{t+1}, which is 7.*

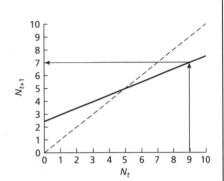

Step 2: *Locate the next value of N_t on the horizontal axis by retracing a line from $N_{t+1} = 7$ to the diagonal, down to the horizontal axis. This pinpoints the new value of $N_t = 7$.*

Step 3: *Find N_{t+2} by repeating steps 1 and 2, starting from $N_t = 7$.*

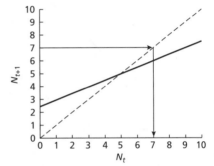

Step 4: *Locate the equilibrium point by continuing to repeat steps 1 and 2 until $N_{t+1} = N_t$, which is at the point where the diagonal and the recruitment curve intersect.*

A *short cut* can be taken by repeating the procedure just described, but by drawing in just the lines that fall between the recruitment curve and the diagonal. These lines summarize the trajectory that will be followed by a population starting at $N_{t=9}$ or $N_{t=1}$.

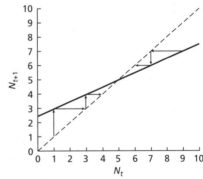

Graph the dynamics of the population by plotting the values you have found for N_t against t.

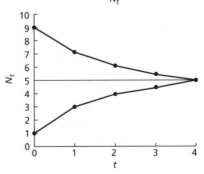

Continued p. 130

Box 5.1 (*continued*)

Other patterns of population behaviour can be seen by cobwebbing recruitment curves where the slope and intercept of the recruitment curve have other values, for example:

Slope = 1.5, $C = -2.5$, the equilibrium point in this population is unstable. For values of $N_t > 5$ the population will increase exponentially, and for values of $N_t < 5$ it will decrease to extinction.

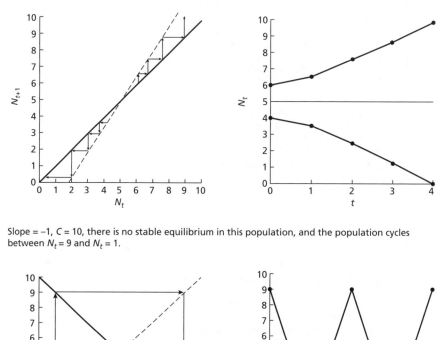

Slope = -1, $C = 10$, there is no stable equilibrium in this population, and the population cycles between $N_t = 9$ and $N_t = 1$.

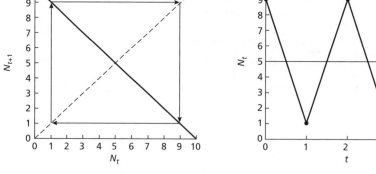

---

the mean seed production s and the exponent b in Eqn 5.10 describing its recruitment curve (Fig. 5.4).

Recruitment curves can be used to investigate how changes in the parameters of Eqn 5.10 will affect a population's dynamics and stability, and to examine the effects of various ecological factors. We shall look at two examples: the effect of self-thinning and the effect of density-independent mortality. Because of its peculiar morphology, the saltmarsh annual *Salicornia europaea* can reach extremely high densities without self-thinning (Ellison 1987a; see also Chapter 4), although some density-dependent mortality may have occurred on the low marsh studied by Jefferies *et al.* (1981). Watkinson and Davy (1985) estimated that the

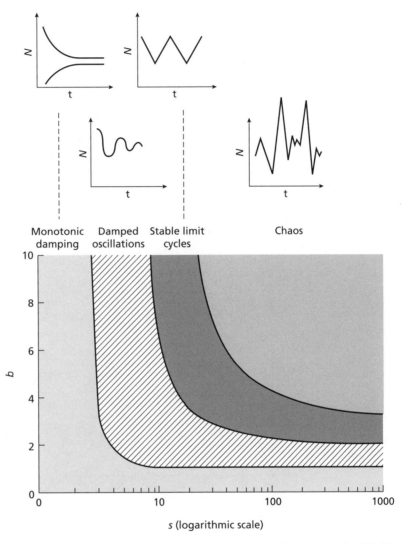

Fig. 5.4 The effect of s and b in Eqn 5.9 on the dynamic behaviour of a population (modified from Hassell *et al.* 1976).

reciprocal of maximum plant density, m, had a value of 8×10^{-5} (Table 5.1). In Fig. 5.5 we have compared the dynamics of this population with and without self-thinning. With self-thinning the population is stable, but without it the population oscillates. This is a particularly interesting finding because it may account for the population cycles that have been reported in a related species *S. patula* in Poland (Symonides 1988).

The level of density-independent mortality that affects a population with a humped recruitment curve may determine whether or not the population cycles. The very regular two-year population cycles of *E. verna* shown in Fig. 5.6(a) were found by Symonides (1984), who studied the species on inland sand dunes

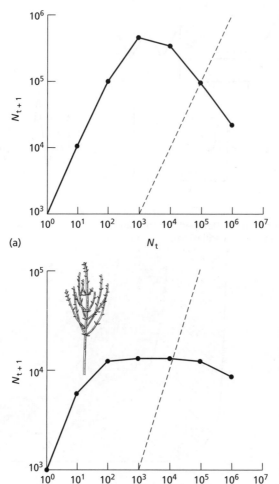

Fig. 5.5 The recruitment curves for *Salicornia europaea* (pictured): (a) without self thinning, $m = 0$; and (b) with self thinning, $m = 8 \times 10^{-5}$ (after Silvertown 1991).

in Poland. She found that parts of the population cycled and parts did not. Symonides *et al.* (1986) analysed this population using a recruitment curve where N_t was the initial seedling density and N_{t+1} was the seed output. The slope of the diagonal that intersects this particular kind of curve represents the proportion of seeds that germinate successfully, or in other words the value p in Eqn 5.10. By varying the slope of the diagonal to represent different rates of germination it was found, when $P = 0.5\%$, that the intersection fell on a part of the curve where no cycles would occur, but if germination success was 1% or greater, cycles similar to those observed in parts of the field population were predicted (Fig. 5.6). Differences in germination success of this order between shaded and unshaded microsites in the *E. verna* population are easy to imagine and would explain why parts of the population cycled and parts did not. Unfortunately, spatial variation in germination success was not measured, but this

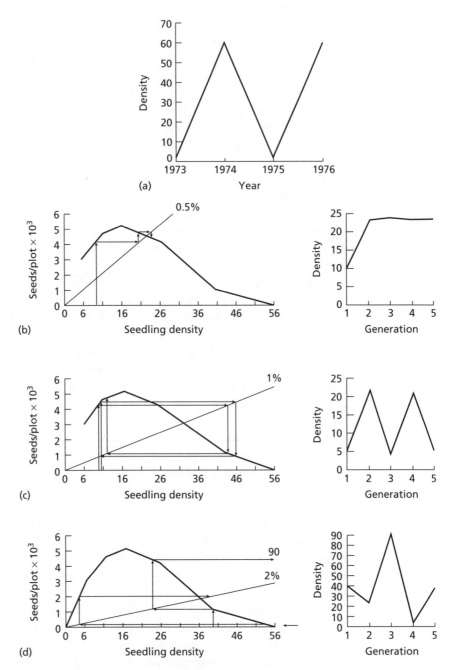

Fig. 5.6 (a) Regular two-year cycles of abundance observed in some parts of a population of *Erophila verna*. (b–d) The dynamics of the *E. verna* population predicted from its recruitment curve when the germination rate (p) is (b) 0.5%, (c) 1% and (d) 2% (from Symonides *et al.* 1986).

population analysis nevertheless illustrates how a population model can help to identify the important parameters that should be investigated in the field. The analysis also demonstrates that quite a small difference in a density-independent factor such as germination can have a major influence on how a population behaves when it interacts with overcompensating density dependence.

The existence of chaotic population dynamics in nature tends to be a very contentious issue, especially in plants, because overcompensating density dependence and humped recruitment curves are undoubtedly rare. The main reason they are rare is because the plasticity of growth in plants usually makes crowded populations obey the law of constant yield (Chapter 4). However, some interesting possible exceptions are to be found among shrubs belonging to the family Proteaceae that are endemic to fynbos heathland in the extraordinarily species-rich Cape floristic province of South Africa. Like all heathlands, the fynbos is a fire-dominated ecosystem and many of the shrub species are killed by fire and then recruit from seed immediately after burning. This creates even-aged populations with generations that do not overlap, so they are demographically just like annual populations with no seed bank. Unexpected population crashes following poor recruitment after fire have been observed in some fynbos shrubs, indicating the possibility of chaotic population dynamics. Bond *et al.* (1995) found evidence that some of these species had humped recruitment curves caused by a peculiar morphology that prevented small, unbranched individuals from flowering at high density. Self-thinning is apparently unusual in these shrubs and thus populations achieving a high density contain many small, sterile plants, produce few seeds in total, and crash after a burn.

In this section we have concentrated on populations with humped recruitment curves because these illustrate a number of important general points. It should not be imagined, however, that curves with this shape are very common, that all plants that have this kind of recruitment curve will cycle, or that all cycles are caused by overcompensation. Pacala and Silander (1985) found a humped recruitment curve for the annual oldfield weed *Abutilon theophrasti* but discovered that in nature the cycles they expected to find were damped by seed dormancy (Thrall *et al.* 1989). Although an annual plant in its growing phase may lack age structure, the existence of a seed bank complicates matters by introducing one.

Population cycles may also be caused by delayed density dependence, for example when the dense litter and dead remains of parent plants interfere with the recruitment of the next generation (Bergelson 1990). A greenhouse study of the annual *Cardamine pensylvanica* that followed experimental populations for 15 generations, found population cycles with a period of 4–5 generations that were caused by delayed density dependence rather than by overcompensation (Crone & Taylor 1996; Crone 1997). The precise mechanism by which the density of the parental generation exercised a time-lagged effect upon the offspring generation was not identified.

Even if it is strongly regulated, a population cannot indefinitely be protected from extinction by density dependence. Populations of sand dune annuals such *E. verna* and *V. fasciculata* are ultimately doomed to local extinction as succession proceeds and perennials usurp their space. Nine years after the start of a study, *V. fasciculata* became extinct in the plots where it was censused originally (Watkinson 1990). Annual species without a seed bank persist in the long term only by colonizing new territory, but this is a very precarious existence. In the particular case of *V. fasciculata* seeds must be buried by wind-blown sand for seedlings to stand a chance of successful establishment. Watkinson (1990) found that *V. fasciculata* had values of $\lambda > 1$ only in plots that were at least 50% covered by sand. For values of $\lambda < 1$, equilbrium population size is, of course, zero. This creates a threshold in the relationship between percent sand cover and equilbrium population size (Fig. 5.7) that would suggest that the density of *V. fasciculata* should be very patchy, and that small changes in the vegetation cover that stabilize sand movement can produce very large changes in the population density of this annual. The behaviour and fate of seeds has important consequences for the dynamics of many plants, so we shall now consider seeds in some detail before looking at the dynamics of such populations in Chapter 6.

5.5 Seeds in the soil

While in the soil only two things matter to seeds: the risk of death, and the chance to germinate. The irony is that chancing germination also risks death. Dormancy is the safest condition for a seed, but a dormant seed bears no offspring. Germination is always risky, but only a plant that reproduces can transmit its genes and multiply its progeny. In Chapter 10 we address some evolutionary strategies that enable plants to resolve this problem. Here we look in some detail at what goes on in seed populations.

Seed populations in soil are usually described as a **seed bank**, a term which gives the erroneous impression that this is a place of safekeeping for plant

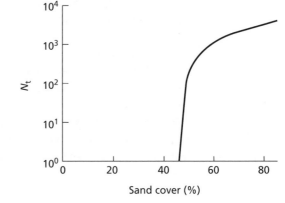

Fig. 5.7 The relationship between percent area covered by sand and the expected equilbrium population density of *Vulpia fasciculata* (Watkinson 1990).

genotypes. In fact, the seed bank is constantly plundered by predators and is highly prone to attack by fungi. Even when they germinate, few seeds make a successful escape. Plants which invest their progeny in the seed bank retrieve only a tiny fraction of them, and each of these is worth (in terms of contribution of genes to future generations) much less than other seeds which have already successfully produced progeny of their own rather than remaining dormant in the soil. In effect, the soil is like a bank that offers little protection from thieves or kidnappers and which guarantees that the value of what is left of a deposit when it is withdrawn will be severely eroded by inflation.

5.5.1 Numbers and distribution

Despite its subterranean perils, the soil beneath many kinds of vegetation is replete with seeds (Table 5.2). The number of seeds in the soil is determined by the rate of input in the seed rain, the rates of loss resulting from fungal disease and predation, and of course losses following germination (Fig. 5.8). Rates and patterns of seed loss through time vary between species (Rees & Long 1993). The fate through the year of cohorts of seeds belonging to two grassland herbs is shown in Fig. 5.9. The buttercup *Ranunculus repens*, studied by Sarukhán (1974) in a field in North Wales, lost half of its seed bank to rodents in the first six months. Few seeds had germinated by the end of the 15 month study, although none of the survivors were innately dormant (see also p.144). Pavone and Reader (1982) studied the legume *Medicago lupulina* in a pasture in southern Ontario, Canada. By contrast with *R. repens*, seeds of *M. lupulina* were dormant when they entered the seed bank. Most deaths occurred during or after germination, and at the end of 9 months the 20% of the initial seed cohort still remaining in the bank was innately dormant.

With the rare exception of apomicts, such as parthenogenetic lineages of the dandelion *Taraxacum officinale*, seeds are always sexual products. Unlike the plants in many populations above ground, every individual in a sexually produced seed population is unequivocally a genet. Above ground, a single genet may in time cover a large area, but below-ground genet density increases with time as seeds accumulate and may reach densities in the tens of thousands per square metre.

In the British and wider European floras, species with persistent seed banks tend to have smaller, rounder seeds than species without (Fig. 5.10) (Thompson *et al.* 1993), and there is significant tendency for seeds of such species to be found more deeply buried in the soil (Bekker *et al.* 1998). Leishman and Westoby (1998) failed to find a relationship like the one shown in Fig. 5.10 in a sample of the Australian flora, possibly because the European data were mainly for herbaceous species and many woody species of fire-prone habitats where the ecology of recruitment is very different were present in the Australian study.

Table 5.2 Numbers of seeds and the predominant species in the seed pools of various vegetation types (from Silvertown 1987a).

Vegetation type	Location	Seeds m^{-2}	Predominant species in the soil
Tilled agricultural soils			
Arable fields	England	28 700–34 100	Weeds
Arable fields	Canada	5000–23 000	Weeds
Arable fields	Minnesota, USA	1000–40 000	Weeds
Arable fields	Honduras	7620	Weeds
Grassland, Heath and Marsh			
Freshwater marsh	New Jersey, USA	6405–32 000	Annuals and perennials representative of the surface vegetation
Saltmarsh	Wales	31–566	Sea rush where abundant in vegetation, grasses
Calluna heath	Wales	17 500	*Calluna vulgaris*
Perennial hay meadow	Wales	38 000	Dicotyledons
Meadow steppe (perennial)	Russia	18 875–19 625	Subsidiary species of the vegetation
Perennial pasture	England	2000–17 000	Annuals and species of the vegetation
Prairie grassland	Kansas, USA	300–800	Subsidiary species of the vegetation, many annuals
Zoysia grassland	Japan	1980	*Zoysia japonica*
Miscanthus grassland	Japan	18 780	*Miscanthus sinensis*
Annual grassland	California, USA	9000–54 000	Annual grasses
Pasture in cleared forest	Venezuela	1250	Grasses and dicot weeds
Forests			
Picea abies (100-yr-old)	Russia	1200–5000	All earlier successional spp.
Secondary forest	North Carolina, USA	1200–13 200	Arable weeds and spp. of early succession
Primary subalpine conifer forest	Colorado, USA	3–53	Herbs
Subarctic pine/birch forest	Canada	0	No viable seeds present
Coniferous forest	Canada	1000	Alder *Alnus rubra*
Primary conifer forest	Canada	206	Shrubs and herbs
Primary tropical forest	Thailand	40–182	Pioneer trees and shrubs
Primary tropical forest	Venezuela	180–200	Pioneer trees and shrubs
Primary tropical forest	Costa Rica	742	Pioneer trees and shrubs

Seed densities are highest in frequently disturbed habitats such as arable fields, and lowest in primary forest (Table 5.2). In all habitats, the species most highly represented in the soil tend to be those with the shortest lifespan in the vegetation (Thompson *et al.* 1998), because these species produce large numbers of small seeds, which are frequently capable of dormancy. Thus, the seeds in the soil

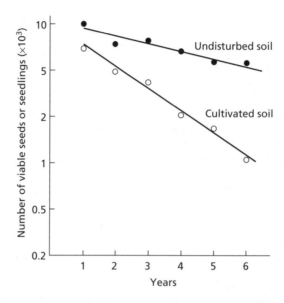

Fig. 5.8 Decay of viable weed seeds in undisturbed soil and cultivated soil (Roberts & Feast 1973).

beneath perennial vegetation such as woodland tend to be unrepresentative of the species above ground (Fig. 5.11a), and consist mostly of the seeds of herbs, and of pioneer trees and shrubs. On the other hand, the seeds beneath arable fields infested with annual weeds may be quite representative (Fig. 5.11b). Debaeke (1988) compared the population sizes of 10 annual broad-leaved weeds with the size of their respective seed banks in fields of winter wheat near Paris, France. Even though no species had a germination rate higher than 14% of the seeds present, the correlation between numbers above and below ground was statistically significant in all of the 10 species. In half of the species, the size of the seed bank accounted for more than 85% of the variance in the number of growing plants. Such correlations can be used as a basis for predicting weed infestations, but are only reliable if the cultivation regime is consistent over a period of years, because different crops and practices favour different weeds (Wilson *et al.* 1985).

The spatial distribution of seeds is inherently patchy because most fall near the parent plant (Fig. 5.12) (Chapter 3). Unfortunately, the patchiness of the seed bank is rarely considered in quantitative assessments, and consequently values such as those in Table 5.2 may be quite inaccurate. Patchy distributions should be studied using many small samples rather than by a few large ones (Roberts 1981; Bigwood & Inouye 1988).

Dispersal agents carry seeds away from the parent, but this does not necessarily make distributions more homogeneous because dispersed seeds are often deposited in clumps. In the Sonoran desert, USA, where occasional sheet flows of water and wind as well as animals transport seeds, Reichman (1984) found extreme densities of seeds concentrated in shallow depressions in the soil surface. Local concentrations of seeds are also created by birds and rodents that cache their food supplies, and ants that carry seeds to their nests (Brown *et al.* 1979; Howe &

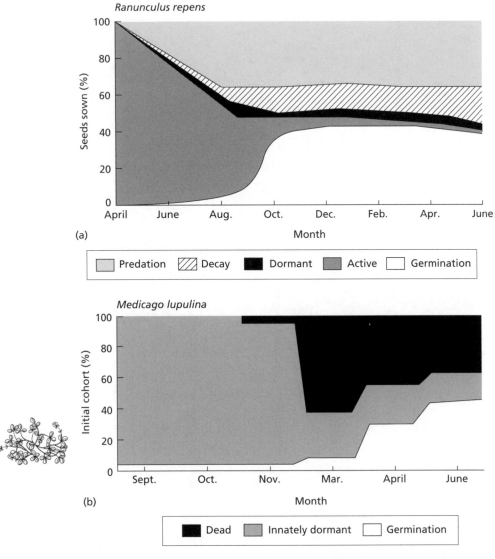

Fig. 5.9 The dynamics of seeds in the soil for two herb species (a) *Ranunculus repens* (from Sarukhán 1974) and (b) *Medicago lupulina* (pictured; data from Pavone & Reader 1982).

Smallwood 1982). Seeds that survive passage through an animal's gut are deposited in clumps with its faeces. Cow pats from a pasture in a forest clearance in Amazonia contained concentrations of grass and sedge seeds 20 times greater than the density of seeds of the same plants in the soil (Uhl & Clark 1983). Grant (1983) found that 70% of the seedlings appearing in grassland plots at a site in North Wales emerged on the site of worm casts. In a Dutch pasture studied by van der Reest and Rogaar (1988), germinable seeds of annual meadow grass *Poa annua* and rush *Juncus* sp. were 15 times more concentrated in worm casts sampled from

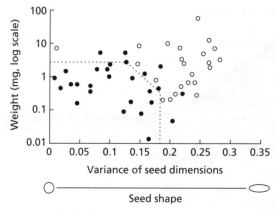

Fig. 5.10 The relationship between seed persistence in the soil for 53 species of British herbs and seed mass and shape. Seed shape was measured as the variance of seed length and width. Solid circles are species whose seeds persist for at least five years in the soil. Open circles are species whose seeds do not. (Thompson *et al.* 1993).

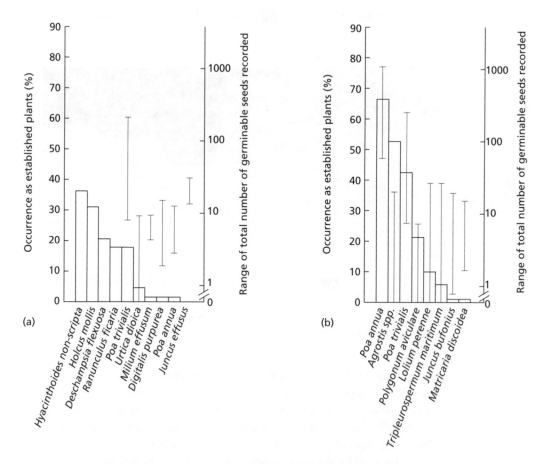

(a)

(b)

Fig. 5.11 The relative abundance of mature plants (histograms) and seeds in the soil (vertical bars) for the major species of (a) the herb layer of a deciduous woodland and (b) an arable field in Britain (from Thompson & Grime 1979).

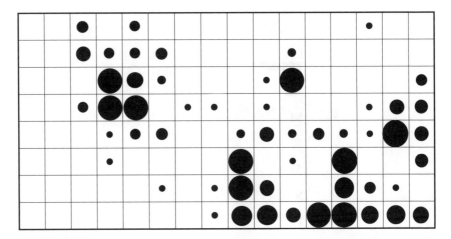

Fig. 5.12 The distribution of seeds of the grass *Danthonia decumbens* in 7 cm × 7 cm, 5 cm-deep blocks of soil taken from a turf of acid grassland on Dartmoor, Devon, England. The smallest dots are one seed, the largest are > 20 (Thompson 1986).

earthworm burrows between 5 and 15 cm below the surface, than in ordinary soil at the same depth.

Animals are important in the burial of seeds and the formation of a dormant seed bank. van der Reest and Rogaar (1988) calculated that earthworm activity transported about 20% of the grassland seed bank from the soil surface to a depth below 4 cm (Fig. 5.13). Although most seeds occur near the soil surface, redistribution by soil animals can carry some seeds to considerable depths. Only those near the surface, or those brought up by earthworm casts, digging animals or uprooted trees stand any chance of successful emergence (Putz 1983). Shallow burial of seeds often increases germination and aids seedling establishment, but deep burial hampers the emergence of germinating seeds (Fig. 5.14).

5.5.2 Dormancy

There are legendary instances of dormant seeds stored in unusual conditions retaining viability for periods of hundreds, and even thousands of years, but these records have little relevance for the normal ecology of seeds in the soil (Baker 1989). Seed dormancy is of selective value to plants because it allows them to time germination to coincide with conditions favourable to seedling survival, not because it allows the occasional seed to sit out the centuries (Section 10.3.3.2).

Seed dormancy is also a tactic against environmental uncertainty, particularly the unreliability of rainfall in arid climates (Section 10.4.2.2). The presence of seed dormancy is correlated with climate and with life history (which are themselves correlated). In the tropics and subtropics the percentage of species with dormancy is negatively correlated with precipitation, being highest in hot deserts. Outside the

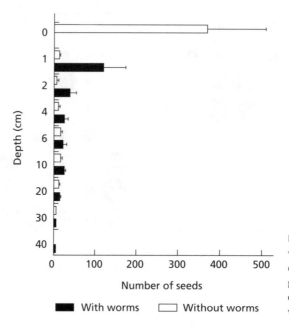

Fig. 5.13 The distribution of viable seeds of *Poa annua* at different depths in the soil at grassland sites with and without earthworm populations (from van der Reest & Rogaar 1988).

tropics, the correlation with climate is more complex (Fig. 5.15). Species that are long-lived as adults often have short-lived seeds. Seeds of the primary tree species of forests tend to lose viability very rapidly. Ng (1983) estimated that the

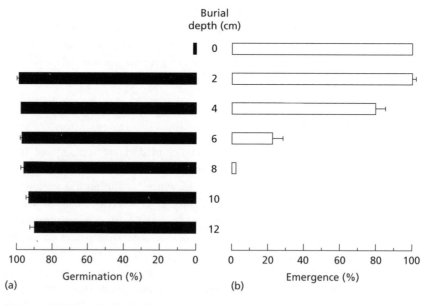

Fig. 5.14 (a) The percentage of seeds of the sand dune grass *Agropyron psammophilum* buried at different depths that germinated, and (b) percentage of germinated seeds that successfully emerged at the surface (data from Zhang & Maun 1990).

viability of seeds in the soil of Malaysian rainforest decayed with an average half-life of only six weeks. In another tropical forest, at Barro Colorado Island (BCI) in Panama, where the rainfall is seasonal, Garwood (1983) found that 80% of species germinated with the arrival of the rainy season following seed dispersal. This involved only a brief period of dormancy for those species that produced seeds outside the rainy season. The rainy season at BCI lasts eight months, but Garwood (1982) found that pioneer trees, lianas and canopy trees, all requiring gaps for germination, appeared only in the first two months of this period, while under-storey and shade-tolerant species germinated throughout the rainy season. Among the gap-requirers, Garwood (1983) found that seedling mortality during the dry season that followed the rains was related to time of emergence. Seedlings that emerged earlier were better able to beat the competition in their gaps and to tolerate the ensuing drought than seedlings that emerged later. No such

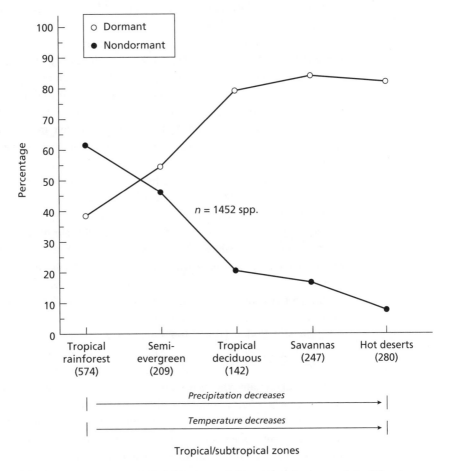

Fig. 5.15 The relative proportions of species with dormant/nondormant seeds in different vegetation types in the tropics and subtropics. Numbers in brackets are sample sizes (no. of species) (from Baskin & Baskin 1998).

143

relationship existed for the shade-tolerant species that germinated throughout the rainy season. The optimal timing of germination for different species at BCI is evidently determined by a combination of seasonally changing climate, biotic interactions, and tolerance of shade.

In seeds that have the capacity for prolonged dormancy, the dormancy state is often changeable. Seeds from different parts of the same plant, or matured at different times, may exhibit different germination behaviours when shed (Silvertown 1984); seeds may change their behaviour with time and in different directions depending upon storage conditions; and all these differences may themselves vary between populations. In many instances the ecological significance of the complex patterns found in laboratory germination tests still requires investigation in the field, but a number of kinds of seed behaviour have apparent adaptive value for the plant. Fresh seeds of knotgrass *Polygonum aviculare* are dormant and will not germinate before winter, but dormancy is broken by chilling in the imbibed state ('**stratification**') which permits germination as the soil warms in the spring. Stratified seeds in the soil that have not germinated by May become dormant again and require another cold period before they will germinate (Courtney 1968; Roberts 1970). The dormancy of fresh seeds prevents germination too early, when seedlings would be exposed to frost, and reacquisition of dormancy in May prevents germination too late when seedlings would probably be at a severe disadvantage to earlier cohorts (e.g. Fig. 4.12). In fat hen, *Chenopodium album*, dormancy is seasonally adjusted in a different manner. Nondormant seeds are produced early in the season and dormant ones in the main crop.

Seeds of many species with long-lived soil populations undergo annual cycles of dormancy induced by seasonal temperature changes, which permit germination at certain times of year and not at others. Both autumn-germinating genotypes of annual thale cress, *Arabidopsis thaliana*, and the spring-germinating annual common ragweed, *Ambrosia artemisifolia*, exhibit dormancy cycles, but these are six months out of phase with each other (Fig. 5.16). Roberts (1986) studied the seasonal patterns of germination in 70 species of annual and perennial herbs of the British flora and concluded that about half of them showed cyclic changes in germination requirements through the year.

Baskin and Baskin (1985, 1998) suggest a scheme of classification for the different kinds of seed dormancy that explicitly allows for changes in dormancy state. The scheme divides seeds into those with **primary dormancy,** which are unable to germinate when shed from the plant, and those with **secondary dormancy,** which acquire dormancy after leaving the parent. Seeds with primary dormancy may become nondormant and then acquire (secondary) dormancy again. Both primary and secondary dormancy are divided into two subcategories: **innate dormancy** and **conditional dormancy**. Innately dormant seeds will not germinate under any normal environmental conditions; seeds with conditional dormancy will germinate in only a narrow range of conditions. When innately dormant seeds alter their dormancy state they usually do so by a gradual change

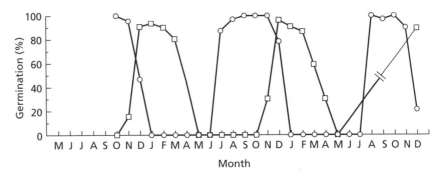

Fig. 5.16 Annual cycles of dormancy in buried seeds of a strict winter annual *Arabidopsis thaliana* (○), and the summer annual *Ambrosia artemisifolia* (□) retrieved from the soil and germinated at their respective temperature optima (from Baskin & Baskin 1983, 1980, respectively).

which takes them through a phase of conditional dormancy. The transition back again from nondormancy to secondary dormancy also passes through a conditional phase (Baskin & Baskin 1987).

An important form of innate primary dormancy caused simply by a hard seed coat is found in many legumes such as *Medicago lupulina* (Fig. 5.9b). With the passage of time in the soil this kind of dormancy may be broken by the decomposition of the testa. Fire or passage through the gut of a vertebrate dispersal agent will also frequently break this kind of dormancy. Conditional secondary dormancy is induced in many species that will normally germinate in the dark but which become light-requiring when exposed to light filtered through a leaf canopy (Silvertown 1980a). Half of a sample of 32 common East African weed species tested by Fenner (1980) showed this behaviour. Other environmental cues may also modulate changes in dormancy. Low humidity caused innate secondary dormancy in seeds of the Nigerian tree *Hildegardia barteri*, which became conditional when humidity was raised to 90% (Enti 1968).

Although seed dormancy is so phenotypically plastic, it is a character with strong effects upon plant fitness. One obvious explanation for this apparent paradox is that there is selection for phenotypic variability in germination behaviour when there is some unpredictability about the best time to germinate (Chapter 10). The fitness of seeds with a particular degree or type of dormancy can be estimated from the survival and fecundity of the seedlings that appear when dormancy is broken. Arthur *et al.* (1973) compared the survival and fecundity of autumn and spring cohorts of seedlings of the poppy *Papaver dubium*, an arable weed of British fields. The autumn seedlings emerging from nondormant seeds were susceptible to heavy winter mortality in some years, but survivors of this cohort produced larger plants and at least 10 times as many seeds as spring cohorts.

There are many examples of annual plants in which seedlings from nondormant seeds have a lower survival but a higher fecundity than later-appearing seedlings from dormant seeds. This pattern has been observed in prickly lettuce *Lactuca serriola* (Marks & Prince 1981) and in charlock (*Sinapis arvensis*) (Edwards 1980) in Britain, between summer and autumn cohorts of *Leavenworthia stylosa* in Tennessee cedar glades in the USA (Baskin & Baskin 1972), and in two closely related annuals *Erodium botrys* and *E. brachycarpum* that occur in annual grass-land in California (Rice 1987). In all instances, variability in climate (rainfall, drought or frost) may favour dormancy in some years and nondormancy in others, causing selection for phenotypic variation. This may arise from phenotypic plas-ticity or from genetic variation in the population. In *Papaver dubium*, seed dormancy has very low heritability, and variation therefore reflects phenotypic plasticity (Arthur *et al.* 1973). The opposite is the case in *Sinapis arvensis* in which dormancy is determined by two alleles at a single locus (Garbutt & Whitcombe 1986). There is also evidence that dormancy has a heritable component in *Erodium* spp. (Rice 1987).

5.5.3 Recruitment and the 'safe site'

Physiological studies of germination in the laboratory can be some guide to the behaviour of seeds in the field, but this behaviour is sensitive to a range of environmental variables and many of these interact with each other to subtle effect. A demonstration of this was provided by Harper *et al.* (1965) in a simple but compelling experiment. A seed bed was sown with equal quantities of seeds of three plantains, *Plantago major, P. media* and *P. lanceolata,* and divided into plots that were given a replicated series of treatments to the soil surface, as described in the legend to Fig. 5.17. The different treatments had markedly selective effects on the emergence of the three species, depending upon the kind of microenvironment each treatment had created (Fig. 5.17). In order to explain the results of this experiment, Harper *et al.* (1965) introduced the term '**safe site**' to describe the specific conditions that allow the seeds of a particular species to emerge success-fully from the soil.

The concept of 'safe site' is similar to the idea of niche (Chapter 8) and shares with it the difficulty that neither of them can be reliably identified until after they have been successfully occupied. Habitats are obviously not uniformly 'safe' for seed germination, but is it more than chance that seedlings appear in some places and not in others? In some instances it clearly is. Decaying logs ('nurse logs') are safe sites for the establishment of seedlings in a number of tree species. In Hawaii there is a tree fern whose name in the native language means 'Mother of Ohia', because Ohia (*Metrosideros collina*) germinates on its fallen trunks and even on living trees.

In the cascade mountains in Oregon almost all of the young western hem-lock *Tsuga heterophylla* found by Christy and Mack (1984) in their 200 m square study plot were growing on decaying logs of Douglas fir *Pseudotsuga menziesii*, although such logs occupied only 6% of the area. By sowing hem-lock seeds onto soil and different kinds of log Christy and Mack (1984) found

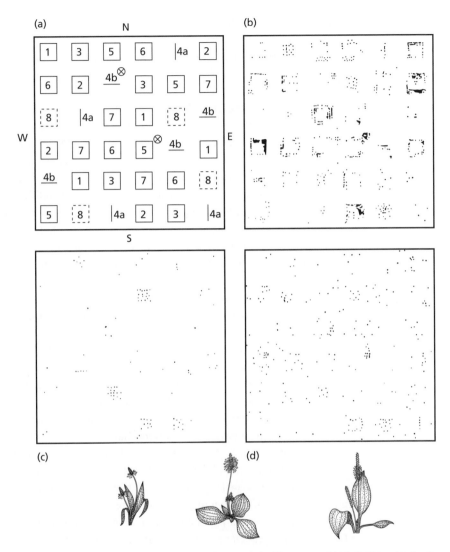

Fig. 5.17 (a) The distribution of various treatments to the *Plantago* seed bed: (1,2) two kinds of depression in the soil surface; (3) a sheet of glass laid on the soil surface; (4a,b) sheets of glass placed vertically in the soil; (5–7) rectangular wooden frames of three different depths pressed into the soil surface; (8) control. ⊗ worm casts. (b–d) The distributions of seedlings of *P. lanceolata*, *P. media* and *P. major* (pictured, left to right; from Harper *et al.* 1965).

that germination was possible on a wide range of surfaces, but that logs with rotten heartwood provided the best substratum for survival. Logs had to be rotten enough to permit hemlock seeds to lodge in them and root, but still sufficiently intact to raise the seedlings above the litter which threatened to bury and kill them.

In a deciduous forest in upstate New York, Handel (1976) found that the sedge *Carex pedunculata* was particularly common on fallen logs, where the species

147

avoided competition from the rest of the woodland ground flora. Like those of many woodland herbs, seeds of *Carex pedunculata* possess an elaiosome (Chapter 1) that attracts ants. *Carex* seeds were carried to rotting logs by ants that nested in them and which discarded the intact seeds there after removing the elaiosome.

In theory, germination safe sites might explain the macrodistribution of species, as well as their microdistribution. Keddy and Constabel (1986) tested this hypothesis in a group of 10 wetland species that are distributed along a gradient of wave exposure on the shores of the North American Great Lakes. The gradient of wave exposure creates a parallel gradient of soil particle size, from silt in sheltered bays to gravel on the most exposed shorelines. Seeds of the 10 species, which ranged in length from 14 mm in *Bidens vulgata* to 0.76 mm in *Lythrum salicaria*, were sown in trays of graded soil. Soil of the finest texture passed a 0.25-mm sieve and the coarsest an 8-mm sieve. When watered well, most species showed no particular preference for soil type, and when watered sparingly all except one established best on the finest soil. Among these plants then, there was no evidence that species have safe site differences related to substratum texture that could explain their distributions along the wave exposure gradient.

An interaction between climate and the germination ecology of the annual weed corn spurrey *Spergula arvensis* appears to explain an intriguing cline in gene frequency in this species in Britain. Three seed morphs are known, determined at a single locus by two alleles. Seeds covered in papillae are produced by one homozygote, the other homozygote produces seeds with a smooth seed coat, and the seeds of the heterozygote are intermediate with half as many papillae as the papillate morph (Fig. 5.18a) (New 1958). All the seeds produced by a particular plant have the same morph because the seed coat is maternal tissue, which makes it possible to score gene frequencies in the field when the plants are in fruit. As the plant is highly selfing, heterozygotes are rare.

New (1958) surveyed populations of *S. arvensis* across Britain in the 1950s and found that the frequency of the papillate morph decreased from the drier regions of the south to the wetter parts of the north and west (Fig. 5.18b). When New (1978) surveyed the cline again in the 1970s she found that there had been no significant change after 20 years. New and Herriot (1981) discovered in laboratory experiments that papillate seeds germinate better than smooth ones in dry conditions, suggesting that natural selection in favour of papillate seeds in dry areas could account for their distribution. What advantage smooth seeds have in wetter conditions is not known, so the full explanation for the cline and its stability is not yet clear, but this appears to be a case where the subtleties of germination ecology affect gene frequencies on a geographical scale. On a more local scale, morph frequencies can vary greatly even within a single field, suggesting a possible role for genetic drift, though local selection cannot be ruled out (Fig. 5.18c).

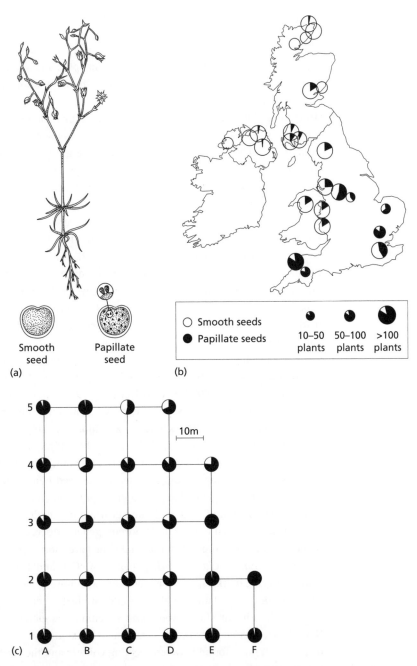

Fig. 5.18 (a) Papillate and smooth seed morphs of *Spergula arvensis* and (b) the cline in frequency of the papillate morph in Britain (from New 1958). The picture of the plant is shown at approximately $\frac{1}{2}$ actual size and the seeds at approximately $\frac{1}{6}$. (c) Variation in seed morph frequency within a single field at Maulden, Bedfordshire (white = smooth morph). The overall frequency of the smooth morph was 12% ($n = 1899$ plants), but there was significant genetic substructuring ($G_{ST} = 0.08$, $P < 0.001$) and morph frequencies varied from 0% to 45% between sampling locations in the field (note the 10 m scale bar) (Silvertown & Gonzalez-Rabanal, unpubl. data).

5.5.4 Genetic consequences of the seed bank

The existence of a seed bank increases effective population size N_e because generation time is increased by the length of time that plants spend as dormant seed in the soil, and because the breeding members of the population may be drawn from different cohorts of seeds. In the absence of selection, the effect of dormancy on N_e depends on the average time spent in the seed bank (Templeton & Levin 1979). The importance of a seed bank in increasing N_e will, of course, depend on which seeds germinate and not simply on the existence of a large dormant population. If most of the seeds in the soil are buried too deeply to germinate, and recruitment occurs mainly from recently produced seeds that are on or near the surface, much of the seed bank may be genetically as well as demographically irrelevant. This is the situation in the neotropical gap-colonizing tree *Cecropia obtusifolia* in which about half of all recruits originate from the seed rain rather than the seed bank (Alvarez-Buylla & Martinez-Ramos 1990; see Section 7.2). In a Mexican rainforest, Alvarez-Buylla *et al.* (1996) found significant genetic differences between the seeds produced by different *C. obtusifolia* trees and between the seeds falling into different gaps in the seed rain (see also Section 3.2.1.3). However, based on F_{ST} values of marker genes, there was no significant genetic differentiation between the seed banks buried beneath different gaps, probably because the seed bank accumulated seeds from many different trees over time. Very small or zero levels of genetic differentiation (F_{ST}) between seed bank samples have also been found in other species (Cabin *et al.* 1998; McCue & Holtsford 1998; Mahy *et al.* 1999).

The seed bank can thus sometimes serve as an important reservoir of genetic variation, buffering the effects of local extinction of genotypes in the adult population caused by selection or drift (del Castillo 1994). If recruitment from a seed bank is important, this should reduce the decay of genetic variability, and retard local subdivision of the population. In a sense, the seed bank can provide a kind of 'memory' that lends genetic inertia to the population. The few comparisons of the genetic composition of seed banks and growing populations that have been made confirm that the seed bank is an important genetic reservoir, especially when it is numerically large and/or when growing plants are short-lived. In the rare endemic Californian annual *Clarkia springvillensis*, McCue and Holtsford (1998) found, using allozyme markers, that the seed bank was more genetically diverse than the above-ground population. Cabin (1996) studied allele frequencies at seven allozyme loci in the short-lived desert perennial *Lesquerella fendleri* at a site in New Mexico. In one of two study years, small but significant genetic differentiation ($F_{ST} = 0.002$) was found between seedling cohorts and the seeds that remained in the top 2 cm of soil after seasonal germination was complete. Differences were also found between the seed bank and established plants in another study of the same species (Cabin *et al.* 1998).

Tonsor *et al.* (1993) found that while allele frequencies at four out of five allozyme loci differed significantly between the seed bank and growing plants in a population of *Plantago lanceolata*, the total genetic diversity within each was similar. The genetic composition of the seed bank and of the above-ground population were similar in a Belgian population of the heathland shrub *Calluna vulgaris*, but this species typically has extremely dense seed banks; Mahy *et al.* (1999) found that a single 10 cm deep soil core of 0.1 L in volume contained seeds with an allozyme diversity similar to that in an above-ground population occupying 800 m². Comparison of the plants germinated from 150-yr-old seed banks of *Luzula parviflora* and *Carex bigelowii* buried by solifluction in the arctic, with plants grown from modern seeds, showed significant differences in many growth characters, suggesting that evolutionary change had taken place in both species (Bennington *et al.* 1991; Vavrek *et al.* 1991). Seed banks can have long memories.

5.6 Summary

Plant populations are **dynamic** and often have high birth and death rates of individuals in them. The simple **difference equation** model introduced to describe population dynamics in Chapter 1 can be used to calculate two related measures of population growth: the **annual rate of increase** λ and the **net reproductive rate** R_0. When these measures have a value greater than one the simple difference equation predicts **exponential** increase. In reality, population increase is limited by **density-dependent** decreases in fecundity or density-dependent increases in mortality which, for an annual plant, can be described by a modification of the yield–density equation used in Chapter 4. The modified equation describes the shape of the **recruitment curve**, which can be used to predict how a population will behave by a graphical method known as **cob-webbing**. Density-regulated populations may show a variety of dynamical behaviours ranging from **monotonic damping** to **chaos**, depending upon the values of the exponent b and the mean seed output s in the equation for the recruitment curve. **Density-independent** mortality may affect population behaviour through its influence on s.

   The behaviour and fate of the seed fraction of a population is often of importance to its overall dynamics, and large numbers of seeds may accumulate in the soil **seed bank**. Seed densities are highest in frequently disturbed habitats and the species most strongly represented in the soil are often those with the shortest life span above ground. The **spatial distribution** of seeds is usually clumped because dispersal distances from the parent tend to be limited and **animal dispersal agents** may concentrate seeds in their faecal deposits. Seed longevity in the soil is prolonged by **dormancy**. Seeds unable to germinate when shed from the plant are said to have **primary dormancy** and those which acquire dormancy after leaving the parent have **secondary dormancy**. The latter may be cyclic. Either of the main kinds of dormancy may be **innate** or **conditional**. The seed bank can serve as an important reservoir of **genetic variation** and it may increase N_e if the recruits from it to the actively growing population represent the full range of genotypes. Seeds often have quite particular germination requirements that restrict the appearance of new plants to **safe sites** that may be limited in number.

5.7 Further reading

Ebert, T. A. (1999) *Plant and Animal Populations: methods in demography*. Academic Press, San Diego, CA.

Baskin, C. C. & Baskin, J. M. (1998) *Seeds*. Academic Press, San Diego, CA.

5.8 Questions

1 Program a spreadsheet to calculate τ, R_0 and λ from a life table. Try out your program on the data given in Table 1.1.

2 Program a spreadsheet (or write a simple computer program) to graph N against time using Eqn 5.9. Explore the effects of varying s_m, a and b in your model. What happens when your introduce a time lag of one year by replacing N_t with N_{t-1}?

3 Design a class exercise to determine the effect of varying the size of samples (volume of soil) and number of samples on the accuracy of seed density estimates. Try out the exercise in the field, or by sampling an artificial array of seeds scattered on a table top.

4 Discuss the concept of 'safe site' in relation to plant distribution. Is it a useful concept, and how can it be investigated in the field?

5 Discuss the importance of seed banks in the dynamics and genetics of plant populations.

Dynamics of age-structured and stage-structured populations

6.1 Introduction

Not all individuals in a population make an equal contribution to its finite rate of increase λ because values of b and d vary with the age, size and stage of an individual (Chapter 4). In this chapter we shall look at how such differences arise and how they affect population dynamics. In populations where recruitment is a frequent event compared to the lifespan of adults, such as in many forest trees, there is an **age structure** (Fig. 6.1a). At any one time, seeds, seedlings, saplings and adults of various ages may all be found in such populations. Younger ages tend to have higher mortality rates than older ones, so a population of 99 seedlings and 1 adult has a quite different future to a population of 1 seedling and 99 adults, even though both have the same total population size.

Plants are highly plastic in their rates of growth and development, so that two individuals of the same age but with different local environments may be at quite different stages of the life cycle. Stunted seedlings of forest trees may persist to 30 years of age on the forest floor without developing to sapling size, while other individuals that germinated at the same time, but happen to have been in a light gap, may already be reproductively mature. Therefore, as well as an age structure, plant populations have a **stage structure** (Fig. 6.1b). The size distributions that we have seen develop in crowded stands are an example (Figs 4.7 and 4.8). Both age and stage structures may influence plant population dynamics, and we will see in this chapter how these are dealt with. However, not all perennial plant populations have an age structure. For example even-aged populations are found in semelparous bamboos (Janzen 1976), and in pines such as *Pinus banksiana* (Kenkel 1988). Whether a plant population has age structure or not depends upon the pattern and periodicity of recruitment.

6.2 Stochasticity, disturbance and recruitment

The world is an uncertain place, especially for sessile organisms that can neither run nor hide. The randomness of events is referred to as **stochasticity**. One of the most important stochastic events for plants is **disturbance**, which we define as the destruction of biomass by any natural or human agency.

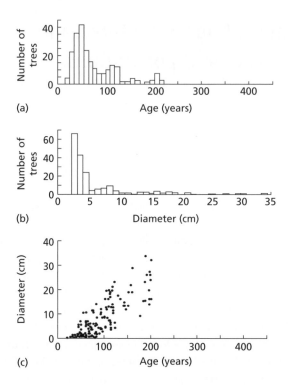

(a)

(b)

(c)

Fig. 6.1 (a) Age structure, (b) size structure, and (c) the relationship between size and age of trees in a population of *Pinus sylvestris* in Sweden (from Ågren & Zackrisson 1990).

Two kinds of stochasticity are usually distinguished. **Environmental stochasticity** refers to random fluctuations in environmental conditions such as the weather. These may affect all the populations in a region, regardless of their size. **Demographic stochasticity** is chance variation in local population structure that affects demographic parameters. This is prone to occur in small populations when, for example, a population happens to consist of only a few old individuals of low fecundity. Kendall (1998) discusses a simple way in which survival data can be tested for the influence of demographic stochasticity.

The age structure of a plant population is often intimately tied up with the periodicity of disturbance. Because large plants suppress the growth of small ones, the entry of new recruits into many populations can occur only when established individuals die or are killed and vacate space. For many, but not all, plants (Grubb 1988) recruitment is coupled with disturbance events which clear space. Populations without age structure inhabit sites where there is periodic, severe and widespread disturbance such as that caused by fire, hurricanes, extreme drought, or tillage of the soil for agriculture. These events remove existing vegetation and provide space for the massive, simultaneous recruitment of new populations, usually from seed.

If the disturbed area is large and recruitment into it is rapid, the populations that become established will be approximately even-aged. If areas of disturbance are small, the population will consist of a mosaic of patches recruited at different times. How the age structure of such a population is described obviously therefore depends upon the scale at which it is studied. Local stands of *Pinus* spp. in Itasca State Park, USA, are even-aged and date from the last fire, but in the Park as a whole the population of this species has an age structure (Fig. 6.2).

Species of pine (e.g. *Pinus banksiana*, Chapter 4), *Banksia* spp. in Australia and Proteaceae of the fynbos heathland of the South African Cape that live in fire-prone habitats, store seeds in sealed (serotinous) cones that accumulate on the tree (Bradstock & Myerscough 1981; Bond 1985). When the heat of a fire unseals them, many years worth of seed production is released at one go onto the newly cleared ground. Serotiny is usually a polymorphic character because fire frequencies are variable in space and time. There is strong selection after a fire against trees lacking serotiny, as well as against those possessing the character in an environment where fire is absent and seeds remain trapped.

In populations of *Pinus resinosa* in the New Jersey pine barrens, USA, Givnish (1981) found that serotiny was most frequent in the most fire-prone areas. Similar correlations have been found in *P. torreyana* and *P. coulteri* (McMaster & Zedler 1981; Borchert 1985). In populations of *Pinus contorta* growing in Montana, the frequency of serotiny was related to the kind of disturbance that had preceded recruitment. Serotiny was common in stands established after fire but uncommon in those initiated by windthrow or disease, suggesting that a single generation of natural selection might be sufficient to radically alter the frequency of genes affecting serotiny (Muir & Lotan 1985).

Serotiny and seed storage in the soil allow the rapid recruitment after disturbance that creates even-aged populations. If recruitment is slower, because no seeds are available at the site of disturbance itself, we can expect the populations that arise to be uneven-aged, and for a succession of species to take place.

The life history characteristics of a species determine whether it may survive over the long term at a site that has a particular frequency and severity of disturbance. Potentially, any species with a seed bank can colonize new space when it becomes available, but not all can complete their life cycle before the next

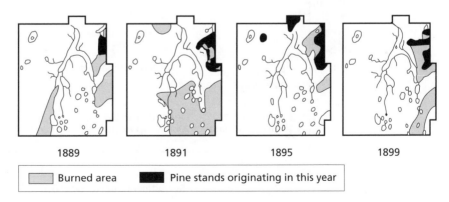

1889	1891	1895	1899

☐ Burned area   ■ Pine stands originating in this year

Fig. 6.2 Stands of pine recruited in Itasca State Park in four different years, in relation to the distribution of fires in the same years (from Frissell 1973).

episode of disturbance occurs. Where it occurs every year, such as in a ploughed arable field or in a desert, annuals are favoured because they can set seed within the year. In herbaceous communities, slightly longer intervals between disturbance events favour semelparous perennials such as teasel *Dipsacus sylvestris*, mullein *Verbascum thapsus* and evening primroses *Oenothera* spp. Populations of these species tend to be evanescent, relying upon a pool of dormant seed to carry them over between one disturbance and the next that occurs at the same spot (Werner & Caswell 1977; Reinartz 1984; Kachi & Hirose 1985).

A useful way to portray the effects of disturbance on population age structure is to graph the average lifespan of adults of a species against the typical periodicity of recruitment (Fig. 6.3). Species falling above the diagonal of this graph have populations with age structure. Those species falling on or below the diagonal tend to form populations that are even-aged, at least locally.

Populations that recruit in a pulse immediately following the death of a cohort lie on the diagonal of the graph. For example, at bottom left on the diagonal are annuals with no persistent seed bank; the various species of serotinous pines fall further up the diagonal as they are, like the mythical phoenix, at one and the same time killed and recruited by fire. Semelparous bamboos (Chapter 1) also fall on the

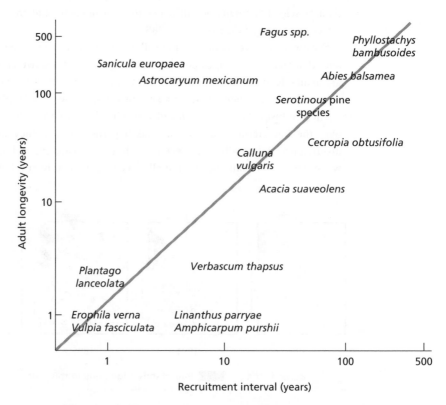

Fig. 6.3 Effect of the relationship between adult lifespan and recruitment periodicity on population age structure, showing some of the species mentioned in Chapters 1–6.

diagonal because their synchronized flowering and death is followed by massive recruitment from seed produced in the terminal act of reproduction. Species falling below the diagonal must survive between the infrequent events that permit recruitment by means of a dormant seed bank. These populations include desert annuals, semelparous perennials and the early colonizers of large forest gaps, such as *Cecropia obtusifolia* and *Piper* spp. in the neotropics. Locally, these species have even-aged populations of adults but if the whole species is considered, including seeds, there is an age structure.

When the period between recruitment events is longer than the lifespan of adults, the finite rate of population increase λ has to be averaged over two phases of the entire life cycle: the phase during which adult plants grow and produce seed, and the dormant phase when the population consists only of seeds in the soil. The best way to compare or combine what happens in these two phases is to measure the size of the population in terms of *seeds* throughout.

Auld (1986a,b; 1987) and Auld and Myerscough (1986) studied the population dynamics of the legume shrub *Acacia suaveolens* that occurs in heathlands and woodlands along the east coast of Australia. The shrub is killed by fire, which is also required to trigger the germination of seeds in the seed bank. As one would expect, populations tend to be even-aged. Adults are short-lived, few reaching 20 years of age. Peak seed production by individuals occurs at age 2 yr; thereafter, the proportion of shrubs producing seeds, and the fecundity of those that do, diminish rapidly with time. This decline in seed production combined with the mortality of adults and predation of seeds in the soil leads to the pulse of seed production that occurs early in the life of a cohort being quickly eroded by exponential decay (Fig. 6.4). It can be seen from Fig. 6.4 that if fires recur with a periodicity of 55 yr or less, the population will survive when the seed bank reaches the replacement level of 1000, although of course numbers of both seeds and adults will cycle. Note that these cycles are driven entirely by fire frequency, not by density dependence as in *Erophila verna* or in fynbos Proteaceae (Section 5.4).

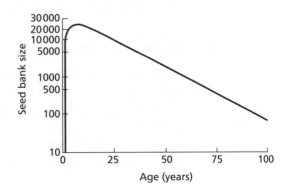

Fig. 6.4 Changes in the size of the soil seed pool of *Acacia suaveolens* for an initial cohort of 1000 seeds (from Auld 1987).

6.3 Population models with age and stage structure

The existence of age and stage structure has to be taken into account in population models. We can treat each age or stage class in a structured population like a population in its own right. If the number of individuals in a class i is n_i, the structure of the population at time t is represented by a list of n_i values called a **column vector**, and given the symbol $\mathbf{n}_t$:

$$\mathbf{n}_t = \begin{pmatrix} n_1 \\ n_2 \\ n_3 \\ \ldots \\ n_i \end{pmatrix}$$

Analogously with the basic equation of population dynamics (Eqn 4.8) we derive $\mathbf{n}_{t+1}$ from $\mathbf{n}_t$ by multiplying $\mathbf{n}_t$ by a coefficient representing the inputs to and outputs from each class. These are gathered together in a matrix called A, so:

$$\mathbf{n}_{t+1} = A\mathbf{n}_t. \tag{6.1}$$

A is a **projection matrix**, or transition matrix, containing all the stage- or age-specific rates of survival and seed production. The projection matrix is square, with i columns and i rows for a population with i classes. For a population with three classes representing, say, seedlings, 1-yr-old rosettes and 2-yr-old flowering adults the projection matrix representing transitions between the classes is:

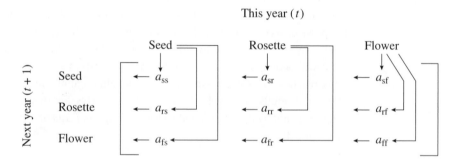

This year (t)

Box 6.1 describes how $\mathbf{n}_{t+1}$ is calculated from $A\mathbf{n}_t$.

6.3.1 Projection matrices

The precise form of the projection matrix will depend upon the details of the biology of the species and how its population structure is represented. If, in our hypothetical plant seedlings, rosettes and flowering plants are true age classes and all 1-year-olds are seedlings, all 2-year-olds are rosettes, and all 3-year-olds flower and then die, then a number of the elements in the matrix will have zero values. You should be able to work out for yourself where these zero values lie.

Box 6.1: Matrix multiplication

Written algebraically, the multiplication of a projection matrix A by a column vector n_t is carried out as follows:

$$\begin{array}{ccccc} A & \times & n_t & = & n_{t+1} \end{array}$$

$$\begin{pmatrix} a_{11} & a_{12} & a_{13} \\ a_{21} & a_{22} & a_{23} \\ a_{31} & a_{32} & a_{33} \end{pmatrix} \times \begin{pmatrix} n_1 \\ n_2 \\ n_3 \end{pmatrix} = \begin{pmatrix} (n_1 a_{11}) + (n_2 a_{12}) + (n_3 a_{13}) \\ (n_1 a_{21}) + (n_2 a_{22}) + (n_3 a_{23}) \\ (n_1 a_{31}) + (n_2 a_{32}) + (n_3 a_{33}) \end{pmatrix}$$

Describing this calculation in words:

- the *first* element of n_{t+1} is the sum of each element in n_t times its corresponding element in the *first* row of A
- the *second* element of n_{t+1} is the sum of each element in n_t times its corresponding element in the *second* row of A
- the *third* element of n_{t+1} is the sum of each element in n_t times its corresponding element in the *third* row of A

and so on, if there are more than three classes.

To find the value of n_{t+i}, the projection matrix is repeatedly multiplied in a process called **iteration**:

$n_{t+1} = An_t$, $n_{t+2} = An_{t+1}$, $n_{t+3} = An_{t+2}$ and so on to n_{t+i}.

Note that A does not change in this process, and that by continually multiplying the population by a constant, iteration produces an exponential increase in population size. The matrix equation $n_{t+1} = An_t$ is the equivalent for an age-structured population of $N_{t+1} = R_0 N_t$ for other populations.

A projection matrix for an age-structured population has non-zero elements *only* in the subdiagonal, which represents annual survival, and in the top row, which represents births. A matrix of this form is called a **Leslie matrix**. As we have discussed already, plant population structure may often be better described by the distribution of individuals among a series of stages. If seeds, rosettes and flowering plants in the hypothetical population have a range of ages and are identified by their size or behaviour rather than by their exact age, we may use a matrix called the **Lefkovitch matrix**. This has non-zero elements in the main diagonal, representing the proportion of seeds that remain seeds, rosettes that remain vegetative, and flowering plants that flower again the next year. The subdiagonal of the Lefkovitch matrix represents growth or development from one stage to the next, not simply an increase in chronological age. Studies of plant populations generally classify individuals by either age *or* stage, although greater accuracy is achieved by using both classifications simultaneously (Law 1983; van Groenendael & Slim 1988). The kind of matrix which classifies individuals by age and stage simultaneously is called a Goodman matrix and, as you can imagine, it is quite complicated, consisting of a matrix whose elements are themselves submatrices.

6.3.2 The stable age/stage distribution and calculation of λ

After a certain number of iterations of Eqn 6.1, the exact number depending upon the projection matrix A, population structure usually attains a stable age-distribution with a fixed ratio between $n_1:n_2:n_3:n_i$. When the stable age-structure has been attained, the finite rate of population increase λ for the population may be found by dividing the size of any age class n_i in one year by its size the previous year. In mathematical terminology λ is the dominant eigenvalue of A. The value of λ depends only upon A, and not the starting conditions n_t. This is a mathematical, not a biological, property of the matrix and is called **ergodicity**.

The stable age distribution derived from iteration of the A matrix can be compared with the actual age distribution of a population in order to determine whether its present composition is stable. Large discrepancies between the real and projected age distributions imply that the age composition of the population will change if present trends continue.

The techniques of matrix analysis described for age-classified populations may also be used for stage-classified ones, although there is an important difference in how age and stage distributions should be interpreted. If an *age* distribution observed in the field has a peak, with many more individuals in one of the older age-classes than in younger classes, this suggests that a strong cohort (or 'baby boom') is passing through and that the age distribution is unstable. A similar peak in the *size* or stage distribution of a population does not necessarily imply departure from a stable stage distribution (Caswell 2000a). Stable stage distributions can have peaks caused by some stages being of longer duration than others.

6.3.3 Sensitivity and elasticity analysis

Density-independent matrix models of the kind described above can be used to determine the sensitivity of λ to small perturbations in the value of the elements a_{ij} in the projection matrix A. Each element a_{ij} of the projection matrix has an associated sensitivity value s_{ij}. These values have two very important applications. First, they tell us the relative importance of different transitions for maintaining population growth rate; thus, if we wish to increase or decrease λ, sensitivity analysis indicates the population's most vulnerable points that may either be attacked or protected accordingly. Secondly, because λ may be used to estimate phenotypic fitness, s_{ij} indicates how changes in a particular transition a_{ij} will affect the fitness of individuals in the population (see Section 1.2.2 for discussion of different fitness concepts). A high s_{ij} value suggests that natural selection should act on characters affecting the transition a_{ij} because small differences between genotypes will affect their fitness.

The sensitivity value s_{ij} for a matrix element a_{ij} is:

$$s_{ij} = \frac{v_i w_j}{<\mathbf{v},\mathbf{w}>}.$$

Mathematically, v and w are, respectively, the left and right eigenvectors of the matrix A and $<v,w>$ is their scalar product. Because $<v,w>$ is not dependent on i or j, it can be ignored for comparisons between values of s_{ij} within a matrix (Caswell 2000a).

Sensitivity values measure the relative effect upon λ of a change in a_{ij} by a small, fixed amount. Values of a_{ij} representing survival probabilities can only range between zero and one, whereas an a_{ij} representing fecundity can take any value. This introduces a difficulty into the comparison of sensitivity values, because a fixed change of say 0.05 to a probability of 0.5 is of much greater significance than a change of the same amount to a fecundity of 100. In order to overcome this problem, de Kroon *et al.* (1986) introduced a refinement to sensitivity analysis. Their modified index of sensitivity, called **elasticity** (e_{ij}), is

$$e_{ij} = \frac{a_{ij}}{\lambda} \times s_{ij}.$$

Elasticities can be used for the dual purposes already described for s_{ij}, and have the additional useful property that they all sum to one within a matrix. This means that e_{ij} represents the proportion of λ resulting from the transition a_{ij} although, of course, we should remember that, to have any value at all, λ requires the transitions of the entire matrix just as a plant cannot skip a stage of its life cycle. Readers wishing to analyse matrices are referred to Ebert (1999) and Caswell (2000a).

6.3.4 Environmental stochasticity

So far we have assumed that the projection matrix A is time-invariant, which is to say values of a_{ij} are not perturbed by environmental stochasticity or altered by density dependence. In practice, stochastic year-to-year variation is found in practically all populations (e.g. Horvitz & Schemske 1995; Valverde & Silvertown 1998). Density dependence undoubtedly occurs in most populations too and is considered in the next section. Analysis of the dynamics of structured populations under conditions of stochastic variation is a complex matter whose full treatment is beyond the scope of an introductory textbook, but some general principles and results can be given. (Further reading on the subject is indicated at the end of this chapter.) The impact of stochastic variation upon λ will generally depend upon four factors: (i) how much different vital rates vary—we can expect some transitions, such as seedling survival, to vary more than others, such as the survival of larger plants; (ii) the sensitivity of λ to those rates that vary most; (iii) the temporal correlation between vital rates (e.g. whether survival and fecundity go up and down together); and (iv) environmental autocorrelation, which is a measure of whether good and bad years occur in runs, or at random (Benton & Grant 1996).

Elasticity (e_{ij}) is a good starting point from which to address stochastic variation because it predicts how much λ will respond when the transition represented by the matrix element a_{ij} is perturbed by a small amount. Although it has been argued that e_{ij} is calculated for a single (or average) value of a_{ij} and may therefore be wrong when

a_{ij} varies stochastically (Ehrlén & van Groenendael 1998), theoretical models have shown that the relative values of elasticities for different a_{ij} are little affected by moderate amounts of stochastic variation (Benton & Grant 1996). Moreover, it has been found that, as a general rule applying to many plant and other populations, the transitions that vary most in field populations are those with the smallest elasticities and, hence, the ones whose variation makes the least impact upon λ (Pfister 1998; Zuidema & Franco 2001). One possible explanation for this is that natural selection has minimized the impact on fitness of life history stages that are vulnerable to stochastic variation. Whatever the explanation for the pattern, it tempts the optimistic conclusion that the existence of stochastic variation need not invalidate elasticity analysis because it most affects those transitions that matter least. However, the reader should note that this matter is not settled (de Kroon *et al.* 2000; Caswell 2000b), and that stochastic variation cannot be ignored for small populations that are vulnerable to extinction. Small populations are considered in Section 7.4.

6.3.5   Density dependence

Matrix models incorporating density dependence can be constructed by replacing the fixed values of a_{ij} in the A matrix with density-dependent functions. Because the value of the elements in the projection matrix then become specific for the particular value of N at each iteration, the matrix is denoted A_n, and Eqn 6.1 becomes:

$$n_{t+1} = A_n n_t.$$

If the survival and fecundity of all age classes are affected equally by density dependence, a function such as we used in Eqn 5.9 can be applied by splitting the density-independent matrix A into two matrices, one called P for the survival probabilities and another called F for fecundities, giving

$$A_n = (1 + aN)^{-1}(F + P).$$

Simple density-dependent models such as this are ergodic, but the assumption that all age classes are affected equally by density is highly unrealistic for plant populations (Chapter 4). For example, Silva Matos *et al.* (1999) found that density dependence acted only at the seedling stage in the neotropical forest palm *Euterpe edulis*. They incorporated density dependence into a matrix model using the function:

$$g_{12} = g_m(1 + aN_1)^{-1},$$

where g_{12} is the transition rate at which seedlings (size-class 1) grew into size-class 2, g_m is the transition rate from size-classes 1 to 2 at low density and N_1 is the initial density of seedlings. Notice the similarity between this expression and Eqn 4.2 describing the effect of density upon individual plant growth. Analysis of matrices for *E. edulis* showed that elasticities for several life cycle transitions were affected by density dependence acting upon seedlings. In general, elasticities are altered more severely by density dependence than they are by moderate amounts of environmental stochasticity (Grant & Benton 2000).

Incorporating density dependence and environmental stochasticity into a matrix model makes it more realistic, but simpler density-independent time-invariant models still have their uses. Matrix models can be used for two distinct purposes which should not be confused: (i) for discovering something (e.g. the value of λ) about the population *at the present time* by **projecting** from its current status represented by A; and (ii) for **predicting** the future dynamics of a population. Density-independent models are useful for projection, but are less reliable for prediction. For example, Bierzychudeck (1999) found that of two populations of the forest herb *Arisaema triphyllum* that she studied and then revisited 15 years later, one had a structure consistent with the predictions of a simple matrix model but the other had a quite different structure, probably because (among other reasons) density dependence had been omitted from the model. We shall see at the end of this chapter that, when used for projection rather than prediction, elasticities derived from density-independent matrices can be useful for making comparisons between species and populations.

6.4 Annuals with a seed bank

The existence of a seed bank confers an age structure on an annual population, although it is one that is mostly buried in the soil. The seed bank is recharged periodically after each successful generation of adults has flowered, and it accumulates a heterogeneous population of seeds that vary in age, depth of burial in the soil and state of dormancy. All of these characteristics affect the probability of a seed germinating, and all change with time.

Annuals with a pool of dormant seed are able to exploit habitats where recruitment opportunities are irregular such as deserts (Kemp 1989; Venable 1989; Venable & Pake 1999), grassland disturbed by animal activity (Rice 1989), and arable land that is periodically left fallow (Cavers & Benoit 1989). Population densities above ground fluctuate wildly from year-to-year, depending upon the variability in weather and soil disturbance (Fig. 6.5). Because these two

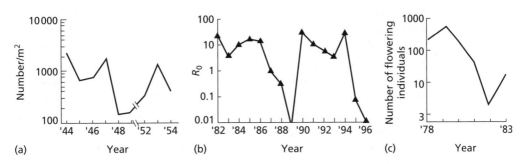

Fig. 6.5 Long-term population dynamics of annuals with a seed pool: (a) density of *Linanthus parryae* (Epling *et al.* 1960); (b) R_0 for *Plantago patagonica* in the Arizona desert (Venable & Pake 1999); (c) flowering plants of *Centaurium erythraea* in calcareous grassland in England (Grubb 1986). Note the log scales on all vertical axes.

163

factors, and the response of dormant seeds to them, are so important to recruitment, the role of changes in seed density in generating the patterns of abundance
seen among adults can be much more difficult to determine than it is for annuals
without a seed bank.

As well as buffering the genetic composition of a population (Chapter 5), a seed
bank also buffers local populations from extinction (Chapter 7) and may affect
spatial pattern (MacArthur 1972). Annuals with no seed bank can only persist
where R_0 is *always* > 0. This is only likely in, or adjacent to, places where the
average value of R_0 is high. We would therefore expect annuals with no seed bank
to be very patchy and that where they do occur they should be abundant. This
constraint is much less rigid for annuals with a seed bank, which may therefore be
less patchy and exhibit a wider range of densities in the years when they are
abundant. In summary then, annuals without a seed bank should have high spatial
patchiness and low temporal variance, while species with a seed bank should have
the reverse characteristics. There is at present no empirical test of most of these
predictions (but see Section 10.4.2.2).

We have seen that for a plant with no seed bank λ is influenced by rates of
mortality and fecundity. For plants with an age or stage structure, the rates of
transitions between classes also influence λ (Box 6.1). In the particular case of an
annual with a seed bank, the survival of seeds in the soil, and the proportion of
seeds which break dormancy each year, which we shall call G, are both transitions
whose value will affect λ. The question is, what contribution do they make to λ,
compared to the contributions of fecundity and survival?

Fluctuations in weed populations from year to year are frequently attributed to
the influence of weather (Freckleton & Watkinson 1998a), but how intrinsically
stable should we expect annuals with a seed bank to be anyway? Rice (1989)
analysed an age-structured model of an annual plant with a seed bank to determine the effect of germination delays upon the **transient dynamics** of such
populations. This analysis counted the number of iterations it takes for a projection matrix to converge on a stable age/stage structure to measure the stability of
the population. Rice found that low values of G, which cause large numbers of
viable seeds to be carried over from one year to the next in the seed bank,
dramatically increase the time it takes for population structure to converge.

The results from models of age-structured seed populations point to the importance of understanding events in the soil better if we are to understand the complicated dynamics of weeds and other annuals with seed banks. Freckleton and
Watkinson (1998b) parameterized one such model for the population dynamics
of the weed *Chenopodium album* growing in crops of sugar beet (*Beta vulgaris* ssp.
cicla). The model was based on a difference equation of the form we met in Chapter
5 (Eqns 5.7–5.10) describing the density-dependent regulation of seed production,
with additional terms for interspecific competition with sugar beet and for the
input, survival and output of seeds in the seed bank. Freckleton and Watkinson
(1998b) used parameter values in their model taken from studies in temperate
regions of Europe, New Zealand and Japan and found remarkable agreement

between localities. Sensitivity analysis of the model suggested that site differences would be much less important to *C. album* population dynamics than variation in the rates of mortality and germination of seeds and control measures used against plants.

Kalisz and McPeek (1992) used a matrix model to determine the importance of seed dormancy and the fate of seeds stored in the seed bank to the demography of experimental populations of the winter annual *Collinsa verna* in deciduous forest in Illinois. Their model contained an age-structured seed bank for seeds that germinated one, two, or three or more years after entering the soil. Elasticity analysis of matrices for two different years revealed that in a year when population growth rate was high ($\lambda = 1.8$) seeds from the seed bank were unimportant to λ, but that in a year of population decline ($\lambda = 0.41$) the germination of 1-year-old seeds made a significant contribution to λ. The contribution of older seeds was negligible in both years. We shall see in Section 10.4.2.2 that temporal variation in reproductive success is crucial to the evolution of seed dormancy.

6.5 Perennials

The distinction between annual and perennial plants is sometimes an arbitrary one. Many species such as annual meadow grass *Poa annua* (Law *et al.* 1977), wild rice *Oryza perennis* (Sano *et al.* 1980), eelgrass *Zostera marina* (Gagnon *et al.* 1980), and even desert 'annuals' such as *Astragalus lentiginosus* and *Tridens pulchellus* (Beatley 1970) have populations in which some individuals are annuals and others perennate. Perennation itself is an important life history trait that influences λ (Chapter 10).

An important reason why so many herbs sit astride the divide between the annual life history and a longer lived one is because perennation is inherent in the modular construction of plants, which grow by the addition of new ramets (Chapter 1). In *Poa annua*, and indeed all grasses, tillers die after flowering. If all tillers flower simultaneously in the first year, the genet is annual and dies after reproduction. If some tillers remain vegetative in the first year, the plant will perennate. In fact, it is generally believed that annual species derive from perennial ancestors and therefore we should really think of annuals as plants that have evolved precocious reproduction, not as plants unable to perennate.

6.5.1 Herbs

Mature individuals of many perennials are composed of collections of a few, to many ramets. The entire demography of such herbs is dependent upon the birth rates and death rates of their parts, because as long as this remains in positive balance the genet is alive and will grow. The genet dies with the death of its last ramet. Perennial herbs vary greatly in their longevity, but this has little if anything to do with the longevity of parts. Tillers of the grass *Festuca rubra*, for example, are quite short-lived, but Harberd (1962) found that most plants of this species in a

90 m × 90 m area in a Scottish population belonged to one of only a few large clones (see also Grant *et al.* 1992). One clone identified at the site had fragments distributed over an area more than 200 m in diameter and must have been quite ancient (Harberd 1961). Molecular genetic markers can be used to identify widely dispersed ramets belonging to the same clone (Widén *et al.* 1994; Harada *et al.* 1997).

Even herbs without the tendency to spread shown by *Festuca rubra* may live to considerable ages. The woodland herbs *Hepatica nobilis* and *Sanicula europaea* were mapped in permanent plots in Sweden by Tamm over a period of nearly 40 years (Inghe & Tamm 1985, 1988). Both species have a rhizome, which slowly advances by tiny increments as the annual shoot dies and a new shoot replaces it. Individuals recorded by Tamm fall into two distinct classes: a group of proven survivors he called the 'old guard' that were present when he first mapped the plots in 1943, many of which were still alive in 1981, and a group of individuals recruited since 1943 that had a much higher mortality rate. By fitting exponential decay curves to the survivorship curves for the old guard, Inghe and Tamm (1985) calculated half-lives (the time taken for 50% of individuals to die) in different plots from 32 years to 360 years for *Hepatica nobilis*, and from 74 to 221 years for *Sanicula euopaea*. These half-lives may underestimate genet survival, because in both species the rhizome can branch. Although branching occurred relatively infrequently, it was sufficient to confer virtual immortality on the old guard in some plots.

As Tamm's study illustrates, there may be significant demographic differences between different parts of the same population. Variation in spatial and temporal demographic parameters occurs in most herbs that have been studied, although few populations have been monitored for as long as Tamm's; thus, it is difficult to know how important demographic variablity is in the long run. The answer should depend upon the longevity of adults and the sensitivity of λ to recruitment, because this tends to be the most vulnerable stage in the life cycle, and consequently also the most variable. Other things being equal, the populations most at risk of extinction from environmental stochasticity should be those with a short lifespan and low population density. How do such populations manage to persist? Rabinowitz *et al.* (1989) investigated this question for four species of prairie grass whose populations are typically sparse (i.e. they occur at low density). Over a nine-year period at Tucker Prairie, Missouri, the fecundity of the four sparse species varied far less than that of three common ones at the same site. The reason was that the sparse species flowered early in the year when rainfall was fairly reliable, but the common species flowered later when rainfall was more uncertain. It appears that early flowering buffered the sparse species against reproductive failure due to drought. The common species were probably better able to survive frequent reproductive failure because they are longer lived and occur at higher population density than the sparse species.

All demographic studies suffer from the problem that data collected over only a few years may be atypical. Also, variability between years may itself play a role in population dynamics and life history evolution that is obscured by using mean values (Chapter 10). Sensitivity analysis can be used to overcome the first problem to some extent, because it can test the robustness of λ to perturbation caused by changes in the value of particular demographic parameters. Moloney (1988) studied a population of downy oatgrass *Danthonia sericea* growing along a soil fertility gradient in a mown field in North Carolina. Plants were mapped three times at yearly intervals, and classified into six size-classes for matrix analysis. No plants in the two largest size-classes died during the study, even though there was a drought in one of the two years. In the smaller size-classes there were significant differences in survival between sites along the gradient, and between years. Elasticity analysis showed that recruitment made little contribution to λ anywhere on the gradient, and that, averaging over all locations and both years, the leading diagonal of the matrix, representing survival of plants in their existing size-class, contributed about 70% and the subdiagonal (growth to the next size-class) most of the remainder. Experimental addition of seeds to cleared and uncleared plots showed that recruitment was largely controlled by the availability of colonization sites (Moloney 1990), competition for which regulated the population. Although Moloney found significant variation from year-to-year in some parameters of the projection matrix, this may not be important in the long run, because the population is regulated and the stages with the largest influence on λ were large plants whose survival was not affected by year-to-year variation. Long-lived adults, like a seed bank, are a cushion against temporal variability.

Natural selection can lead to evolutionary divergence in demographic parameters, causing them to differ on a large or small spatial scale (Chapter 3). This has been elegantly demonstrated in ribwort plantain, *Plantago lanceolata*, by van Groenendael (1985), who compared populations in a dry dune grassland and a wet meadow in the Netherlands. The dune grassland was nutrient poor, subject to drought and had been grazed by cattle and horses for over three centuries. In contrast the wet meadow was on waterlogged peat, was mown once a year and was more fertile. Substantial demographic differences were found between the populations of *P. lanceolata* on the two sites: (i) juveniles in the dry grassland had much higher survival and adults much lower survival than in the wet meadow; (ii) fecundity in the dry site was higher and seed size was half that in the wet meadow; (iii) plants of the dry grassland first reproduced at one year of age, while in the wet meadow this occurred at three years; and (iv) dry grassland plants produced vegetative side-rosettes while wet meadow plants did not (van Groenendael & Slim 1988). Elasticity analysis of Goodman matrices for the two populations pointed to some underlying demographic similarities between them, as well as confirming the significance of their differences. Although there was a significant seed bank at the dry site, this was of little importance in either population. Fecundity was subsidiary to establishment and survival in both populations, although it was most important in the dry grassland (van Groenendael & Slim 1988). Timing of germination was different in the two populations and significantly influenced λ. At the dry site, most seeds germinated in spring, while at the wet site most germinated in autumn.

By raising plants from the two populations in a uniform greenhouse environment, van Groenendael (1985) showed that a number of the differences observed in the field had a genetic basis. Seeds, juveniles and adults were also transplanted between habitats to determine the extent to which adaptation to the native habitat created a handicap in a different environment. After two years in alien habitats, transplants showed significantly decreased survival compared to natives (van Groenendael 1985). Seedling transplants of *P. lanceolata* were also made by Antonovics and Primack (1982) between different habitats in North Carolina. In this instance the fate of transplants was different between sites, but there was no detectable genetic component to how seedlings of different origin survived in different environments. Other examples are given in Chapter 3, Table 3.4.

6.5.2 Forest trees

The recruitment of new individuals into tree populations is controlled by disturbance of the forest canopy and the appearance of gaps. Forest trees fall into two broad categories: **pioneers** and **primary species**, both of which depend upon gaps for recruitment, though in different ways (Whitmore 1989).

Pioneers such as *Betula* spp. in northern temperate forests and *Cecropia* spp. in the neotropics appear soon after gaps are created, germinating from seeds already in the soil or from seeds brought in by wind or animals. Pioneers are fast growing, short-lived and produce copious, regular crops of small seeds that recharge a dormant seed bank in the soil. These seeds are unearthed by the upheaval of roots when a tree falls (Putz 1983; Whitmore 1983). Seeds of *Cecropia obtusifolia* in Mexican rainforest are triggered into germination by the light conditions in gaps (Vazquez-Yanes & Smith 1982), and in French Guiana *C. obtusa* actually establishes best on the mounds of fallen trees (Riera 1985; Nuñez-Farfán & Dirzo 1988). Germination of another neotropical pioneer *Heliocarpus donnell-smithii* is stimulated by the diurnal temperature fluctuations characteristic of gaps (Vázquez-Yanes & Orozco-Segovia 1982). The pulsed recruitment typical of pioneer species generates local patches of even-aged individuals.

In temperate regions at high elevations, and where disturbance such as fire is recurrent, pioneers may replace themselves (Whitmore 1988), but these trees are relatively intolerant of shade as juveniles and are generally succeeded by primary species that are slow growing, but more shade-tolerant as well as longer-lived. This was the pattern found by Brokaw (1985) in gaps in the seasonal tropical forest at Barro Colorado Island, Panama (BCI).

Seed production by primary species such as the oaks *Quercus* spp. and beeches *Fagus* spp. of the northern temperate zone or the dipterocarps of SE Asia is typically irregular (Nilsson & Wästljung 1987; Ashton *et al.* 1988). They produce huge crops of seeds every 5–10 years, but few or none in the intervening periods. The seeds are often large and do not remain viable for long. This, along with the heavy losses resulting from animal predation, prevent the formation of a dormant seed bank. Although irregular, fruiting is usually synchronized within a species, and sometimes between species, over large geographical areas. This pattern of seed production is called **masting** and appears to reduce losses to predation by swamping seed consumers with food in mast years and starving them in the intervening years of scarcity (Section 10.3).

With no dormant seed bank from which recruits can germinate when a gap opens, but with the ability to tolerate shade, primary tree species recruit from seeds that germinate under a closed canopy. These produce stunted juveniles that linger around the photosynthetic compensation point, waiting for a gap to throw light down upon them. Like the character Oskar in *The Tin Drum* by Günter Grass, these juveniles age but do not grow. Until, that is, a gap appears admitting light. The forest floor in temperate and tropical forests alike may be littered with **oskars** (ageing juveniles), many of them decades old. In beech–maple forest in New York State, Canham (1985) found that mature sugar maples had gone through several cycles of suppression and release before reaching the canopy. Hemlock *Tsuga canadensis*, another primary species of forests in eastern North America, typically shows signs of successive bouts of suppression and release in the width of growth rings in the trunks of mature trees (Henry & Swan 1974). Oskars of *Aglaia* sp. growing in Malaysian dipterocarp rainforest also needed several openings of the canopy before reaching maturity (Becker & Wong 1985).

Most forest gaps are small (Fig. 6.6), caused by branch fall, and these are rapidly filled by the growth of the surrounding canopy. Slightly larger gaps are created by trees that die on their feet (called snags) and oskars of primary species may be 'released' by these for a brief period, but pioneers usually require larger gaps. Large gaps occur when trees are blown over, often taking their neighbours with them. At BCI, Brokaw (1985) found that the pioneer *Trema micrantha* was strictly confined to large gaps. In several North American forests studied by Runkle (1982, 1998) larger, longer lasting gaps were colonized by more species than smaller ones that offered a briefer opportunity for recruitment. Slower growing saplings

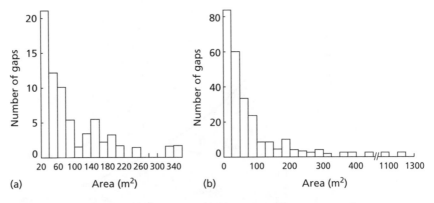

Fig. 6.6 The size distribution (m^2 surface area) of forest gaps: (a) 66 gaps occurring over a 5–year period in 13.4 ha of tropical forest on Barro Colorado Island, Panama (from Brokaw 1982); (b) 243 gaps occurring over a two-year period in 34.7 ha of temperate beech forest at Tillaie, Fontainebleau, France (from Faille *et al.* 1984).

required larger gaps or more recurrent openings to reach the canopy than faster growing ones (Fig. 6.7).

Gap-dependent regeneration creates a mosaic of patches of different species and age composition (Fig. 6.8). At BCI, Hubbell and Foster (1986b) found that the abundance of shade-intolerant species was correlated with the abundance of canopy gaps. Sarukhán *et al.* (1985) devised an ingenious way to map the mosaic age structure in a 5-ha permanent plot of tropical rainforest in Mexico, using the understorey palm *Astrocaryum mexicanum*, which was abundant there, and can be aged from leaf scars on its trunk. Tree and branch falls in the forest knock over any slender *Astrocaryum* stems beneath them. Palms prostrated in this way are rarely killed in the process, but continue growth by turning the apical shoot to the vertical again. This produces a kink in the stem of the tree that can be used to calculate how long ago the palm was knocked down by counting leaf scars back from the apex. The map produced by this means showed that a large proportion of the forest had experienced some form of disturbance in the previous 100 years (Fig. 6.8b). Gaps occurred particularly frequently on steep slopes, and their annual rate of formation was correlated with rainfall (Martínez-Ramos *et al.* 1988).

The demography of *Astrocaryum mexicanum* was studied by Piñero *et al.* (1984), who found that its population in the study area as a whole was at equilibrium. However, λ was most sensitive to the growth rate of trees and, although it is a primary species that is tolerant of shade, adult *A. mexicanum* grew faster and produced more fruit in gaps than under a closed canopy. This created local variation in the value of λ, depending upon canopy conditions (Martínez-Ramos *et al.* 1989).

It may seem a paradox that forests where parts of the canopy are continually being torn down and regrown are at equilibrium, but this alternation between disturbance and stability is the very essence of how forests work. A study of tropical rainforest at La Selva in Costa Rica, for example, found that over a 13-year period

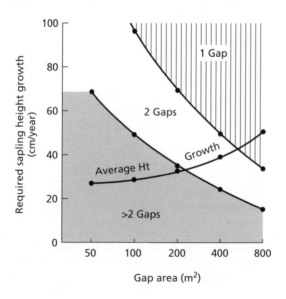

Fig. 6.7 Rates of height growth required by saplings to reach the canopy for gaps of different size and frequency of recurrence in old-growth forests of the Great Smoky Mountains National Park, USA (from Runkle & Yetter 1987).

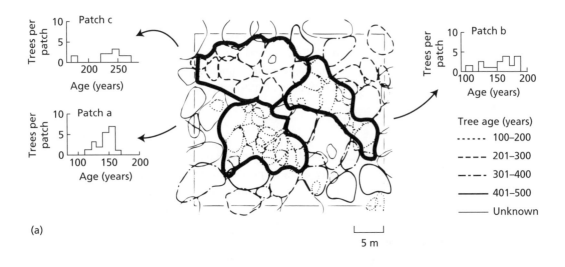

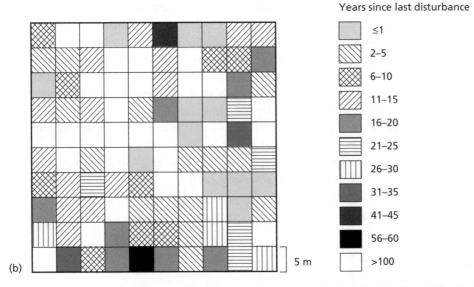

Fig. 6.8 (a) The age-structure mosaic of a 30 m × 30 m plot in a subalpine Tsuga-Abies-Picea forest in Honshu, Japan; trees in each patch are similar but not identical in age because they began life as oskars (from Kanzaki 1984). (b) The age-structure mosaic in a 50 m × 50 m plot of tropical rainforest at Los Tuxtlas, Mexico (from Sarukhán *et al.* 1985).

there was an annual loss of 2% of all tree stems ≥ 10 cm d.b.h., nearly a third of which fell, and a quarter of which died standing (Lieberman *et al.* 1985a). The same annual rate of mortality among trees of this size has also been found at BCI (Condit *et al.* 1995). At La Selva this remarkably high rate of loss was compensated *exactly* by recruitment. There were 105 deaths ha^{-1} over the period of study and 104.3 recruits ha^{-1} (Lieberman & Lieberman 1987). This demonstrated numerical equilibrium in the forest as a whole, but there was also evidence that populations of individual species were at equilibrium too. For a sample of 44 tree species at La

Selva, Lieberman *et al.* (1985b) found that the rate of recruitment was negatively correlated with the longevity of species (Fig. 6.9), or in other words mortality and recruitment rates were positively correlated for individual species as well as for the forest as a whole. Density–dependent population regulation is also prevalent among the trees at BCI (Wills *et al.* 1997) (See Section 8.4.5)

6.5.3 Overview

It is easy to get lost in the details of demographic studies and to lose sight of the bigger picture. Elasticity analysis provides a tool that can be used to compare populations and species with one another, affording an overview. The technique makes use of the fact that elasticities always sum to unity across an entire matrix and that stage projection matrices can be divided into regions that represent different processes (Fig. 6.10) (Silvertown *et al.* 1993). For example, the cells in the leading diagonal of a matrix represent the probabilities of individuals remaining in the same stage between two census intervals (called **stasis**) and the sum of the elasticities for these cells tells us how important stasis is to λ. The impact on λ of small changes in fecundity is given by the sum of elasticities in the top row of a typical matrix (excluding cell a_{11}) and growth between size-classes is represented by the cells that lie below the leading diagonal. By dividing a matrix into elements that represent each of the processes growth (*G*), stasis (*L*) and fecundity (*F*), and summing the elasticities relating to each of these, we obtain measures of the proportional impact of *G*, *L* and *F* upon λ that can be compared with other populations and species.

Populations can be compared with each other by plotting their locations in *G*–*L*–*F* space. The space is triangular because $G + L + F$ always $= 1$, and is referred to

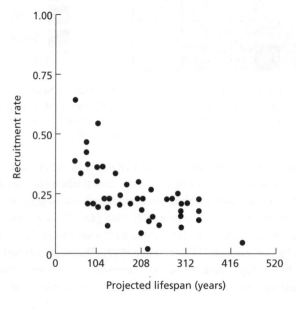

Fig. 6.9 Relationship between recruitment rate and longevity for 44 species of trees at La Selva (from Lieberman *et al.* 1985b).

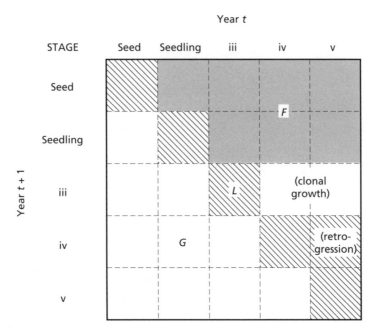

Fig. 6.10 An ideal stage projection matrix illustrating regions that represent growth (G), survival (L) and fecundity (F) (after Silvertown *et al.* 1996).

as the **demographic triangle**. This simple procedure has produced some useful insights. First, it is seen that semelparous perennial species fall mainly in the top corner of the triangle where growth and fecundity are more important than stasis (Fig. 6.11a). Iteroparous perennials tend to lie nearer the L-axis, with woodland herbs nearer to this axis than herbs of open habitats (Fig. 6.11b,c). Woody species lie closest to the L-axis and the longest-lived ones cluster in the $L = 1$ corner (Fig. 6.11d). There are mathematical constraints on where woody species and certain other types of life history may fall in the demographic triangle (de Matos & Silva Matos 1998), but demographic and environmental processes still influence their precise locations. So, for example, populations of the palm *Euterpe edulis* move along a trajectory of increasing importance of L and decreasing importance of G and F as population density increases (Fig. 6.12a).

Bullock *et al.* (1994) found that different populations of the semelparous thistle *Cirsium vulgare* all lay in a small, upper segment of the demographic triangle, but that increasing grazing pressure reduced λ and moved populations towards higher values of L and lower values of F (Fig. 6.12b). Grazing upon six species of South African perennial savannah grasses studied by O'Connor (1993) had a similar relative effect upon elasticities (Fig. 6.12c). Silvertown and Franco (1993) suggest that both within-species and between-species patterns of distribution in the demographic triangle are consistent with the overall successional trajectory shown in Fig. 6.12(d).

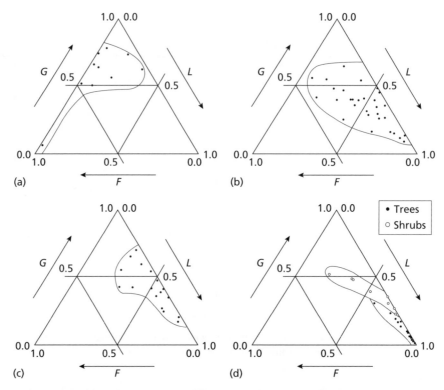

Fig. 6.11 Location in the demographic triangle of: (a) 10 semelparous herbs of open habitats; (b) 32 iteroparous herbs of open habitats; (c) 18 iteroparous forest herbs; and (d) 24 woody plants comprising eight shrubs (open symbols) and 16 trees (closed symbols) (from Silvertown *et al.* 1996).

6.6 Summary

Recruitment in plants tends to follow **disturbance**, which is an important source of **environmental stochasticity** for plant populations. **Demographic stochasticity** is chance variation in local population structure that affects demographic parameters. Populations in which recruitment is a frequent event by comparison with the lifespan of individual plants develop an **age structure**. Even among plants of uniform age, size may vary considerably and create a **stage structure**. A population structured by age or size may be represented by a **column vector**, which lists the number of individuals in each age- or size-class at a particular time. This vector may be used to project the numbers in each class at a later time by multiplying the column vector by a **projection matrix** which contains the rates at which individuals move from one class to another during the appropriate time interval.

Annuals with a **seed bank** in the soil have an age structure that provides a demographic buffer against extinction and a genetic buffer against genetic drift and natural selection. The importance of the seed bank, or indeed transitions involving

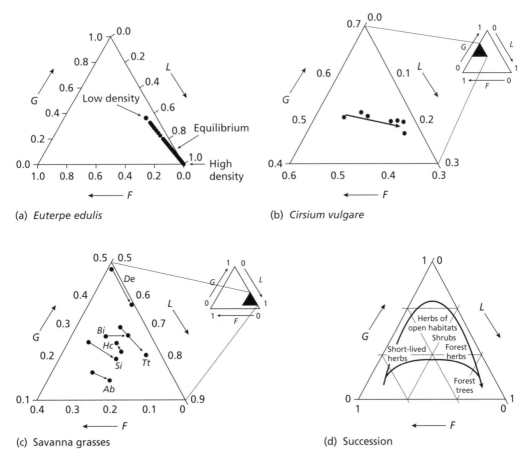

(a) *Euterpe edulis*

(b) *Cirsium vulgare*

(c) Savanna grasses

(d) Succession

Fig. 6.12 Effect upon location of populations in *G–L–F* space of: (a) population density in *Euterpe edulis* (Silva Matos *et al.* 1999); (b) sheep grazing intensity in populations of *Cirsium vulgare*; (c) grazing upon South African perennial savanna grasses; and (d) succession (b–d from Silvertown & Franco 1993).

any component of the life cycle, to the value of λ can be calculated by **sensitivity** and **elasticity analysis** of the projection matrix. Because λ can be used as a measure of fitness as well as a measure of population growth, elasticity analysis indicates the contribution of different parts of the life cycle to **fitness**. The ramets of perennial herbs usually have a relatively brief lifespan. The genets of these herbs may reach large size and live to advanced age, though survival rates tend to vary a great deal in space and through time. Spatial variation in mortality rates can produce **disruptive selection** that leads to **genetic differentiation** between populations. Forest trees fall into two broad categories: **pioneers** and **primary species**. Pioneers are usually the first trees to appear in treefall **gaps**, but are generally replaced by primary species that are able to recruit under a closed canopy. However, primary species also usually require small canopy gaps to permit recruits to reach the canopy. The effect of canopy gaps on recruitment creates a mosaic pattern of age structure and species composition in forests. Demographic differences between plant populations are reflected in the relative value of elasticities for three different components of their respective projection matrices, known as G, L and F. Populations and species may be compared by plotting these values in the **demographic triangle**.

6.7 Further reading

Bond, W. J. & van Wilgen, B. (1995) *Fire and plants*. Chapman & Hall, London.

Caswell, H. (2000) *Matrix Population Models*, 2nd Edn. Sinauer Associates, Sunderland, MA.

Ebert, T. A. (1999) *Plant and Animal Populations: Methods in Demography*. Academic Press, San Diego, CA.

Heppell, S., Pfister, C. & de Kroon, H. (2000) Elasticity analysis in population biology: methods and applications (Special Feature). *Ecology*, 81, 605–708.

6.8 Questions

1 Distinguish between: (a) age structure and stage structure; (b) environmental and demographic stochasticity; (c) a column vector and a projection matrix; (d) Leslie and Lefkovitch matrices; (e) prediction and projection from matrix models; and (f) pioneer and primary species of forest tree.

2 Use your own observations and/or any information you can find in the literature to estimate the longevity and recruitment interval for some plant species that grow in your locality. Do observed population age structures for these species fit the prediction of Fig. 6.3? If there are deviations from expectation, how can you explain them?

3 Use the information given in Box 6.1 to program a spreadsheet to perform matrix multiplication for a Leslie matrix representing an age-structured population. Then modify your program to calculate λ for the stage-structured population of *Dipsacus sylvestris* depicted in Fig. 1.5(c).

4 Use a spreadsheet to construct a matrix model for *Chenopodium album*, including one stage for the seed bank (see p. 164), but ignoring competition from sugar beet. Parameters used by Freckelton & Watkinson (1998b) in their model were: probablity of seed germination $= 0.10$ per annum; probability of seed survival in the seed bank $= 0.70$ per annum. Use Eqn 5.9 to calculate the seeds produced in each iteration of your model and parameterize it with the following values from Freckelton & Watkinson (1998b): $s_m = 1.5 \times 10^4$, $a = 0.10$, $b = 1$.

5 Use your model to perform transient analysis (p. 164) for *Chenopodium album*. Explore how the dynamics of *C. album* change when the values of various parameters are altered one at a time. For example, what happens when $b = 2$, or when seed survival is halved? If you wish to incorporate competition from sugar beet into your model, you need to know that Freckelton & Watkinson (1998b) found the effect of one crop plant on *C. album* to be the same as the effect of one weed plant.

7 Regional dynamics and metapopulations

7.1 Introduction

So far in our exploration of population dynamics we have implicitly assumed that the principal demographic processes that affect numbers are birth and death, and that immigration and emigration are of little numerical importance, even though migration may be significant enough to create gene flow (Chapter 3). In some circumstances this omission is certainly wrong: immigration rates in particular affect the size and fate of local populations. To consider this kind of question we need to go up in scale from the local to the **regional dynamics** of populations.

The spatial distribution of species is patchy at many scales (Fig. 7.1, and Section 3.2.3); it tends to be particularly fragmentary near range boundaries and for many species it is becoming increasingly fragmented by habitat destruction, which is often concentrated in the middle of species' geographical ranges (Channell & Lomolino 2000). Patchiness at a local scale often reflects past disturbance. In herbaceous vegetation, microhabitats such as gopher mounds in Californian serpentine annual grassland (Hobbs & Mooney 1985), badger mounds in tallgrass prairie (Platt 1975), molehills and old dungpats in pastures (Jalloq 1975, Parish & Turkington 1990a,b), and ant hills in British chalk grassland (King 1977a), are all colonized by species rare or absent in the surrounding vegetation. The mosaic structure of forest composition reflects past spatial patterns of disturbance and recruitment (e.g. Fig. 6.7). Should each of these island-like patches be regarded as separate populations, or as fragments of a whole? The answer must depend upon how regionally interdependent patches are (i.e. how much migration occurs between populations, see Chapter 3).

Three kinds of regional dynamics that intergrade with one another can be distinguished (Eriksson 1996), depending upon the relative importance to local populations of between-population processes in the form of net migration (i–e) and within-population processes in the form of net births (b–d) (Thomas & Kunin 1999). First, a **metapopulation** is a network of local populations connected by dispersal, whose persistence depends upon *reciprocal* but unsynchronized migration between local populations (Fig. 7.2a). Second, if migration is consistently one-way, from source populations where $R_0 > 1$ to recipient populations where $R_0 < 1$, **source–sink** dynamics prevail (Pulliam 1989) (Fig. 7.2b). Third, **remnant**

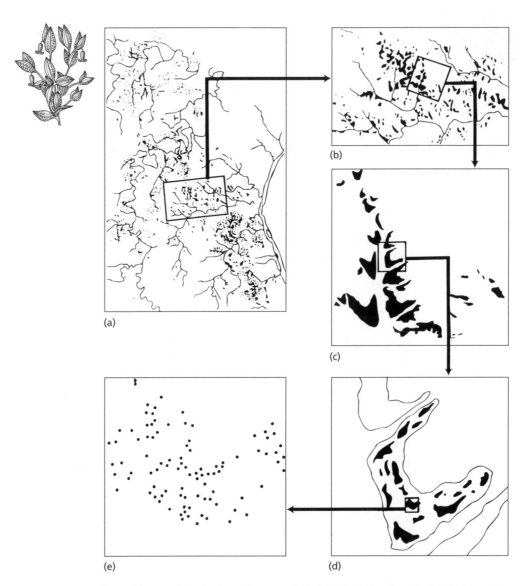

Fig. 7.1 The spatial distribution of the perennial herb *Clematis fremontii* var *Riehlii* (pictured) at a range of scales within its geographical range, which was found by Erickson (1945) to be limited to an area of 1129 km² in Missouri. Only about 1.5% of this area provided habitat for the plant, which was confined to rocky outcrops called 'barrens' where the broad-leaved forest cover was broken, forming open glades. (a) The geographical range; (b) one of four major regions within the range, each of which contained on average 50 glade clusters; (c) a glade cluster which contained on average 30 glades; (d) a glade containing patches of the plant; and (e) individual plants within a patch (from Erickson 1945).

populations may be able to persist without immigration through periods that are unfavourable to recruitment, because they are long-lived as seeds in the soil, oskars (Chapter 6) or adults (Grubb 1988, Eriksson 1996) (Fig. 7.2c).

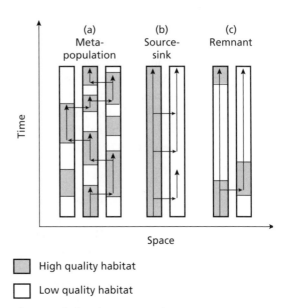

Fig. 7.2 A comparison between three types of regional dynamics based upon seed flow between patches and persistence within them. (a) Metapopulation dynamics; (b) source–sink dynamics; and (c) remnant population dynamics (after Eriksson 1996).

7.2 The classical metapopulation model

Just as the size of a local population is determined by the birth rate and death rate of individuals, the size of a metapopulation is determined by the rates of establishment and extinction of component populations (Levins 1970). Imagine a species that requires a particular type of site for successful establishment and reproduction and that these sites have a scattered distribution. The plants of badger or gopher mounds, and forest gaps, are examples. We will call the density of populated sites at time t P_t and the density of vacant sites V_t. Populated sites become extinct at a rate x per population per time interval and vacant sites are colonized at a rate c per population per time interval. The number of populated sites per unit area at time $t + 1$ is then:

$$P_{t+1} = P_t + cP_t V_t - xP_t, \tag{7.1}$$

where $cP_t V_t$ is the number of new sites that are colonized and xP_t the number of sites where existing populations become extinct. This simple equation makes some interesting predictions about how metapopulations will behave (Levins 1970). At equilibrium $P_{t+1} = P_t$ so from Eqn 7.1:

$$cP_t V_t = xP_t$$

thus,

$$V_t = x/c, \tag{7.2}$$

which means that if a species is to spread ($P_{t+1} > P_t$), the density of vacant sites V_t must exceed the **relative extinction rate** x/c. Only very slight changes in extinction or colonization rates are needed to cross the **extinction threshold** from a situation where $V_t > x/c$ and a species is spreading, to a violation of these conditions that will cause extinction of the metapopulation when $V_t < x/c$.

A little manipulation of Eqn 7.1 reveals how sensitive the regional abundance of populations can be to the relative extinction rate. Let the total number of sites $V_t + P_t = T$, and the *proportion* of sites that are populated $P_t/T = p$. Then $V_t/T = 1 - p$, and from Eqn 7.2 we can replace V_t with the relative extinction rate x/c to give:

$$p = 1 - (x/cT). \tag{7.3}$$

We can apply this equation to the question of why some species are rare. Local populations of rare species are generally few but, with some exceptions (Rabinowitz 1981), individuals in them are abundant. This suggests that the causes of this kind of rarity may lie in the regional dynamics that determine population number rather than in the local dynamics that determine abundance within populations. Rare species that occupy only a few sites have low values of p. For a rare species with $p = 0.01$, a decrease of only 10% in the value of x/cT, (from 0.99 to 0.90) is all that is required to increase the proportion of occupied sites tenfold, to $p = 0.1$ (Levins & Culver 1971). It is thus easy to imagine how slight ecological differences between species in the traits that influence x, c and T could be difficult to detect and yet radically affect their commonness or rarity.

The classical metapopulation model of Levins (1970) unrealistically assumes that patches are of equal size and are equally connected to each other by migration. Hanski (1985) created a simple extension of the Levins model to allow for two sizes of patch (large and small) and found that the fraction of populated sites (p) could be much greater than predicted by the Levins model (Eqn 7.3) because, when strong enough, migration created a **rescue effect** by turning small patches that were vulnerable to extinction into large ones that did not become extinct. The rescue effect operates only when the fraction of populated sites is large, because when many sites are vacant migrants from large populated sites rarely land in small populated sites. This and other refinements and alternatives to the classical metapopulation model are discussed in detail by Hanski (1999).

All metapopulation models based upon Levins' original conception have in common the idea that abundance at the metapopulation level results from a balance between local population extinction and colonization rates. For these models to be relevant to plants, it must be established that natural populations do have turnover (i.e. do go locally extinct), that local population dynamics are not synchronized across populations (or all might go extinct simultaneously), and that seed dispersal between populations, rather than the seed bank, is the chief source of colonists in vacant sites (otherwise the remnant population model would apply, see pp. 177–8). We will consider these questions for a number of examples in the next Section and then consider the causes of local extinction (Section 7.4), the genetic and evolutionary consequences of regional dynamics (Section 7.5) and the link to the geographical scale (Sections 7.6–7.8).

7.3 Regional dynamics of plants

Unequivocal examples of pure metapopulations are very hard to find in plants, and indeterminate or intermediate cases appear to be so common that it is impossible to place most species into one of the three categories of regional dynamics. By default, we therefore discuss examples broken down by life history rather than by type of regional dynamics.

7.3.1 Annuals and semelparous perennials

Extinction and colonization rates are often high, and thus easiest to quantify, in short-lived plants. The water hyacinth *Eichhornia paniculata* is found in temporary pools and water-filled ditches in NE Brazil where it behaves as an annual. Through repeated censuses Husband and Barrett (1998) found that on average 36% of local populations of *E. paniculata* went extinct each year, apparently without leaving behind a viable seed bank. Local population dynamics were not synchronized between patches within a region. Comparing different regions, there was a positive correlation between the density of available sites (T_t) recorded along transects and the number occupied (P_t), with a lower threshold value of T_t of around 0.2 sites km^{-1} (Fig. 7.3), suggesting that metapopulation dynamics and an extinction threshold governed the regional distribution of *E. paniculata*. The regional dynamics of two short-lived semelparous perennials was studied by van der Meijden *et al.* (1992) in a sand dune area in the Netherlands. Ragwort (*Senecio jacobaea*) presented a complicated picture with local populations at some sites repeatedly going extinct, but others persisting throughout 17 years of observation. Vacant sites were recolonized about equally from a seed bank and by dispersal from persistent refugia, suggesting that the regional dynamics of the species lay somewhere between the remnant population and source–sink types. Another species at the site, *Cynoglossum officinale*, showed unsynchronized local

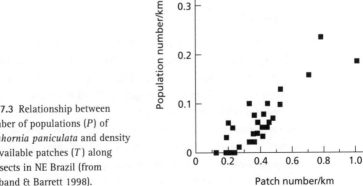

Fig. 7.3 Relationship between number of populations (*P*) of *Eichhornia paniculata* and density of available patches (*T*) along transects in NE Brazil (from Husband & Barrett 1998).

dynamics, and extinctions were more frequent than colonization events during 9 years of monitoring, perhaps reflecting regional dynamics nearest the remnant population type.

In open habitats, high-density plant patches may act as sources of seeds that disperse into nearby sink populations where plants can establish but not maintain $R_0 > 1$. The landward population of *Cakile edentula* on the transect studied by Keddy (1981) (Fig. 4.19) was a sink maintained by seed migration from a source at the seaward end of the transect (Watkinson 1985). In the savannah woodlands of NW Australia seeds migrating from dense patches of the grass *Sorghum intrans* sustain sink populations of lower density in poor quality patches (Watkinson *et al.* 1989). Kadmon and Shmida (1990) experimentally assessed the relative contributions of seed dormancy, seed dispersal and *in situ* seed production to λ in two populations of the annual desert grass *Stipa capensis* in Israel. The seed pool made only a minor contribution to either population, but in one it was estimated that 90% of plants arose from seed dispersed from elsewhere. Venable and Pake (1999) conducted similar experiments with six annual species growing under shrubs and in open habitat in the Sonoran Desert at Tucson, Arizona. They found that seed dispersal was of minor importance to most populations in either type of habitat, even though the direction of seed flow between shrub and open-habitat populations differed between species.

Watkinson and Sutherland (1995) caution that, if density dependence is ignored, some populations may appear to be sinks when in fact they could sustain themselves at a lower density without immigration. They called these **pseudo-sink** populations. To test for this possibility in desert annuals found beneath shrubs and in adjacent open areas in Israel, Kadmon and Tielbörger (1999) isolated plots of each type from external sources of seed and monitored the densities of annuals compared to non-isolated controls over four years. Effects were not clear-cut because natural abundances of species varied a good deal and were often low, but *Ifloga spicata*, one of the most consistently abundant species, did decline under shrubs to about one quarter of its density in control plots. Only 8% of recruits in this species came from the seed bank, so it may have formed pseudo-sink populations under shrubs.

7.3.2 Iteroparous perennials

In a study of a group of five perennials that colonize soil disturbances made by badgers in tallgrass prairie, Platt (1975, Platt & Weiss 1977, 1985) describes a community in which metapopulation dynamics appear to be responsible for between-species differences in spatial distribution and abundance. The study site at Cayler Prairie, Iowa, was on sloping ground where the density of ground squirrels was highest at the top and decreased progressively downslope. The density of badger disturbances, made by these animals digging for ground squirrels, followed the same pattern and decreased downslope, while soil moisture increased in the same direction. None of the five plant species studied had a persistent seed pool, so they all relied on seed dispersal to reach new disturbances. The species were segregated along the density gradient, of disturbances in a remarkable way (Fig. 7.4). At one extreme of the density gradient, only *Mirabilis hirsuta* occurred, because other species could not colonize the very dry sites that were found there. At the other extreme of the density gradient, only *Apocynum sibiricum* was present, because no other species had sufficiently good dispersal to reach remotely spaced sites (Platt & Weiss 1977). Why then didn't all species occur at intermediate disturbance densities? The answer seems to be that rates of colonization and extinction for each species were negatively correlated. The well-dispersed species, such as *Asclepias syriaca*, had high rates of colonization, but low rates of establishment and reproduction when they arrived at sites where other species were present (Platt & Weiss 1985). A high rate of extinction thus kept them out of areas where disturbance density was great enough for more competitive species to colonize. The more poorly dispersed species were confined to areas of high disturbance density, because their colonization rates were too small in areas of low disturbance density. Their competitive superiority over other species ensured low rates of extinction where colonization was possible.

Extinction and colonization rates are difficult to observe directly in long-lived perennials, but can be estimated indirectly when populations occupy successional sites such as forest gaps that can provide a chronosequence of populations of different age. Conceptually, the spatial structure of a metapopulation of perennial plants can be represented by a series of life-cycle graphs, one for each local population, linked together by paths that represent migration between their appropriate stages (Fig. 7.5), and can be analysed using an extension of the matrix methods already applied to age- and stage-structured populations in Chapter 6 (Hanski & Ovaskainen 2000).

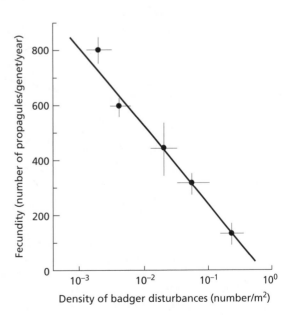

Fig. 7.4 Relationship between mean plant fecundity and mean density of sites occupied by five species of plant colonizing badger disturbances (modified from Platt & Weiss 1985).

The European woodland herb *Primula vulgaris* (primrose) grows in light gaps, and as the canopy closes over these gaps and light falls to below about 2% of full sun, λ drops below the replacement rate of one (Valverde & Silvertown 1995, 1997a, 1998). The coupling of primrose population dynamics to forest canopy dynamics makes it possible to calculate the expected time to extinction for local primrose populations and to model the dispersal rate from established populations that is required to colonize enough new gaps to maintain a metapopulation (Valverde & Silvertown 1997b,c). *Primula vulgaris* has no long-term seed bank, so new gaps must be colonized by dispersal. Most seeds that disperse out of a local population never reach a new gap, and thus the level of dispersal required to maintain a metapopulation significantly lowers the value of λ in local populations (Valverde & Silvertown 1997b,c). However, the depressive effect of dispersal on local values of λ might have been overestimated because no density dependence was incorporated into models of local population dynamics. Similar results have been obtained for the tropical forest herb *Calathea ovandensis* in Mexico (Horvitz & Schemske 1986) and for other temperate species (Cipollini *et al.* 1993, 1994). The metapopulation dynamics of the pioneer tree *Cecropia obtusifolia* colonizing treefall gaps in rainforest at Los Tuxtlas, Mexico, was analysed by Alvarez-Buylla

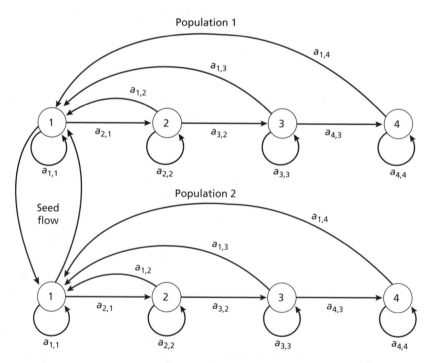

Fig. 7.5 Life cycle graphs for a metapopulation of two local populations of *Primula vulgaris* linked by seed dispersal. The populations each have four stage-classes (1–4, where class 1 are seeds). Within each population the values $a_{1,x}$ are fecundities; values $a_{x,x}$ are the rates of survival for plants that do not change stage class and values $a_{x+1,x}$ are the rates at which plants move up a stage class (compare with Fig. 1.5b).

and García-Barrios (1991) (Fig. 7.6). Although this species has a significant pool of dormant seed in the soil, these are relatively short-lived, and seeds brought into new gaps by birds were the source of about half of all recruits (Alvarez-Buylla & Martinez-Ramos 1990).

7.4 Extinction

Viewed on a sufficiently long timescale, local extinction is the ultimate fate of every population, so it is the *rates* and causes rather than the fact of local extinction that should concern us. Extinction can be caused by **deterministic processes** from which there is no escape such as advancing glaciers or bulldozers, and by **stochastic processes** (Section 6.2) that increase the random probability of extinction. The two kinds of process are not independent of one another. For example, populations are made smaller and more vulnerable to stochastic extinction by the fragmentation of habitats. Chance variation in local population structure caused by demographic stochasticity can lead to extinction. Like genetic drift, the effects of demographic stochasticity increase in importance as population size decreases, and can be retarded by immigration. Population projection models

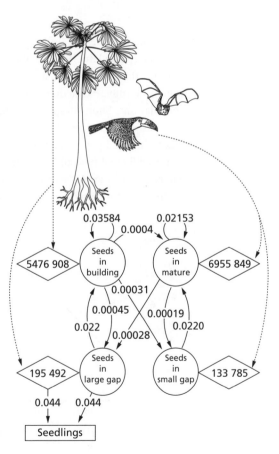

Fig. 7.6 A metapopulation model for *Cecropia obtusifolia* based upon rates of seed production and transport in four types of forest patch (circles). *C. obtusifolia* only recruits successfully in large gaps. Diamonds enclose numbers of seeds falling into each patch type. Arrows show probabilities of seeds in one patch type being transferred to another by dispersal, and by a patch changing from being of one type to another as a consequence of treefall or canopy development (from Alvarez-Buylla & García-Barrios 1991).

(Chapter 6) can be used to estimate how environmental stochasticity will affect demography and influence the probability of extinction (Menges 2000).

Direct estimates of extinction rates in the field are few because they require repeat censuses of large numbers of populations. Bräuer *et al.* (1999) revisited 359 sites in the German state of Lower Saxony where population sizes of eight rare, short-lived plant species had been estimated 10 years previously. Twenty-eight percent of the original populations had become extinct over the 10-year period, and in seven out of eight species the probability of a population surviving was size-dependent (Fig. 7.7). These extinctions were not the consequence of habitat destruction or other obvious deterministic causes. If it is assumed that past trends will continue, then the observed relationship between population size and survival can be used to estimate how big a population has to be to have any given probability of surviving for another decade or more (Brook *et al.* 2000). This population size is often referred to as the **minimum viable population** (MVP). In the case of the biennial *Gentianella germanica* in Lower Saxony the MVP for 90% survival over 10 years estimated by Bräuer *et al.* (1999) was 750 (Fig. 7.7). In *G. germanica* and two other species, including one with an MVP as high as 1300, all of the surviving populations were smaller than the MVP. Populations of *G. germanica* that are extinct above ground might be restored from their seed banks with appropriate conservation management (Fischer & Matthies 1998c).

Extinction rates are influenced by life history traits. Stöcklin and Fischer (1999; Fischer & Stöcklin 1997) analysed extinctions among more than 150 grassland species that were recorded at 26 sites in the Swiss Jura Mountains in 1950 and again in 1985. Extinction rates over the 35-year interval were higher in smaller populations. Short lifespan, the absence of a seed bank and strict habitat specificity also each independently increased the extinction rates of species with these traits. In small populations, species with an outcrossing mating system may suffer reduced fecundity resulting from pollen limitation of seed set (Ågren 1996). This is an example of an **Allee effect**, which occurs when individual reproductive success increases with population size or density, and which can increase extinction rates

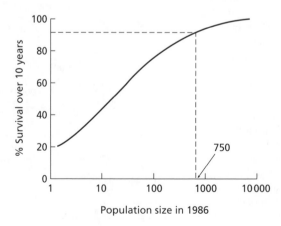

Fig. 7.7 Relationship between original population size in 1986 and probability of survival over 10 years for *Gentiana germanica* in Lower Saxony. The dotted line indicates the population size (*n* = 750) required to ensure a 90% probability that a population would survive 10 years (from Bräuer *et al.* 1999).

of small, or low density, populations (Groom 1998). Pollination failure is probably one of the commonest deterministic threats to small plant populations, second only to habitat destruction, with which it is often associated (Lamont *et al.* 1993; Nason & Hamrick 1997; Ghazoul *et al.* 1998; Curran *et al.* 1999).

A separate, genetic consequence of small population size is inbreeding depression through selfing or mating with relatives (Oostermeijer *et al.* 1994; Heschel & Paige 1995; see also Section 2.8.2). In combination or on their own, Allee effects and inbreeding depression may create a vicious circle, or **extinction vortex**, in which the deleterious fitness consequences of small population size lead to a further reduction in population size and so on to extinction (Gilpin & Soulé 1986). It has been argued that the deleterious genetic consequences of small population size are usually likely to be slight by comparison with the much greater risks posed by stochastic processes (Lande 1988), but potential genetic effects should not be forgotten.

Fischer and Matthies (1997, 1998a,b) compared a number of genetic and demographic parameters for *G. germanica* populations in Switzerland ranging in size between 40 and 5000 plants, and found that smaller populations had significantly lower growth rates. Smaller populations also had lower seed set per fruit and lower total fecundity, lower genetic variation at RAPD loci, and produced progeny that survived less well when grown from seed in a common garden. Given that *G. germanica* has a mixed-mating system ($S \approx 0.66$), and that experimental exclusion of pollinators significantly reduced seed set, low fecundity in small populations might be the result of pollinator limitation, and poor offspring survival in these populations might result from inbreeding depression. However, two other factors could explain away or nullify the relationship between population size and plant performance, and should always be considered. First, large and small populations may well differ in habitat quality, which could explain the correlation with plant performance without any genetic effects. Secondly, density dependence could decouple population growth rate from individual plant performance, because poor survival in inbred progeny could be compensated for at the population level by better survival or higher fecundity among the remaining plants. Neither of these explanations applies in the case of *G. germanica*. Moreover, the likelihood of inbreeding depression is supported by the fact that progeny from experimental selfing and matings between plants 1 m metre apart had less than two thirds the survival in a common environment of progeny from matings between parents 10 m apart.

While it is clear that inbreeding poses a potential danger to the survival of some small plant populations, it is less clear whether this actually matters in comparison with the threats posed by Allee effects and stochasticity. Newman and Pilson (1997) attempted to answer this question for the annual *Clarkia pulchella* by comparing survival in the native prairie habitat of experimental populations established with either 12 unrelated plants (a high N_e treatment) or three sets of four full sibs plants (a low N_e treatment). Mean fitness was 79% lower in the more inbred population than in the high N_e treatment, as a result of poorer germination and survival. After four generations and exposure to some severe weather conditions, 75% of populations in the high N_e treatment were still in existence while only 31% of populations in the low N_e treatment survived. This suggests that inbreeding can significantly increase extinction rates, over and above the effect of environmental stochasticity.

Another approach to evaluating the sensitivity of extinction rates to inbreeding depression is to include its effects in a stochastic population model. Burgman and Lamont (1992) found that inbreeding had no detectable effect in simulations of *Banksia cuneata*, an Australian shrub that recruits after fire, although their estimates of inbreeding depression used as input in the model were based largely on guesswork. Oostermeijer (2000; Oostermeijer *et al.* 1994) found significant inbreeding depression affecting survival and pollinator limitation affecting fecundity in small populations of the perennial herb *Gentiana pneumonanthe*. Stochastic matrix projection models suggested that in some circumstances the combined effects of inbreeding and pollinator limitation, but not one operating without the other, could cause extinction when stochasticity alone would not (Oostermeijer 1996).

7.5 Genetic and evolutionary consequences of regional dynamics

In a system where there is population turnover, the manner in which new populations are founded has important genetic consequences. The effects have been studied using theoretical models in which colonization and extinction occurred (equivalent to the metapopulation case) and with models without population turnover (equivalent to the remnant population case). Colonization and extinction cause the number of ancestors of the species as a whole to be lower than the observed number of individuals, because individuals may trace back to common ancestors quite recently, if turnover is rapid. The entire species thus behaves as if it has gone though a bottleneck, which decreases its overall genetic diversity (H_T; see Section 3.3.6). The diversity within any population (H_S) will also tend to be low, and to be similar for all populations (Pannell & Charlesworth 1999). At the level of the common measure F_{ST} between local populations (see Section 3.3.1), the net effect may be either an increase or a decrease, depending upon the relatedness of the individuals founding new populations (Wade & McCauley 1988; Whitlock & McCauley 1990). When colonists are closely related to one another (the **propagule pool** model), as seeds dispersed within a single fruit are, metapopulation dynamics generally increase F_{ST}, provided that the size of the colonizing group, k, is smaller than the local **carrying capacity** (i.e. the maximum size of a local population). In a contrasting scenario, colonists founding new populations are drawn at random from all established populations (the **migrant pool** model). In this model, the colonization process more strongly homogenizes the subpopulations over the species range, and may thus decrease F_{ST}. If k is small by comparison with gene flow, as it may be when established populations exchange pollen, this can create conditions in which extinction/colonization dynamics reduce F_{ST} by more than gene flow alone would in a remnant population system.

A study has been made of the genetics and regional dynamics of the short-lived, perennial dioecious herb *Silene alba* along roadsides in Virginia (McCauley *et al.* 2001). Populations have a high turnover rate, with new ones typically appearing within 160 m of existing populations and consisting of five or fewer individuals. The smallest and most isolated populations have the highest extinction rate (Richards 2000). McCauley (1994; McCauley *et al.* 1995) compared the genetic structure of 11 old populations with 12 recently founded populations of *S. alba*. The theoretical models suggest that if metapopulation dynamics raises F_{ST}, then this value should be higher in recently founded populations than in old ones, while the opposite result would suggest that conditions are such that extinction/colonization dynamics lower F_{ST}. Values of F_{ST} calculated for allozyme data were found to be 50% higher for young populations than for old ones. F_{ST} values calculated for variation in cpDNA, which is maternally inherited in *S. alba* and therefore only transmitted by seed, were five times greater than values based upon allozyme variation, which is biparentally inherited. Both of these results indicate that the propagule pool model is probably the relevant scenario for this system, and that new populations are founded by small numbers of close kin.

A follow-up study of the 12 young populations of *S. alba* conducted five years after the initial census found that some had increased in size while others had decreased. Populations that had increased also contained numbers of allozyme alleles, presumably acquired through gene flow. Populations that had decreased in size had lost alleles, probably through genetic drift. In contrast to allozyme diversity, which changed with changes in population size, cpDNA haplotype diversity decreased in all populations to the point where, only three years after the initial census, seven out of eight populations were monomorphic (McCauley *et al.* 2001). This suggests that no new seed immigrants had arrived and that genetic drift had a powerful influence upon haplotype diversity.

Do the genetics of small populations add to the extinction risk for *Silene alba*, as they appear to do in some other herbs (Section 7.4)? Richards (2000) found strong inbreeding depression in *S.alba* that reduced seed germination by 60% in the offspring of full-sib matings in greenhouse conditions. In the field in Virginia, experimental crosses between individuals within small, isolated populations produced seeds with a low germination rate typical of that found in greenhouse full-sib crosses. Experimental crosses made between plants in small, but less isolated, field populations produced seeds with germination rates typical of outcrossing. Other field experiments with *S. alba* showed that such populations receive proportionately more gene flow than larger or more remote populations (Richards *et al.* 1999). Richards (2000) suggests that gene flow into small populations can rescue them from inbreeding depression and from consequent extinction, which is the fate of more remote populations that do not receive pollen from established populations. He termed this **genetic rescue**, by analogy with the demographic rescue effect observed in some metapopulation models (Section 7.2).

7.6 Geographical range limits

Any stable geographical limit to a species' range is, by definition, the point at which it goes locally extinct. The majority of plant species' distribution limits appear to correlate with geographical or topographical trends in climate, but trends in temperature and rainfall tend to be less abrupt than plant distribution limits. This discrepancy can be explained if a species' regional dynamics conform to the metapopulation model, because site occupancy (p) is very sensitive to relative extinction rate x and to site density T (Eqn 7.3), and hence values of these parameters need be altered only slightly by changes in climatic conditions to tip a metapopulation into extinction (Lennon *et al.* 1997). Because the cause of such boundary limits lies in metapopulation dynamics and not in the performance of individual plants or populations, there need be no measurable deterioration in the growth of individual plants at, or even beyond, the range boundary.

Carter and Prince (1981, 1988) suggested that the metapopulation model could explain why some short-lived species, such as the wild lettuce *Lactuca serriola* that they studied in England, have very abrupt distribution limits. When introduced experimentally, such plants typically grow and set seed quite successfully beyond the boundaries of their natural distribution limits, but the species still fail to establish self-sustaining populations. Seeds of the annual *Phlox drummondii*, which is confined to sandy, nutrient-poor sites, were sown by Levin and Clay (1984) at stations along a transect crossing an abrupt part of the species' boundary that coincided with a change in soil type near San Antonio, Texas. Sites just beyond the boundary were able to support self-sustaining populations ($R_0 > 1$) where the soil was locally poor, but further away such sites were rare and competition from other plants made establishment impossible (Fig. 7.8). Pockets of sandy soil habitable by *P. drummondii* did lie well beyond the species' boundary, but Levin and Clay concluded that colonization rates there were too low, and extinction rates too high, for them to remain occupied.

The metapopulation model of range limits is more likely to apply to short-lived plants than to long-lived perennials with remnant type regional dynamics. With the dramatic exception of invaders, considered in the next section, most range boundaries are constant through time unless there is a climate change. An interesting question is why natural selection acting on individuals does not drive all range boundaries ever outwards. One of several possible explanations is that populations near range boundaries tend to be sinks. For a sink population to expand its range it must first adapt sufficiently to local conditions to raise R_0 to a value above one, but this may be impossible because gene flow from source populations swamps it with genotypes that are not locally adapted and because the evolutionary change required to overcome this effect is too great (Holt & Gomulkiewicz 1997; Kirkpatrick & Barton 1997). Evolution of reproductive isolation from the source of gene flow is necessary before local adaptation can occur. Reproductive isolation may evolve in a single generation by polyploidization or though the evolution of apomixis, which is often correlated with polyploidy. This is

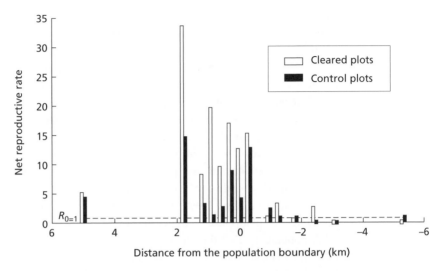

Fig. 7.8 Net reproductive rate R_0 of experimental populations of *Phlox drummondii* sown in plots cleared of vegetation and in control plots along a transect crossing the species' boundary in Texas. The boundary of the natural population is at zero (drawn from data of Levin & Clay 1984).

perhaps why polyploids and apomicts are often found at the edge of the ranges of the diploids from which they have arisen (Bierzychudek 1987a; Peck *et al.* 1998).

7.7 Invasions

Like the end of the rainbow, the tail of the seed dispersal curve (see Section 3.3) is impossible to reach. The occasional seed is carried quite extraordinary distances by chance events, but these seeds are so few that we can only ever know where they end up when they attract attention by starting a new population in an alien site. Invasions are a lesson in the sometimes overwhelming significance of very rare events. Alien plant species are constant hitchhikers on international trade, and botanists with an eye for the exotic have monitored these 'adventive' floras for nearly a century (Crawley 1987). We know from such records that, in Britain at any rate, about 20% of vascular plant introductions have become established and widespread, and of these about 25% are considered pests (Williamson 1996).

There is often a significant time lag between a species' first arrival and its subsequent spread. For example, the Brazilian tree *Schinus terebinthifolius* was first introduced into Florida in the 19th century, but it was not until the 1950s that it began an explosive spread across South Florida (Ewel 1986). Such a time lag could be more apparent than real if it represents the early phase of exponential population increase (Fig. 5.2), or it might be an actual delay, reflecting the existence of a threshold constraining the establishment of a metapopulation (Section 7.2). Most invasive plant species cannot invade closed vegetation, and human disturbance generally provides them with the opportunity to spread. The annual cheatgrass *Bromus tectorum* invaded NW USA with great rapidity at the start of the 20th century, permanently replacing native perennial bunchgrasses when they were overgrazed (Mack 1981). Similar invasions must have occurred in the history of the California annual grasslands that are composed almost entirely of aliens (Jackson 1985), but this replacement happened earlier than in NW USA and the events were not chronicled.

Once plants begin to spread they can do so with incredible rapidity. The initial population of an invading weed typically spawns new colonies at a distance (e.g. Forcella & Harvey 1988), and it is these small outliers that push the range of an invading species forward so rapidly. Such invasions are a metapopulation phe-nomenon in which the rate of colonization consistently exceeds the rate of extinc-tion. Moody and Mack (1988) have pointed out that control measures against invading alien plants usually target the largest, oldest foci of infection and ignore small populations that they term 'nascent foci'. However, numbers of scattered nascent foci make a bigger contribution to spread than a single large population of the same total area; thus, control programmes that ignore nascent foci because they are small and inconspicuous are likely to fail.

Most of the plants of the northern temperate zone have spread to the northern-most parts of their ranges in geologically very recent time following the retreat of the ice at the end of the last glaciation (e.g. *Picea abies*, Fig. 1.1). The rates of this spread have been calculated from pollen maps (e.g. Fig. 1.1a) for the major tree genera and turn out to have been remarkably fast—of the order of 7 or 8 km per generation for oak *Quercus* and beech *Fagus* in North America, for example (Johnson & Webb 1989). In a pioneering paper modelling dispersal, Skellam (1951) found that after the retreat of the ice, oak spread into Britain at a rate far too fast for the process to have happened by diffusion along a single advancing front. Long-distance dispersal is required to account for such rapid spread (Clark 1998). Today, the large seeds of species such as beech and oak in both North America and Europe are dispersed by jays and pigeons that may well have been responsible in the past for creating small nascent foci well beyond the trees' main front of advance, thus accelerating their spread and allowing them to move as fast as species with much smaller, wind-dispersed seeds (Bossema 1979; Johnson & Webb 1989). The passenger pigeon, which existed in North America in such huge flocks that they are said to have darkened the sky as they flew overhead, may have been important long-distance dispersers of seeds until the sudden extinction of the species under pressure from hunting at the beginning of the 20th century (Webb 1986).

Pollen maps do not have sufficient spatial or temporal resolution to reveal the structure of postglacial metapopulations, but there are gaps in some historical or present-day distributions that suggest long-distance dispersal did take place. Webb (1987) suggests that an isolated population of *Fagus grandifolia* in eastern Wisconsin that lies 70 km beyond the current main distribution limit of the species in Michigan may have arisen by a very rare long-distance dispersal event over Lake Michigan or the Illinois Prairie Peninsula. Birks (1989) found that Scots pine (*Pinus sylvestris*) arrived in Scotland and alder (*Alnus glutinosa*) arrived in Wales

8000 years BP, well ahead of other areas nearer the source of colonists for these species, and he suggests that these populations may also have been founded by long-distance dispersal. Lodgepole pine (*Pinus contorta*) may also have spread up the NW coast of Canada by this means (MacDonald & Cwynar 1991).

7.8 Phylogeography

Plant invasions leave not only a fossil imprint upon the pollen record of the territory that they colonize (Fig. 1.1a), but also a living footprint impressed upon the DNA of present-day populations. In plants, this footprint is most clearly seen in the chloroplast genome because this is usually inherited uniparentally, multiplies clonally and does not recombine (Chapter 2). There is thus an unbroken, untangled lineage between the cpDNA sequences found in modern populations and those of their long dead ancestors. Chloroplast DNA mutations that occurred in one individual plant at some remote time and place will be carried wherever the descendants of that plant may go. The paths of those migrations may be traced by reconstructing the evolutionary tree, or phylogeny, that links all modern populations through their (and their chloroplasts') common descent, and overlaying the result on to a geographical map (Fig. 7.9). This combination of intraspecific phylogenies with geographical distribution is called **phylogeography** (Avise 2000), and has successfully reconstructed the migratory history of some species

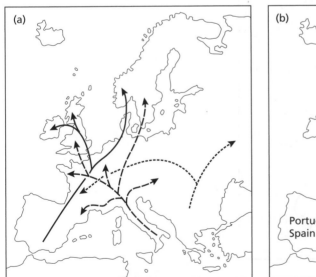

Fig. 7.9 (a) Phylogeographic reconstruction of the migration routes of *Quercus* spp. in Europe, showing three separate lineages identified by their cpDNA haplotypes; (b) an unrooted tree showing the phylogenetic relationships among geographical populations in Europe. These relationships apply within *Quercus*, *Fagus sylvatica* (beech) and within animal taxa including brown bear, shrews, newts and grasshoppers (modified from Taberlet *et al.* 1998).

(e.g. Hewitt & Ibrahim 2001; Petit *et al.* 2001). For example, in the Pacific NW of North America, many species of plants and animals have continuous distributions up the coast from the Sierra Nevada of northern California along the Cascade Range and into British Columbia and Alaska. Some of these show genetic discontinuities between northern and southern populations (Brunsfeld *et al.* 2001). Within each species, the northern and southern populations form separate clades, suggesting that postglacial recolonization by these species took place from two geographically distinct refugia—one leading to the southern clade and one to the northern one (Fig. 7.10). A general note of caution should be sounded here because isolation by distance (Chapter 3) can cause divergence between populations too. Ideally, independent evidence that migration has taken place is needed.

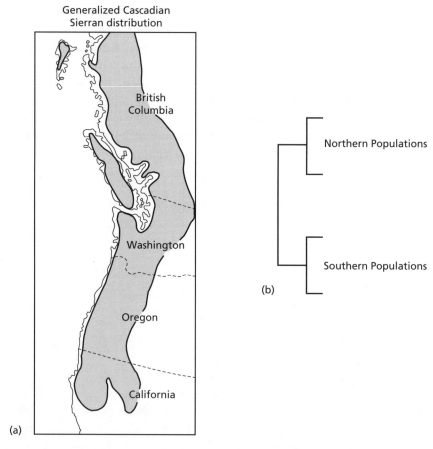

Fig. 7.10 (a) A generalized distribution of plants and animals found in Pacific NW North America across the junction of the Sierra Nevada of northern California and the Cascade Range; (b) simple phylogeny illustrating the relationship between northern and southern populations within many species that show the Cascadian–Sierran distribution. These include the herb *Tiarella trifoliata*, the tree *Alnus rubra*, the shrub *Ribes bracteosum*, the fern *Polystichum munitum* and animals including a moth, rainbow trout, salamanders, a frog and a shrew.

In Europe, the story told by the genes corroborates that which can be read in the pollen record of tree migrations in the last 12 000 years, but gives much more detail. For example, maps of cpDNA haplotypes in white oak populations in France show that trees in whole areas of 100 km^2 or more have the same haplotype, although at a slightly larger scale the different haplotypes form a mosaic pattern (Petit *et al.* 1997). This picture suggests that a few original colonists must have been dispersed over great distances—a scenario consistent with our knowledge of recent invasions (Section 7.7). Long-distance dispersal would have created a bottleneck that reduced haplotype diversity and also permitted genetic drift to fix different haplotypes in different places, rather as was observed to happen on a smaller scale in *Silene alba* (Section 7.5). Chloroplast genes are especially prone to drift because, being haploid and inherited uniparentally, they have a lower effective population size than nuclear genes (Avise 2000).

Even more remarkable than the fixation of haplotypes over large areas is the discovery that the *same* haplotypes are fixed in populations of *different* oak species where they occur together (Dumolin Lapegue *et al.* 1997, 1999; Petit *et al.* 1997). cpDNA is maternally inherited in oaks. The geographically local sharing of haplotypes between species therefore implies that the original seed colonists in such an area belonged to only one of the two species, and that genes in pollen of the other species invaded by hybridizing with the first colonists after their arrival. In the case of *Quercus robur* and *Q. petraea* in France, for example, these two oaks fill slightly different ecological niches. Once hybrids had appeared, disruptive selection must therefore have acted on the hybrid progeny, re-creating two distinct populations that differed in their nuclear genomes but still shared neutral markers in their cpDNA. Co-occurring species that share local cpDNA haplotypes are also known in wild species of cassava (*Manihot* spp.) from Amazonia (Schaal & Olsen 2000), in Mediterranean species of *Senecio* (Comes & Abbott 1999) and *Alnus* (King & Ferris 2000), in *Eucalyptus* in Australia (Jackson *et al.* 1999) and in *Penstemon* spp. (Wolfe & Elisens 1995) and *Ipomopsis* spp. in North America (Wolf *et al.* 1997). Many more cases are likely to be found.

7.9 Summary

Regional dynamics describes the changes affecting *groups* of populations. Three types of regional dynamics are distinguished.

1 **Metapopulation** dynamics are dominated by the rates of **colonization** and **extinction** of component populations. The balance between colonization and extinction generates an **extinction threshold** that can sometimes determine where the boundary of a species' **distribution** lies. Migration from large populations may **rescue** small populations from extinction by increasing their size.

2 **Source–sink** dynamics describes the situation where some populations (sinks) are not self-sustaining and depend entirely upon seed flow from high-density

source populations. **Pseudosinks** are self-sustaining populations whose equilibrium size is significantly increased by immigration.

3 **Remnant populations** persist over long periods without immigration, but may depend upon an *in situ* seed bank for recruitment.

Examples of true metapopulation dynamics and source–sink dynamics appear to be few in plants, with remnant populations probably being typical.

Extinction is the ultimate fate of all populations, so it is the *rates* and causes of extinction that matter. Causes may be divided between **deterministic** and **stochastic processes**, though the two may be interdependent. Life history traits, genetic structure and, above all, population size are known to influence extinction rates. Stochastic models incorporating these and other relevant variables can be used to estimate **minimum viable population (MVP)** sizes for desired survival probabilities over specified periods of time. In outcrossing species there is increasing evidence that in small populations reproductive failure, sometimes resulting from an **Allee effect**, and inbreeding depression may create an **extinction vortex**, which causes them to spiral into extinction. Gene flow from large populations into small ones may prevent this by a **genetic rescue effect**.

At the edge of species' ranges gene flow can, at least in theory, prevent local adaptation and hence create a barrier to geographical spread. When the climate changes, or new species are introduced to an area by humans, plant distributions may advance rapidly towards new limits. The speed of these advances is increased by the long-distance dispersal of a few pioneer individuals that establish new populations. The genetic footprint of this process may still be seen in the distribution of cpDNA haplotypes many thousands of years after the event. **Phylogeography** uses cpDNA haplotypes and other neutral genetic markers to match phylogenies with geographical distributions. It has identified Pleistocene refugia and reconstructed the routes by which plants reinvaded the Northern temperate zone from these.

7.10 Further reading

Akçakaya, H. R., Burgman, M. A. & Ginzburg, L. R. (1999) *Applied Population Ecology using RAMAS Ecolab.* Sinauer Associates, Sunderland, MA.

Avise, J. C. (2000) *Phylogeography.* Harvard University Press, Cambridge, MA.

Hanski, I. A. & Gilpin, M. E. (Eds) (1997) *Metapopulation Biology, Ecology, Genetics and Evolution.* Academic Press, San Diego, CA.

Hanski, I. (1999) *Metapopulation Ecology.* Oxford University Press, Oxford.

Silvertown, J. & Antonovics, J. (Eds) (2001) *Integrating Ecolog and Evolution in a Spatial Context.* Blackwell Science, Oxford.

7.11 Questions

1 What are the three types of regional dynamics? How do they differ from one another? What kinds of plants would you expect to find fitting each type of dynamics?

2 Why are small populations vulnerable?

3 What effects can gene flow (or its absence) have upon the distribution of plant populations at the regional scale?

4 When is a knowledge of local population dynamics alone an insufficient basis for managing plant populations for conservation?

5 How and why does the distribution of cpDNA haplotypes reflect the dispersal and migratory history of plant populations? Would you expect nuclear genetic markers to tell the same story?

8 Competition and coexistence

8.1 Introduction

Plants usually live in mixtures of species. Because plants all require water, nutrients, light and space, competition for these resources is their dominant, but not their only, mode of interaction with each other. This chapter is about the two sides of a coin: how plants in species mixtures compete on the one side, and how competing species are able to coexist and form stable mixtures on the other. We will look at competition and other interspecific interactions first because these can determine whether stable coexistence is possible or not. In the section on coexistence (Section 8.4), we shall examine the apparent paradox that mixtures of species are the rule rather than the exception in nature, even though most plants compete with other species that they could potentially exclude or be excluded by.

Definitions of competition vary among plant ecologists, depending upon the relative degree of emphasis some place upon the mechanism and others place upon the outcome of interaction. Grime (1979) epitomizes the mechanistic camp, defining competition as 'the tendency of neighbouring plants to utilize the same quantum of light, ion of a mineral nutrient, molecule of water, or volume of space' while, at the other extreme, a definition based entirely upon demographic outcome would describe interspecific competition as an interaction between species in a mixture in which each lowers the net reproductive rate R_0 of the other. These different definitions can be reconciled with one another, as we shall see below, but the approach used in this book emphasizes outcomes.

8.2 The variety of interactions between plants

Based upon outcomes alone, there are three possibilities for the individual species in a mixture, each of which may gain (+), lose (−), or be unaffected (0) by the presence of the other. In a two-species mixture both populations may be depressed (in shorthand this is a − − interaction); one may increase and the other lose (+−); both may benefit (++); or the interaction may be neutral for one species and beneficial (+0), or detrimental (−0) to the other. These combinations define five kinds of interaction:

Competition $--$

Parasitism $+-$

Mutualism $++$

Commensalism $+0$

Amensalism -0.

Goldberg (1990) has argued that most interactions between plants occur through some intermediary such as light, nutrients, pollinators, herbivores or microorganisms. She suggests that we should dissect interactions between plants into the **effect** of each species on the abundance of the intermediary, and its **response** to an increase in the abundance of the intermediary. So, for example, a grass population limited by nitrogen availability will deplete that resource ($-$ effect) and will increase in abundance if the nitrogen supply increases ($+$ response). The various permutations of positive and negative effect and response define interactions between plants based upon the net result for the intermediary. In this example the net result is negative ($-$ effect times $+$ response $=$ negative result), and two species interacting via nitrogen experience resource competition (Table 8.1). Goldberg's approach reveals how the resource-based and demographic definitions of competition can be linked to each other. This link between mechanism and outcome reveals that not all $--$ interactions need be the result of straightforward competition and not all $++$ interactions need be the result of mutualism. Natural enemies may generate **apparent competition** and **apparent mutualism**.

Proven examples of apparent interactions are hard to find, partly because investigators of competition in the field rarely investigate natural enemies as well, but possibly also because they may be genuinely rare phenomena (Holt & Lawton 1994). A possible example of apparent competition that illustrates what to look for was reported by Reader (1992), who planted seedlings of three herb species in an experiment in an abandoned pasture in Ontario, Canada. Experimental treatments of neighbour removal, herbivore exclusion (using 6-mm mesh cages) and both combined were used and the survival of seedlings was monitored over 16 weeks. With herbivores present, seedling survival in all three species was adversely affected by the presence of neighbours, suggesting that competition

Table 8.1 Types of interaction between plants that involve an intermediary. The net result is the product: effect × response (after Goldberg 1990).

Type of interaction between species	Intermediary	Effect on intermediary	Response to intermediary	Net result (effect × response)
Resource competition	Resources	$-$	$+$	$-$
Apparent competition	Natural enemies	$+$	$-$	$-$
Allelopathy	Toxins	$+$	$-$	$-$
Positive facilitation	Resources	$+$	$+$	$+$
Negative facilitation	Resources	$-$	$-$	$+$
Apparent facilitation	Natural enemies	$-$	$-$	$+$

occurred. However, seedling survival in the caged treatment was unaffected by neighbours, indicating that herbivores (probably slugs) and not competition may have been responsible for the observed effect of neighbours upon seedlings, perhaps by providing shelter for the herbivores. Although this experiment suggests that apparent competition might have occurred, we do not actually know whether the apparent negative effect of neighbours upon seedlings was reciprocated, which, strictly, it would have to have been to qualify as a genuine (– –) interaction (apparent or otherwise).

Given that competitive interactions between plants are often highly asymmetric or even one-sided (larger plants affecting smaller ones but not *vice-versa*; Section 4.4) it might be sensible to accept asymmetry in cases of apparent competition too (Connell 1990; Huntly 1991). If the definition of apparent competition is relaxed in this way, then more examples are known (e.g. Bartholomew 1970; Atsatt & O'Dowd 1976; Rice & Westoby 1982; Callaway 1995). Among such examples is the case of the small cactus *Opuntia fragilis*, which Burger and Louda (1994) found grew better in the open than when shaded by grass, not because of release from competition but because attacks by larvae of the cactus moth borer (*Melitara dentata*) were less frequent in open conditions than in shade.

Interactions between plants often first come to the attention of ecologists when positive or negative patterns of association are noticed in their spatial distribution. However, patterns can be quite misleading about the processes that generate them. The annuals *Minuarta uniflora* and *Sedum smallii* that are found together in shallow depressions in granite outcrops in North Carolina and Georgia, are positively associated because of similiar habitat requirements. At a finer spatial scale these annuals are segregated along a gradient of soil depth within their rock depressions as a consequence of competition between them (Sharitz & McCormick 1973). Positive associations between shrubs and smaller plants are frequent in deserts where shrubs act as **nurse plants** to juveniles of other species, particularly cacti (Yeaton 1978; McAuliffe 1984; Arriaga *et al.* 1993; Suzan *et al.* 1996). In the Tehuacán Valley in Mexico, juveniles of the columnar cactus *Neobuxbaumia tetetzo* benefit from the shade cast by nurse plants of the legume shrub *Mimosa luisana* (Valiente-Banuet & Ezcurra 1991), but the cactus inhibits the growth of its nurse and eventually replaces it (Flores-Martinez *et al.* 1994). In a subtropical savanna habitat in southern Texas, Barnes and Archer (1999) found that *Prosopis glandulosa* nurse trees grew and survived better when associated understorey shrubs were experimentally removed from beneath them. The beneficial influence of nurse plants upon their associates is often cited as the commonest example of positive (or **facilitative**) interactions between plants (Hunter & Aarssen 1988; Callaway 1995) but if, as is the case for *Neobuxbaumia* and *Mimosa*, the benefits are all one-way then it would be more accurate to describe such relationships as parasite/host interactions (+–).

Whether one-sided or not, the study of positive interactions between plants has been neglected until recently, although it is now becoming clearer that facilitation (where at least one species benefits) as well as competition operates in many plant communities (Callaway & Walker 1997). For example, in middle zones of salt marshes at Rhode Island, USA, the presence of the rush *Juncus gerardi* reduces soil salinity and increases soil aeration, with positive consequences for other species (Hacker & Bertness 1999); however, whether the effect of other species on *J. gerardi* is detrimental (and hence the interaction is +−) or neutral (+0) is not clear. Competition excludes some species from elsewhere on the marsh. In other environments there is a recently discovered mechanism for positive interactions that may turn out to be quite widespread. Plants whose roots span a large vertical gradient in water potential passively transport significant quantities of water at night from depth to the surface where the water leaks out of their roots. Emerman and Dawson (1996) estimated that a mature sugar maple tree transported 100 L of water in a single night. This process of **hydraulic lift** may benefit not only the plants that lift the water but also neighbouring species (Caldwell & Richards 1989; Caldwell *et al.* 1998). Our understanding of the range of types of interaction between plants is improving. Yet, although there is more variety in these interactions than was once appreciated, competition remains central to understanding communities and mixtures.

8.3 Competition

The composition of mixtures is dynamic because populations are dynamic. Competition changes the composition of mixtures, or sometimes stabilizes it. The starting-point and the end-point of an experimentally contrived interaction can therefore be marked by the starting densities and the finishing densities of the species (allowing for differences in size). A simple way to describe the composition of any mixture of two species is to plot the density of one against the density of the other on a two- dimensional graph called a **joint-abundance diagram** (Fig. 8.1a). Any point on this graph can be described by a pair of co-ordinates N_i, N_j denoting the density of species i and the density of species j in the mixture. The total density is, of course, $N_i + N_j$, and this total is constrained by the same limitations that constrain the density of a monoculture (Chapter 4). With increasing density, mixtures reach an upper limit just as monocultures do. If we assume, for the moment, that the precise ratio of the two species doesn't affect the upper limit on total density, then $N_i + N_j$ is a constant, which we will call K. This is the maximum sustainable density, or the **carrying capacity**, for any kind of mixture. All possible values of $N_i + N_j = K$ are represented on the graph by a straight line of slope -1 (Fig. 8.1a). This line intersects the horizontal axis at $N_i = K$ and intersects the vertical axis at $N_j = K$. Remembering that any straight line of negative slope can be defined by an equation of the form $y = a - bx$, we can write the equation for the line $N_i + N_j = K$ as:

$$N_i = K - \alpha N_i, \tag{8.1}$$

where K is the intercept, and α is the slope of the line. Ecologically, the coefficient α represents the number of species j that must be lost to increase species i by

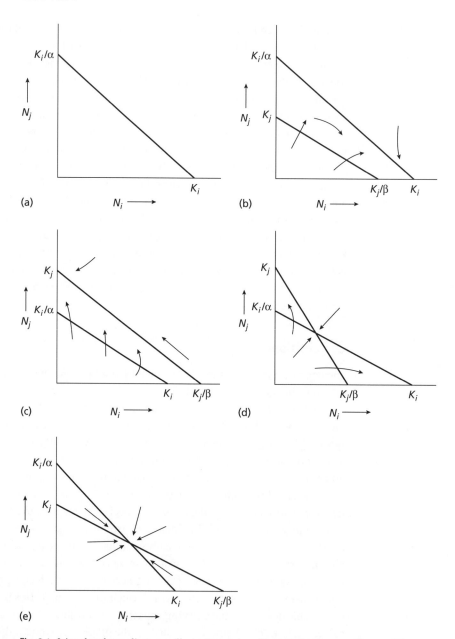

Fig. 8.1 Joint abundance diagrams, illustrating various theoretical outcomes of competition between two species: (a) an isocline for one species, described by Eqn 8.1; (b) competitive exclusion of species j by species i; (c) competitive exclusion of species i by species j; (d) an unstable equilibrium; and (e) a stable equilibrium.

a single individual. When we made the assumption that the ratio $N_i:N_j$ doesn't affect K we implicitly decided that the two species were equivalent, so $\alpha = 1$ in this example. The line $N_i + N_j = K$ can also be expressed in terms of N_j:

$$N_j = K - \beta N_i. \tag{8.2}$$

Here, the coefficient β represents the number of individuals of species i that must be lost to increase species j by a single individual. Again, because we have assumed that the competitors are equivalent to one another, its value is also one. The coefficients α and β are called **competition coefficients** and are fundamental to most theoretical formulations of competition between species in population biology. The model is a version of the **Lotka–Volterra** competition model.

   Although in the simplest of examples $\alpha = \beta = 1$ and the intercepts on both axes are the same (K), in most mixtures the competing species will not be completely equivalent. In such cases α and β will have different values, and the carrying capacities (K_i, K_j) that determine the intercepts will also be different. When α and β, and K_i and K_j are different, Eqns 8.1 and 8.2 describe different lines (e.g. Fig. 8.1b). The lines, which are called **zero isoclines**, pass through the points on the graph that represent mixtures at carrying capacity. The lines are crucial boundaries between mixtures below the line, where populations have net reproductive rates > 1 (and will therefore increase), and points above, where net reproductive rates are < 1 (where populations will decline).

   Using the isoclines as a tool, we can now explore how mixtures will behave, according to the relationships between the competitors that are described by the competition coefficients and carrying capacities of the species. This exploration is best tackled using the graphs in Fig. 8.1(b)–(e). First, note that now we have abandoned the assumption that the competitors are equivalent, then Eqns 8.1 and 8.2 must be rewritten:

$$N_i = K_i - \alpha N_j \tag{8.3}$$

$$N_j = K_j - \beta N_i. \tag{8.4}$$

You should be able to see how the intercepts of the lines, shown in Fig. 8.1(b)–(e) can be calculated from these two equations. The arrows drawn on the graphs indicate how the joint abundances of the competing species change, depending upon their starting densities. Figure 8.1(b) illustrates the conditions for the competitive exclusion of species j by species i. This will occur when $K_i > K_j/\beta$ and $K_i/\alpha > K_j$. When these two inequalities are reversed, species j will exclude species i (Fig. 8.1c). In the relationship shown in Fig. 8.1(d) either species can exclude the other, depending upon starting conditions, or upon chance perturbations (inevitable, sooner or later) of the 50/50 mixture where there is an unstable equilibrium. The conditions for this situation are $K_j > K_i/\alpha$ and $K_i > K_j/\beta$. Reversing these two inequalities creates a stable equilibrium mixture at the intersection of the two isoclines, as shown in Fig. 8.1(e).

   In what ecological circumstances are the conditions for stable coexistence of competitors fulfilled? For stable coexistence to occur, the inequalities $K_j < K_i/\alpha$ and $K_i < K_j/\beta$ must be satisfied. If the species are ecologically equivalent, then $K_i = K_j$ and $\alpha = \beta = 1$. By inserting these values into the inequalities you will see that the conditions for stable coexistence are violated. From this we draw the important conclusion that ecologically equivalent species cannot coexist. This is sometimes called the **competitive exclusion principle** or **Gause's principle**.

The simplest way for both inequalities to be met is for the maximum mono-culture densities K_i and K_j to be similar, but $\alpha < 1$ and $\beta < 1$. This implies that each species is more sensitive to its own density than to that of its competitor. This can happen if there is some kind of **niche separation** (see Section 8.4.2). For example, Berendse (1981, 1982) found that competition between two grassland species that commonly grow together, the grass *Anthoxanthum odoratum* and *Plantago lanceolata*, was ameliorated by *P. lanceolata* having deeper roots than the grass. When separation of the rooting zones was experiment-ally prevented in pot or field experiments, the grass had a significantly greater negative effect upon its competitor than in controls where deep rooting by *P. lanceolata* was possible. There was also some evidence that the presence of *A. odoratum* caused *P. lanceolata* to develop deeper roots than it did when growing by itself. Competitive relationships have to be elucidated through experi-ments of this kind that compare plants growing in mixtures with plants in mono-cultures.

8.3.1 Two-species competition experiments

There are four basic designs of competition experiment used to explore the behaviour of mixtures. The essential features of each can be described by plotting the mixtures they use on a joint-abundance diagram. In the **partial additive** design a constant density of one species is combined with a range of densities of another (Fig. 8.2a). Experiments that determine the effect of neighbours on a single target plant are the simplest form of this design. The total density of the mixture and the relative proportions of the species vary in parallel in the partial additive design, which makes it impossible to disentangle their separate effects. This problem is partly alleviated by the **replacement** series design, which maintains the *total* density of mixtures constant and varies the ratio of species to each other (Fig. 8.2b). This design has the drawback that it cannot be assumed that the proportions in a mixture have the same influence at different densities. Only the **additive series** and **complete additive** (Fig. 8.2c,d) designs explore the effects of varying density and proportions of species independently of each other. The relative merits of the different designs for competition experiments are reviewed in greater detail by Gibson *et al.* (1999).

Competitive interactions first influence the performance of plants – that is their size or their yield – and as a consequence of this may later affect density by altering survival or fecundity and thus R_0. Depending upon the time over which an interaction is observed, changes in either performance or density may be chosen to measure the effects of competition. In agricultural mixtures of crops and weeds, performance is the usual choice.

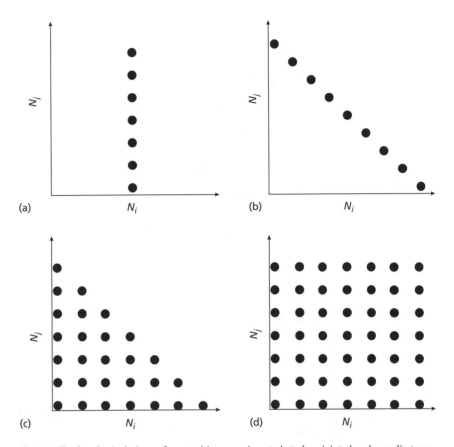

Fig. 8.2 The four basic designs of competition experiment plotted on joint abundance diagrams for two species: (a) partial additive; (b) replacement series; (c) additive series; and (d) complete additive.

8.3.1.1 *Partial additive designs*

This is the design often chosen to determine the effect of weeds on a crop, for example the effect of sicklepod *Cassia obtusifolia* or redroot pigweed *Amaranthus retroflexus* on cotton yield (Fig. 8.3a). The partial additive design is also used to look at the effect of different numbers of neighbours of a competitor on a single individual of a 'target' species planted in their midst (Goldberg 1987; Goldberg & Fleetwood 1987; Miller & Werner 1987; Pacala & Silander 1990). The results of such **target-neighbour** experiments are used to parameterize individual-based competition models (Section 8.3.2.2). Goldberg and Fleetwood (1987) used the design to compare the effect of different neighbour species upon five annuals. The results of these, and similar experiments, typically show a hyperbolic relationship between the weight of the target plant and the density of its neighbours (Fig. 8.3b). This indicates that the first few neighbours have a large effect upon the target, but additional ones have proportionately less effect. The hyperbolic response curve of a target plant to the density of competitors around it has essentially the same

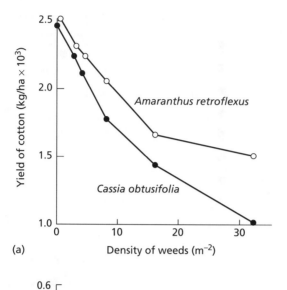

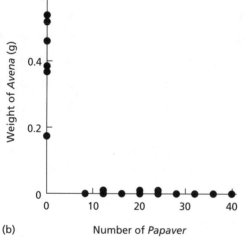

Fig. 8.3 Results of partial additive competition experiments between: (a) cotton *Amaranthus retroflexus* and sicklepod *Cassia obtusifolia* (data of Buchanan *et al.* 1980); and (b) *Papaver rhoeas* and *Avena sativa* (Goldberg & Fleetwood 1987).

shape as the response of yield per plant w_i to density N seen in monocultures (Fig. 4.1c). The equation (4.2) we used to describe this in Chapter 4 was:

$$w = w_m(1 + aN)^{-1},$$

where w_m is the maximum weight of an isolated plant and a is the area needed for a plant to achieve w_m. Equation 4.2 can describe the effect of competitors on a target of another species if we substitute N with an equivalent density of competitors αN_j. Recall that α is the competition coefficient that measures the equivalence of the two species in their effects on the target species:

$$w_i = w_{mi}(1 + a_i \alpha N_j)^{-1}. \tag{8.5}$$

The extra subscripts in the terms w_i, w_{mi} and a_i indicate that these measures apply to species i. They would, of course, have different values for species j.

One particular situation in which the partial additive design is useful is in separating the effects of root and shoot competition.

8.3.1.2 Root and shoot competition

There are a variety of ways to test for the effects of competition for resources above and below ground. Water or nutrients may be added and the effects observed, and estimates of rates of utilization can be made (Eissenstat & Caldwell 1988). Alternatively, physical partitions can be used to separate competitors above or below ground, and to compare their performance with controls that are interacting fully, or not at all (Snaydon & Howe 1986). Groves and Williams (1975) used an additive design to look at the effect of root and shoot competition between subterranean clover *Trifolium subterraneum* and skeleton weed *Chondrilla juncea*, which is a weed of cereal crops in SE Australia. *Chondrilla juncea* is able to persist through the fallow period between cereal crops when fields are used as pasture, so suppressing it with subterranean clover during this period could aid in its control as a cereal crop weed. Strains of the rust fungus *Puccinia chondrilliana* have been used as biological control agents of *C. juncea*, so this was included as a treatment in the competition experiment (Fig. 8.4a).

Trifolium subterraneum did not suffer significantly from competition with *C. juncea*, but the latter was suppressed by both root and shoot competition. Rust-free plants were reduced to 65% of control weight by root competition, to 47% of control weight by shoot competition, and to 31% ($= 65\% \times 47\%$) by the combination of the two. An identical synergism between the effects of competition above and below ground occurred in the infected plants (Fig. 8.4a). An experiment using the replacement series design was used by Martin and Field (1984) to look at root and shoot competition between white clover *Trifolium repens* and perennial ryegrass *Lolium perenne* (Fig. 8.4b). *Lolium perenne* greatly benefitted in this interaction, and its suppressive effects on *T. repens* showed synergism between root and shoot competition (Fig. 8.4b). Such interactions between root and shoot competition are not uncommon, but are typically sensitive to the availability of resources, depending upon light above ground and nutrients and water below (Casper & Jackson 1997).

Wilson (1988) surveyed the results of 23 studies in which investigators had used partitions to compare root and shoot competition. In the majority of cases (68%) root competition had a greater adverse effect on growth than shoot competition. Some of the competition experiments were repeated at a range of densities using the partial additive design, and in over 70% of these the effects of root competition intensified more rapidly as density increased than the effects of shoot competition. All the species studied were agricultural crops (including pasture grasses) or weeds, so competitors would have been approximately equal in stature, limiting the potential for one species to shade the other and making it more likely that root competition would be detected. Wilson and Tilman (1991) grew three species of grass in field plots along a nitrogen gradient and compared their growth when competing only below ground with growth when competing both above and below

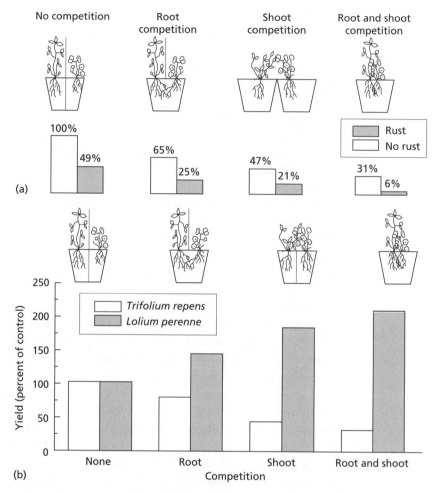

Fig. 8.4 Competition experiments separating the effects of root and shoot using two different designs: (a) effects on skeleton weed of root and shoot competition with subterranean clover (redrawn from Groves & Williams 1975); (b) effects on white clover and perennial ryegrass of root and shoot competition between them (Martin & Field 1984).

ground. In all three species, root competition from neighbours was relatively more important at lower levels of soil nitrogen than at higher levels. Root competition may therefore be even more important in infertile natural communities than experiments with agricultural species suggest. This conclusion is supported for forest communities by a review of 47 experiments that isolated plants in experimental plots from the roots of neighbours by a trench or other below-ground barrier. Coomes and Grubb (2000) found that 40/47 of these experiments resulted in improved growth in isolated plots compared to controls. The exceptions were mainly in moist forests on fertile soil where below-ground resources were less limiting than light, which had to penetrate a dense forest canopy.

Different rooting depths can promote niche separation and coexistence between species, so this might also occur between genotypes, under certain conditions. Ennos (1985) showed in white clover with different rooting depths that length of root had a high heritability. He evaluated the significance of this trait for plant fitness by selecting plants with either relatively long (L) or short (S) roots and planting them in a replacement series competition experiment. Although the *total* number of stolons produced in monocultures of L or S was the same as in mixtures, L in mixtures produced 22% more stolons than S. Under conditions of drought, though, the dry matter yield of a 50:50 mixture was significantly greater than that of either monoculture. Depending on the occurrence of drought then, the exploitation of different soil layers by the two genotypes could promote their coexistence and help maintain genetic diversity.

8.3.1.3 *Replacement series*

This design, which was first advanced by de Wit (1960), has been used very widely to compare the performance of species in mixtures with their performance in monoculture using their relative yields:

$$\text{Relative yield of species i} = \frac{\text{yield per unit area of species i in mixture}}{\text{yield per unit area of species i in monoculture}}$$

$$\text{Relative yield of species j} = \frac{\text{yield per unit area of species j in mixture}}{\text{yield per unit area of species j in monoculture}}$$

The sum of the relative yields for the two species in a particular mixture (usually 50:50) is called the **relative yield total** (RYT). Unfortunately, its value has often been misinterpreted. In fact, RYT can only be interpreted under some rather restrictive conditions (Jolliffe 2000). The calculation of relative yields is based on the assumption that yields per unit area in monocultures do not change across the range of single-species densities used in the experiment, or in other words that these densities lie within the range that generates constant final yield (Chapter 4). *On condition* that this is true, an RYT > 1 indicates that there is a yield advantage in the mixture, *at that particular density and proportion* of the species. This may, or may not apply to other densities and proportions, even if they meet the condition of constant final yield (Taylor & Aarssen 1989).

Validly interpreted, an RYT > 1 may indicate that there is some niche separation between the species (e.g. *Plantago lanceolata* vs. *Anthoxanthum odouratum*), but this may not occur at other densities (Willey 1979; Connolly 1986), so one cannot conclude from RYT that mixtures of the two species meet the general conditions for coexistence that we derived earlier (Inouye & Schaffer 1981). Despite these handicaps, the replacement design can be useful for examining the effects of herbivory, disease or some other agent on competition.

For many purposes it is desirable to be able to determine when mixtures of crops produce a greater yield than they would in an equivalent area planted with separate monocultures. Because of the limitations of RYT as a measure of the performance of mixtures in relation to monocultures, Connolly (1987) proposed an alternative index called the **relative resource total** (RRT) to answer this kind of question. This requires yield–density curves for the two monocultures to be available. RRT is derived as follows. The yields per individual of two species in mixture are w_i and w_j, when they are growing at densities N_i and N_j. For species i there will be some monoculture density N_{i0} that produces the same yield w_i as i in the mixture. The area occupied by a plant in this monoculture is a measure of the resources needed to produce a plant with yield w_i, and is given by the reciprocal of the monoculture density, $1/N_{i0}$. To produce the equivalent yield of N_i individuals with yield w_i requires an area N_i/N_{i0}. The corresponding quantity for the other species is N_j/N_{j0}, and

$$\text{RRT} = \frac{N_i}{N_{i0}} + \frac{N_j}{N_{j0}}.$$

If RRT > 1, the mixture is more efficient at turning resources into yield than are the monocultures. In effect, this measures the extent to which the species utilize different resources, and is a measure of niche separation. If RRT > 2, the species are yielding as if the other species in the mixture didn't exist! Note that RRT > 1 does *not mean* that the mixture necessarily yields better than both of the monocultures. One of the monocultures may still perform better than the mixture (Connolly 1987).

The Land Equivalent Ratio (LER) is another index, based upon a similar idea to RRT, which has been used a great deal in evaluating mixtures of agricultural crops. For species i and j producing yields per unit area in mixture that are w_iN_i and w_jN_j:

$$\text{LER} = \frac{w_iN_i}{w_{i0}N_{i0}} + \frac{w_jN_j}{w_{j0}N_{j0}}$$

where $w_{i0}N_{i0}$ and $w_{j0}N_{j0}$ are the yields of i and j in the highest yielding monocultures, regardless of density. As with RRT, an LER > 1 implies that there is a benefit in combining the species, though this might not produce a yield greater than one of the monocultures. Like RYT, the LER of a mixture also has the disadvantage that its value is dependent upon the particular monoculture chosen as a standard of comparison. Riley (1984) describes a method for comparing LERs that have been calculated using different monoculture yields in the denominators.

8.3.1.4 *Additive series and complete additive designs*

Additive series and complete additive competition experiments are similar in design and effectiveness, so we will refer to both by the shorthand 'additive'. Additive experiments vary N_i and N_j independently of each other, and can therefore be used to map the isoclines in a **joint abundance diagram**. One way to do this with a minimum number of experimental mixtures is to measure the starting and finishing densities (or seed outputs for annuals) of each species in each mixture and to use regression to fit a population model to the data that incorporates terms for the effects of species upon each other. We can start with the model we used for a single species in Chapter 5 (equation 5.9):

$$N_{t+1} = s_m N_t (1 + aN_t)^{-b}$$

in which s_m is the maximum seed output of a plant and the exponent b accounts for the effect of density dependence upon the shape of the recruitment curve. Freckleton *et al.* (2000) modelled the interaction between two grasses (an agricultural species *Lolium rigidum* and a weed *Vulpia bromoides*) and a legume (*Trifolium subterraneum*) growing in rangelands in New South Wales, Australia. They found that they could describe the dynamics of experimental mixtures of the three annuals by adding one term for each competing species inside the brackets of Eqn 5.9. Using subscripts i, j, k to denote terms applying to different species, the population size for species i at time $t + 1$ was given by:

$$N_{i(t+1)} = s_{mi} N_{i(t)} (1 + \alpha_{ij} N_{j(t)} + \alpha_{ik} N_{k(t)} + \alpha_{ii} N_{i(t)})^{-1}, \tag{8.6}$$

where α_{ij} and α_{ik} are, respectively, coefficients describing the effect of each individual of species j and species k upon species i, and α_{ii} is the effect of species i on itself ($= \alpha$ in Eqn 5.9). (Note that in these experiments the value of the exponent was $b = 1$ and that to standardize the coefficient α_{ij}, say, to the form used for competition coefficients in Eqns 8.1–8.4 it must be divided by α_{ii}.) After parameterization using data from two- and three-species experimental mixtures, equations of the form of Eqn 8.6 produced the joint abundance diagrams for *Lolium*, *Vulpia* and *Trifolium* shown in Fig. 8.5.

Figure 8.5(a) and (b) show that mixtures of either of the grasses with the legume (*Trifolium*) are stable, but that invasion by *Vulpia* creates an unstable situation from which either *Lolium* or *Vulpia* may emerge the winner, depending upon intitial densities when sown. The zero isoclines for the legume with either grass are nearly at 90° to each other indicating that they scarcely influence each other's densities, possibly because there is niche separation of nitrogen resources. Droughts are frequent in the climate of New South Wales and these greatly increase the density of *Vulpia* in rangelands there, making it almost inevitable that it will eventually reach densities at which it displaces *Lolium* (Freckleton *et al.* 2000).

If we are interested in yield rather than population dynamics, then intra- and interspecific effects of the species in a mixture can be handled in a manner analogous to the way in which we developed above the three-species competition model from the one-species case. Starting this time with Eqn 8.5, which describes the additive effect of one species upon another, intraspecific effects may be incorporated simply by inserting $a_i N_i$ into the right-hand side:

$$w_i = w_{mi} [1 + a_i (N_i + \alpha N_j)]^{-1}. \tag{8.7}$$

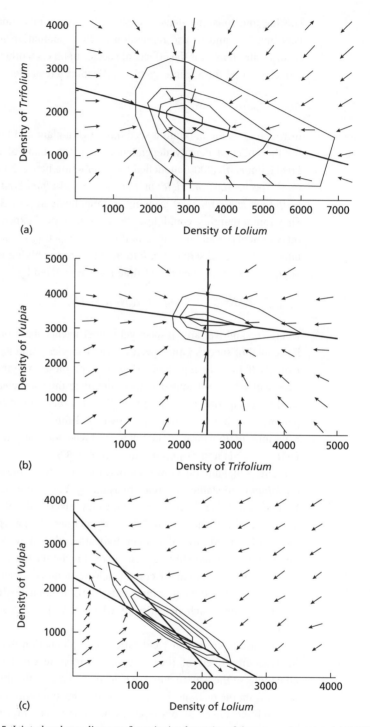

Fig. 8.5 Joint-abundance diagrams for pairwise dynamics of three annual species: (a) *Trifolium subterraneum* and *Lolium rigidum*; (b) *T. subterraneum* and *Vulpia bromoides*; and (c) *L. rigidum* and *V. bromoides* (modified from Freckleton *et al.* 2000).

For species j, the equivalent expression is:

$$w_j = w_{mj}[1 + a_j(N_j + \beta N_i)]^{-1}.$$

The yield–density response (Eqn 4.2) from which we have ultimately derived this equation assumes that yield approaches an asymptote. However, particularly if the yield we are interested in is some plant part such as the seed, the yield–density response may have a different shape. As we saw in Chapter 4, this may easily be accommodated by replacing the exponent -1 with a variable $-b$, which may assume a different value for each species. Pantone and Baker (1991) fitted Eqns 8.7 and 8.8 to data from a competition experiment between crop and weed varieties of rice, *Oryza sativa*, grown in Louisiana. Response surfaces for the grain yields per plant of the crop and the weed (called red rice) are shown in Fig. 8.6 and the equations for the two species were for grain yield per crop plant (w_c):

$$w_c = 149[1 + 0.29(N_c + 3.37N_r)]^{-0.88};$$

for grain yield per weed plant (w_r):

$$w_r = 218[1 + 0.61(N_r + 0.30N_c)]^{-0.85},$$

where N_c and N_r are, respectively, the densities of the crop and red rice (weed) varieties. Note from the values of parameters in these equations that isolated plants of the weed were much bigger than crop plants ($w_m = 218$ g vs. 149 g), occupied a greater area per plant ($a = 0.61$ m² vs. 0.29 m²) and that weeds had 3.37 times the effect on crop plant yield as crop plants had on each other ($\alpha = 3.37$). Exponents in both equations were significantly smaller than unity ($b_c = 0.88$, $b_r = 0.85$).

If the exponents of the yield density responses are unity, then a simpler model than those in Eqns 8.7 and 8.8 can be used to calculate competition coefficients. This is the **reciprocal yield model** of Wright (1981) and Spitters (1983). Using the reciprocal of yield per plant ($1/w$) instead of yield per plant (w) as the dependent

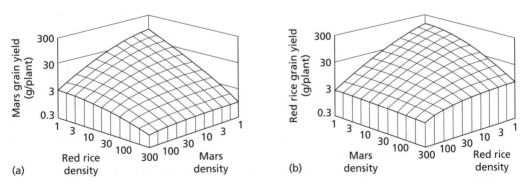

(a) (b)

Fig. 8.6 Response surfaces for the grain yield per plant in competition experiments of (a) *Oryza sativa*, cultivar 'Mars' and (b) a weed variety of *O. sativa* known as red rice (Pantone & Baker 1991).

variable produces a more friendly looking formulation of yield–density relationships that has found favour in agricultural research (Radosevich & Rousch 1990). In a monoculture:

$$1/w = B_{i0} + B_{ii}N_i, \tag{8.8}$$

where B_{i0} is $1/w_m$ and B_{ii} measures intraspecific competition (the effect of species i on itself). Comparison of this equation with Eqn 4.2 should convince you that they are equivalent. Equation 8.8 can readily be expanded to two or more species. For species i in a mixture with species j:

$$1/w_i = B_{i0} + B_{ii}N_i + B_{ij}N_j, \tag{8.9}$$

where B_{ij} measures the effect of j on i and N_j is the density of j. The corresponding equation for species j is:

$$1/w_j = B_{j0} + B_{jj}N_j + B_{ji}N_i. \tag{8.10}$$

These two equations are the equivalents of Eqns 8.7 and 8.8, but separate the effects of intraspecific and interspecific competition more clearly. The ratio of interspecific effects to intraspecific effects in Eqns 8.9 and 8.10 gives us the competition coefficients of Eqns 8.7 and 8.8:

$$\frac{B_{ij}}{B_{ii}} = \alpha \text{ and } \frac{B_{ji}}{B_{jj}} = \beta \tag{8.11}$$

The reciprocal yield model has been fitted to a number of weed–crop mixtures. For example Rejmánek et al. (1989) used the model to describe the results of an additive competition experiment between the tomato cultivar 'Ace VF55' and Japanese millet Echinochloa crus-galli var. frumentacea, which is a close relative of barnyard grass, one of the world's worst weeds. The model fit the yields of both components of the mixtures well, and showed that tomato was more sensitive to interspecific competition than to intraspecific competition, while the reverse was the case for Japanese millet (Table 8.2). The competition coefficients show that each millet plant had an effect on tomato equivalent to 3.7 other tomatoes. Tomato was equivalent to only 0.14 millet plants in its effect on millet. In another experiment, with wheat and annual ryegrass, wheat had a greater effect on the weed (ryegrass), than vice versa (Table 8.2).

Analysing the results of competition experiments is very much an empirical exercise in finding which models best fit the data (Cousens 1985; Connolly 1987; Freckleton & Watkinson 2000a). This can make it difficult to generalize about plant competition, although there is also a lesson to be learned from the very difficulty itself. The outcome of interspecific competition can be highly contingent upon the conditions of the experiment, including the density and proportions of the species in a mixture. In an additive experiment with Vulpia fasciculata and Phleum arenarium, Law and Watkinson (1987) found that the model that best fitted their data allowed competition coefficients to vary with frequency and density. Equations such as the reciprocal yield model were inadequate to describe

Table 8.2 Parameter values for the reciprocal yield model fitted to two additive competition experiments, between (a) tomato and Japanese millet and (b) between wheat and annual ryegrass.

Species	Variable				
	Intraspecific effect (B_{ii})	Interspecific effect (B_{ij})	Competition coefficient (B_{ij}/B_{ii})	Fit of the model (R^2)	Source
(a) Tomato	0.003	0.011	3.67	0.94	Rejmánek *et al.*
Japanese millet	0.001	0.00014	0.14	0.95	(1989)
(b) Wheat	1.18	0.17	0.14	0.90	Concannon, *in*
Annual ryegrass	3.21	4.51	1.41	0.43	Radosevich & Rousch 1990

the interaction between the two species across the full range of values of $N_i N_j$ that their experiment explored. In a similar competition experiment between two other annuals, Connolly *et al.* (1990) found that competition coefficients varied with total density and time.

Most competition experiments have measured competitive effects rather than outcomes and have often used seedlings. It is also important to caution that competition coefficients by themselves do not predict the outcome of competition or the relative fitness of the competitors. This depends upon the configuration of the isoclines in the joint-abundance diagram (Fig. 8.1), and competition coefficients are not the only variable determining these.

8.3.2 Multi-species competition

The competition experiments and models we have discussed so far have nearly all involved mixtures of just two species, but most communities contain more than this. How, then, do we scale up from the results of competition experiments between two species to models of more realistic communities? As a first step to answering this question we need to know how competition affects species in field conditions.

8.3.2.1 *Field experiments*

The simplest test for interspecific competition in the field is to remove individual species from a community and to monitor the response of those that remain (Goldberg & Scheiner 1993). The majority of such **removal experiments** reveal some degree of interspecific competition (Aarssen & Epp 1990; Goldberg & Barton 1992; Gurevitch *et al.* 1992), but removal experiments also demonstrate that the behaviour of a plant in a mixture of two species may be quite different from its behaviour in more diverse mixtures. For instance, when plantain (*Plantago lanceolata*) was removed from field plots in a grassland community in North Carolina, the abundance of winter annuals increased. However, this only happened if sheep's sorrel (*Rumex acetosella*) was absent from the experimental plot. Where

Rumex was present and *Plantago* was removed, *Rumex* and not winter annuals benefited from the removal (Fowler 1981). Hence, in the field situation the relationship between specific pairs of species is contingent upon the presence or absence of other species.

Competition between several species at once can lead to the apparently paradoxical result that a species may decrease in abundance when another species is removed because of the effect this removal has on a third species. Such an effect was found in the study of a community of desert annuals by Davidson *et al.* (1985) and was also observed by del Moral (1983) when he removed *Carex spectabilis* from plots where it was dominant in a subalpine meadow community in Washington State. The grass *Festuca idahoensis* increased and produced a significant decrease in four other species. Effects of this kind plainly depend upon which species occupy a particular experimental plot. The pattern of plant distribution is important. Fowler (1981) found that up to 67% of the variance in the response of a species to the removal of another from her plots was the result of differences in plant distribution between plots. On the whole, field experiments involving species removal demonstrate very few specific interactions between species that cannot be accounted for by the disposition of individual plants before removal was carried out. In the short term, the species to respond most strongly to the local removal of another may be whichever one happens to be in the gap that has been created. In the longer term, gaps may be colonized by plants regenerating from seed, and on this timescale the size of gap may well determine which species appears in it (Bullock *et al.* 1994, 1995).

Competing plant species are rarely evenly matched in the field. The relative size of a plant, and whether it or its competitors are seedlings, juveniles or larger, usually determines the outcome of competition (Goldberg & Werner 1983). This was the case in a greenhouse experiment in which Goldberg and Landa (1991) grew seven species in all pairwise combinations in partial additive experiments that combined nine target individuals of one species with 0–356 neighbours of another. The effects of varying numbers of neighbours upon targets fitted the reciprocal yield model (Eqns 8.9 and 8.10), which could then be used to estimate competition coefficients for all the pairwise interactions among species (Eqn. 8.11). Freckleton and Watkinson (2000b) found that these coefficients were predictable from the difference in maximum size between two competing species. Field experiments also indicate relative size to be important. For example, del Moral (1983) found that adult transplants survived better than seedling transplants and that survival of both depended upon the productivity of different sites, and hence the size of competing plants, within his subalpine meadow community. Grace (1985) found that the outcome of competition between two cattail *Typha* species was different in mixtures of juveniles raised from seed and in mixtures of mature plants.

If competitive relationships among groups of species are determined by their relative sizes, then species should fall into linear hierarchies or pecking orders of competitive effect (e.g. species $A > B > C$, etc.). In such cases competitive relationships are said to be **transitive**, and the handful of experimental tests of the kind Goldberg and Landa (1991) performed usually demonstrate near-complete transitivity (Grace *et al.* 1993; Shipley 1993; Silvertown & Wilson 2000). The sum of evidence from field and greenhouse experiments points clearly to three simple conclusions with far-reaching consequences: (i) interspecific competition among plants is ubiquitous; (ii) it is confined mainly to immediate neighbours; and (iii) the relative size of neighbours influences the outcome of interactions. These are the premises upon which individual-based spatial models of competition are built.

8.3.2.2 *Individual-based neighbourhood models*

Spatial models of the kind used to describe intraspecific interactions among neighbours in Chapter 4 may be extended to incorporate two or more species competing with each other. The first step in obtaining the data to parameterize such a model is to map all plant locations, and then to use multiple regression to determine the effect of each neighbour on the growth, survival and fecundity of target individuals. Neighbour identity (species), time of germination, size, distance from the target and angular dispersion around the target may be included in the model, although not all variables may prove to have a significant effect (Bergelson 1993; Lindquist *et al.* 1994; Garrett & Dixon 1998; Wagner & Radosevich 1998). The simplest version of this approach is to use a partial additive competition experiment that varies only the number of neighbours. Typically, most of the competitive effect on the target is accounted for by only the first few neighbours (e.g. Fig. 8.3) and plants lying outside a limited radius can be ignored. This radius defines the **neighbourhood area** (also termed the **area of influence** in weed science) for the target species, and its optimum value in a particular community can be determined by multiple regression. For example, in greenhouse mixtures of the annuals *Abutilon theophrasti* and *Amaranthus retroflexus* grown by Pacala and Silander (1987), the optimum neighbourhood radius for the species' effects upon each other's biomass proved to be 70–75 mm.

In tree communities where competition is chiefly for light, neighbours below a certain height, as well as those beyond a certain radius from the target, may have negligible effects; thus, vertical as well as horizontal spatial dimensions must be considered. For Douglas fir (*Pseudtsuga menziesii*) seedlings competing with shrubs in Oregon, Wagner and Radosevich (1998) found that the regression models which best fitted the height growth of targets ignored shrubs with a relative height less than 1.25 (height of target = 1). The effect upon *P. menziesii* seedlings of shrubs taller than this increased rapidly with height (Fig. 8.7a). In the horizontal plane, the optimum neighbourhood radius was zero because the shrubs with the greatest effect upon *P. menziesii* growth were those whose crowns intersected with the stem of a seedling. The effect of shrubs upon seedlings decreased rapidly with the distance between the target's stem and the edge of a neighbour's crown (Fig. 8.7b).

The ultimate neighbourhood model of interspecific competition describes the complete dynamics of a community and must include seed dispersal and all stages

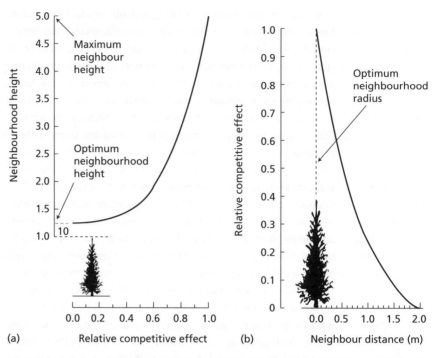

Fig. 8.7 Relative competitive effect of shrubs upon the height growth of Douglas fir *Pseudotsuga menziesii* seedlings in relation to (a) relative height of neighbour (height of target = 1) and (b) distance of neighbour from the target's stem (modified from Wagner & Radosevich 1998).

of the life cycle of all competing species. Such spatial population dynamics models are complex, but can be built from several simpler regression equations or **predictors** that describe how the germination, survival, fecundity and seed dispersal of an individual varies with the properties of its neighbourhood. In Pacala and Silander's (1987) experiment with *Abutilon theophrasti* and *Amaranthus retroflexus* already referred to, they found that individual plant fecundity correlated with individual plant biomass, enabling them to create a fecundity predictor for incorporation into a more complex model. The full model based upon empirically calibrated neighbourhood predictors of germination, survival, fecundity and dispersal successfully predicted the dynamics of mixtures over a four-year period but so too, it turned out, did a **mean field** model (Pacala & Silander 1990). Mean field models ignore, or average out, spatial variation between neighbourhoods. The critical factor in determining whether an individual-based spatial model is more accurate than a mean field model appears to be whether there is variation in intraspecific aggregation at the scale of the neighbourhood (Garrett & Dixon 1998), which in the case of Pacala and Silander's (1990) experiments there apparently was not. The scale of aggregation in relation to the size of sampling units (e.g. quadrats) also influences the accuracy with which removal experiments can measure competition coefficients (Freckleton & Watkinson 2000c).

In contrast to Pacala and Silander's (1987) experiment with annuals, an individual-based neighbourhood model designed to simulate the dynamics of broad-leaved forest in north-east North America, proved to be much more successful than its mean field equivalent. The model, called SORTIE and developed by Pacala *et al.* (1993, 1996), simulated the population dynamics of nine tree species including American beech *Fagus grandifolia*, eastern hemlock *Tsuga canadensis*, sugar maple *Acer saccharum*, white ash *Fraxinus americana* and red oak *Quercus rubra*. Measurements of sapling growth and environmental factors in the neighbourhood of individual trees showed that light was a limiting resource for all species and that there was a negative correlation between a species' rate of growth under high light and its rate of survival under low light conditions, defined as 1% of full sunlight (Fig. 8.8) (Pacala *et al.* 1994; Kobe *et al.* 1995). Neighbourhood predictors of growth and mortality for each species were derived from the functional relationship between these variables and local light availability. Two other parameters important in SORTIE were also correlated with light: (i) species with high growth in high light, at what was effectively the shade-intolerant end of the continuum,

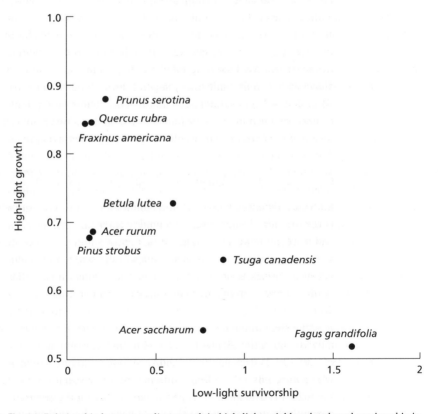

Fig. 8.8 Relationship between sapling growth in high-light neighbourhoods and survivorship in low light for nine tree species (adapted from Kobe *et al.* 1995). Both variables are plotted on relative scales.

cast less shade with their canopies and (ii) had better seed dispersal than species at the more shade-tolerant end.

Individual trees in a SORTIE simulation competed with one another solely through the intermediary resource, light. Each tree received an amount of light determined by its canopy size. The size, identity and location of its neighbours were then used to 'predict' its growth and survival accordingly. Fecundity and seed dispersal were then predicted from tree size and location, and the success of recruits was determined by light availability in the neighbourhood in which seeds germinated. Pacala and Deutschman (1995) compared the performance of SORTIE with an equivalent nonspatial (mean field) model and found that the nonspatial model predicted a standing crop biomass for the forest only about half that actually observed, while SORTIE got the biomass right. Standing crop biomass is the result of a balance between the death of large trees and the growth of younger ones. The mean field model failed so badly because the forests in question contain variation in light availability between sapling neighbourhoods that only a spatial model can reproduce. The significance of this variablity for the prediction of standing crop biomass lies in the fact that sapling growth in the forest is a concave function of light availability, increasing disproportionately with the percentage of full sun (Pacala *et al.* 1994). In a nonspatial model all saplings receive an average amount of light, but in a spatial model, as in a real forest, some saplings receive enough light to grow disproportionately better than the average. This raises the total rate of biomass accumulation among saplings above what can be achieved by saplings all experiencing only the average neighbourhood. What is surprising is that this mechanism is strong enough to shift the balance between biomass accumulation and loss to twice the value to be expected for the same species growing in a spatially uniform light environment.

8.3.3 Effects of other organisms on competition between plants

Most experiments on plant competition are carried out in pots where the outcome is constrained. Just as when two boxers enter the ring, there has to be a contest, and there has to be a victor. In the field, plants may pass each other by, analogous to the boxers passing on opposite sides of the street. The outcome of competition between species is probably also much less deterministic than pot experiments imply. One reason for this is that populations confronting each other in the field do not do so alone. To continue the analogy, there are other plants present that may interfere in the fight; herbivores that can turn a heavyweight contestant into a flyweight; pathogens that build up locally and create hostile conditions (Bever *et al.* 1997; Mills & Bever 1998); and symbiotic microorganisms that can subvert the whole match by channelling nutrients from one species to another.

Grazing can influence competitive outcome, but the reverse may also occur and competition may influence grazing damage. This was demonstrated by Cottam *et al.* (1986) who found that the beetle *Gastrophysa viridula* significantly reduced leaf area and weight of *Rumex obtusifolius* compared to ungrazed controls, when the plant was growing in a grass sward, but not when it was growing free from competition. Herbivory exercises the most obvious influence over the outcome of competition between plants, but the influence of pathogens, mycorrhizal symbionts and nitrogen-fixing bacteria can also be strong. Their effects are less obvious than those of insect and vertebrate herbivores that chomp holes in leaves or defoliate plants, so their importance can easily be underrated.

The balance between competing plant species may be determined by their relative palatability to herbivores. In many grasses, such as perennial ryegrass *Lolium perenne* and the fescues *Festuca* spp. that are normally palatable to herbivores, infection by endophytic fungi reduces their palatability to vertebrates and invertebrates because the fungi produce toxic alkaloids (Clay 1990). The relationship between endophytic fungi and the plants they infect ranges from the parasitic to the mutualistic, depending upon the fungi involved and the prevalence of herbivory. Heavy grazing favours infected plants that increase at the expense of uninfected individuals. Clay and Holah (1999) found that endophytic infection of *Festuca arundinaceae* increased its dominance and decreased the species richness of successional communities when compared with control plots containing uninfected *F. arundinaceae*.

The most important group of microbial symbionts that infect plants are the mycorrhizal fungi. The roots of a majority of plant species are colonized by **endomycorrhizal fungi** (or AMF, short for Arbuscular Mycorrhizal Fungi) or, in temperate region trees by **ectomycorrhizal fungi**, that have large effects upon nutrition and hence upon competitive success. The effects on nutrition are of two types: (i) improved acquisition of resources from the soil, particularly phosphorus; and (ii) transfer between plants of mineral nutrients, and perhaps carbon (Simard *et al.* 1997; Robinson & Fitter 1999), through the network of fungal hyphae in the soil. The consequences of these processes for particular interspecific interactions between plants depend upon plants' relative dependence upon mycorrhizal fungi, the directionality of transfers and the specificity of the symbiosis.

In a review of interspecific competition experiments that contained treatments with and without mycorrhizal infection, Allen and Allen (1990) found that the result was altered by infection in most cases, although in only one case out of eight was the outcome actually reversed. Hetrick *et al.* (1989) grew two grasses of tallgrass prairie together in pots, with and without an innoculum of the AMF *Glomus etunicatum*. With the fungus present, the warm-season grass *Andropogon gerardii* supressed the cool-season grass *Koeleria pyramidata*, but in the absence of mycorrhizal infection *A. gerardii* grew so poorly that *K. pyramidata* was unaffected by competition. Although both grass species were infected by *Glomus*, and its presence was essential to *A. gerardii*, the mycorrhizal association was of no nutritional importance to *K. pyramidata*. Field experiments confirmed that eliminating AMF by applying fungicide to plots in tallgrass prairie in Kansas caused a decline in *A. gerardii* and other dominant, warm-season grasses that were also highly dependent on mycorrhizas, leading to a significant increase in total species diversity over five years (Hartnett & Wilson 1999; Smith *et al.* 1999). In the field, the warm- and cool-season grass species appear to coexist because their main growth periods are separated phenologically. Although suppressed by competition from *A. gerardii* when mycorrhizal fungi are present (the natural situation), labelling experiments with ^{32}P have shown that cool season species up to half a metre away do receive transfers of phosphorus in varying amounts from *A. gerardii*, although mycorrhizal links may not be the only route by which this occurs (Walter *et al.* 1996).

Dominant species are not always the ones to benefit most from mycorrhizal infection, possibly because resources may move from source (well-resourced plants) to sink (poorly growing plants). Grime *et al.* (1987) grew experimental mixtures of 18 herb species in treatments with and without AMF and with and without clipping to simulate herbivory. Both AMF infection and clipping reduced the dominance of the most abundant species in the mixture (the grass *Festuca ovina*), causing other species to increase in biomass as a consequence of competitive release. Transfers of radiolabelled carbon occurred between plants with mycorrhizal infection and not between those without, but a *net* transfer was not established. The effects of AMF on this experimental community could be accounted for by a direct nutritional benefit conferred selectively on some species, and the release from competition of others caused by reduced growth in the dominant grass. The role of mycorrhizal *links* between species remained an open question (Bergelson & Crawley 1988).

Particularly in natural habitats, but less so in agricultural ones (Helgason *et al.* 1998), soils contain large numbers of AMF taxa (identification to species level is very difficult). Although individual plants may be infected by 10 or more AMF taxa, it appears that plant species differ from each other in which AMF benefit them most (van der Heijden *et al.* 1998a). van der Heijden *et al.* (1998b) grew separate experimental mixtures of species from European calcareous grasslands and North American oldfields, each in a series of treatments that varied the number or identity of AMF taxa present. Out of 11 plant species in the calcareous grassland mixtures, only the grasses *Bromus erectus* and *Festuca ovina* and the nonmycorrhizal sedge *Carex flacca* grew well in the absence of any AMF, and altering which AMF taxa were present altered the composition of the experimental community. In the oldfield communities increasing the number of AMF taxa in steps from 0 to 14 steadily increased plant species diversity. It is remarkable that while the relative size of neighbours appears to be the key to understanding interspecific competition among plants in the rather artificial conditions of pot experiments (Section 8.3.1–2), it looks as though in the field mycorrhizal interactions may restore the importance of the identity of species and of the presence of plants outside the immediate neighbourhood.

Although mutualists tend to be much less host-specific than pathogens, some microbial genotypes are more infective or more beneficial to some hosts than others. Chanway *et al.* (1989) studied the relationship between three genotypes of white clover *Trifolium repens* sampled from different parts of a Canadian pasture,

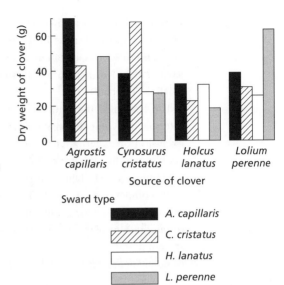

Fig. 8.9 The dry weight of plants of *Trifolium repens* from a permanent grassland sward, sampled from patches dominated by four different perennial grasses and grown in all combinations of mixture with the four grass species (Turkington & Harper 1979).

and three genotypes of *Rhizobium* and perennial ryegrass *Lolium perenne* growing with them. When all combinations of the *Trifolium*, *Rhizobium* and *Lolium* genotypes were grown in a glasshouse, genotype combinations from the same source outyielded others by 27%. Strangely, this effect was almost entirely the result of a specificity between *Rhizobium* and *Lolium* genotypes, and it made little difference with which *Trifolium* genotype they were combined. A bacterium, *Bacillus polymyxa*, associated with *Lolium* was found to improve the yield of *Trifolium* in mixtures, especially when the bacterial strain used was the one native to the *Lolium* genotype in the experiments (Turkington 1996).

These results throw some interesting light upon field studies by Turkington and Harper (1979), Aarssen and Turkington (1985) and Evans *et al.* (1985), who all found that yields of *T. repens* in experimental mixtures were highest when clover genotypes were grown with the grass species or *L. perenne* genotype occurring with them in the source field (Fig. 8.9). It now seems likely that these specificities are due to two-, three-, or even four-way interactions involving microbial as well as plant genotypes.

8.4 Coexistence

The richest plant communities contain very large numbers of species: at Yanamomo in the upper Amazon forest of Peru, Gentry (1988) counted all trees 10 cm in diameter or larger in a 100 m × 100 m (1 ha) plot and found that every second tree he encountered added a new species to his list which finally totalled 283. The total for plants of all species would certainly have been much larger. At Barro Colorado Island (BCI), which is one of the best studied areas in the tropics, 1316 plant species have been recorded in the 15 km² of tropical forest there. This is roughly the number of native vascular plant species in the whole of the British

Isles, which has an area of $314\,375\,km^2$, but on a different scale there are high diversities in small areas in the temperate zone too. In British calcareous grasslands, up to 30 species can be found coexisting in $1/8th\,m^2$, and in pine-wiregrass savannah in North Carolina 42 species may be found in plots of the same area (Walker & Peet 1983).

The challenge that these facts present to theory is immense. In the model of interspecific competition introduced in Section 8.3 coexistence and competitive exclusion are alternative states and changing parameters can flip species from coexistence to extinction. The model predicts that species should coexist in a community only when each inhibits its own population growth more than that of its competitors (Fig. 8.1). One way this can happen is if species have different **niches** (see below for further explanation). Can this really explain the coexistence of hundreds of plant species when all share a requirement for water, light and the same mineral nutrients? In this section we shall see how this model and a variety of other theories match up to explaining coexistence in real plant communities.

8.4.1 An overview of coexistence mechanisms

We will discuss, in different degrees of detail, six mechanisms that have been proposed to explain how plant species are able to coexist (Table 8.3). Each mechanism has been analysed using theoretical models, and although only the conclusions from these models will be presented here, it is important to remember that a mathematical demonstration that a mechanism can work is as important as field data to deciding what is really going on. A necessary and sufficient criterion for whether a theoretical model will promote coexistence is whether it predicts that species will increase when they are rare. A similar criterion applied to allele frequency is used in population genetics models to predict whether a polymorphism will be stable (Section 3.5). In ecological situations, coexistence through negative frequency-dependence can be achieved by any mechanism that provides species with a **refuge** from competition when at low abundance. Models differ mainly in how the refuge from competition comes about.

The Lotka–Volterra competition model, which requires that coexisting species have different niches, was the first formalization of how competitors interact and is also the conceptual basis for many of the coexistence mechanisms that have been proposed as alternatives to niche separation. The Lotka–Volterra model can justifiably be called the 'classical' model of coexistence, and is a useful benchmark against which to compare other models. These other models divide into two groups: (i) those in which *coexistence* is an equilibrium condition; and (ii) those in which *competitive exclusion* is the equilibrium condition, but in which coexistence occurs because some mechanism prevents equilibrium being attained. The latter are usually called 'nonequilibrium' models of coexistence. The equilibrium models variously add space and dispersal, time, age structure and natural enemies to the brew of the classical model (Table 8.3). Nonequilibrium models add temporal fluctuations in recruitment or density-independent mortality, or make

Table 8.3 The six coexistence mechanisms discussed in the text, giving the conditions that must be met for the mechanism model used to derive these conditions (numbers refer to Sections of Chapter 8).

	Necessary conditions for coexistence	
Equilibrium mechanisms		
8.4.2 Resource partitioning (niche separation)	Intraspecific > interspecific competition due to niche differences	
8.4.3 Spatial segregation	Within neighbourhoods:	Among neighbourhoods:
	$\dfrac{\text{interspecific competition}}{\text{intraspecific competition}} <$	$\dfrac{\text{concentration of intraspecific competition}}{\text{concentration of interspecific competition}}$
8.4.4 Recruitment limitation	Dispersal/competitive-ability trade-off	
8.4.5 Pest-pressure	Species-specific natural enemies, limited dispersal	
Nonequilibrium mechanisms		
8.4.6 Storage effect	Overlapping generations, asynchronous recruitment	
8.4.7 Density-independent mortality	Density-independent mortality disproportionately kills competitive dominants	

assumptions that slow the progress of competitive exclusion to an infinitesimal rate (Shmida & Ellner 1984; Hubbell & Foster 1986a).

Several different coexistence mechanisms share some important conditions that both add realism and facilitate coexistence by reducing the strength of interspecific competition without always being sufficient, on their own, to prevent competitive exclusion. The first of these conditions, which appears in several models, is that there is a phase of the plant life cycle when individuals show little or no response to interspecific competition. This **stage-specific refuge** can occur in the seed bank or, for example, once established plants have reached sufficient height not to be overtaken by later-germinating neighbours. This is especially important to the storage effect (Section 8.4.6) and to recruitment limitation (Section 8.4.4). The second condition is that species are not randomly distributed in space. Models such as the classical Lotka–Volterra model, that ignore the spatial distribution of individuals, are **mean field** models and frequently produce quite different predictions from those that take account of dispersal and assume some **aggregation** of distribution (Dieckmann *et al.* 1999). Two mechanisms that are particularly dependent upon spatial effects are recruitment limitation and spatial segregation (Section 8.4.3). Most spatial models treat competition as a **lottery** in which an element of chance, usually weighted by abundance or competitive ability, influences which of the species germinating in any particular microsite capture it from the others. This also favours coexistence.

8.4.2 Resource partitioning

The theoretical conditions for coexistence derived from the classical model (that intraspecific competition > interspecific competition) may be met if the population growth rate of each species is limited by a different resource. The coexistence of *n* species requires *n* different limiting resources. Each available resource can be represented by an axis on a graph or, if it is a nutrient, a gradient of concentration. These **resource axes**, define a **niche space**. Two axes, for example one for nitrogen and one for phosphorus, define a two-dimensional niche space (Fig. 8.10). Resource axes are often referred to as **niche dimensions** and there is no theoretical limit to their number. Hutchinson (1957) described the niche as an '*n*-dimensional

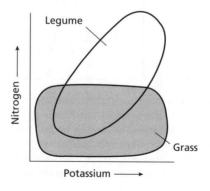

Fig. 8.10 A two-dimensional niche space showing the hypothetical use of two mineral resources by a legume and a grass.

hypervolume'. The resource use of competing species can be mapped onto a graph of niche space if we know which resources limit population growth rate, and where the limits of each species' consumption of each limiting resource lie. For the coexistence conditions to be satisfied, competing species must be sufficiently different in their use of resources to reduce overlap in niche space. There is a substantial literature discussing how niche overlap should be measured (reviewed by Neet 1989) and what degree of overlap is compatible with coexistence (reviewed by Abrams 1983), but both questions are moot unless the relevant niche axes can be identified, and therein lies the real problem.

Resources can be partitioned between the species in a community in a variety of ways. First, a community can usually be divided up into **guilds**, or groups of species that exploit resources in a particular way. For example at BCI (Chapter 6) plants fall into groups that live in different horizontal strata of the forest, and into a group that regenerates in gaps and a group that regenerates in a closed-

Table 8.4 Life forms of plant species at Barro Colorado Island, Panama (Croat 1978).

	Number of species	Percentage
Cryptogams		
Epiphytes	41	3.1
Hemi-epiphytes	1	0.1
Aquatics	6	0.5
Vines	4	0.3
Other terrestrials	47	3.6
Tree ferns	5	0.4
Total cryptogams	104	7.9
Phanerogams		
Trees > 10 m tall	211	16.0
Trees < 10 m tall	154	11.7
Shrubs 2(−3) m tall	93	7.1
Epiphytic or hemi-epiphytic trees and shrubs	16	1.2
Parasitic shrubs	7	0.5
Total arborescent spp.	481	36.6
Lianas or woody climbers (including 10 climbing trees)	171	13.0
Vines	83	6.3
Epiphytic or hemi-epiphytic vines	11	0.8
Total scandent spp.	265	20.1
Epiphytic herbs	135	10.3
Aquatic herbs	54	4.1
Herbs of clearings	197	15.0
Forest herbs	75	5.7
Parasitic herbs	1	0.1
Saprophytic herbs	4	0.3
Total herbs	466	35.4
Total native plant species	1316	100

canopy environment (Table 8.4). This kind of partitioning of physical space is relatively easy to see, but cannot explain how 93 shrubs or 171 lianes coexist within their respective guilds.

Plants of species-rich grasslands can generally also be divided into a guild that colonizes disturbances and a guild of 'matrix-forming' species (Grubb 1986). Grasses and legumes can be considered separate guilds because the former tend to be limited by nitrogen, while legumes, which have nitrogen-fixing symbionts, tend to be limited by phosphorus (Fig. 8.10). At first sight the scope for further partitioning of mineral elements is greatly limited because all plants share a requirement for a relatively small number of nutrients. Braakhekke (1980) and Tilman (1982) suggested a mechanism by which small-scale spatial patchiness in the relative concentrations of essential elements could permit more than two species to coexist on only two essential resources. The **resource ratio hypothesis** requires that species switch from limitation by one resource to limitation by another, depending upon the local ratio of resource concentrations, but a test of the hypothesis in grassland at Cedar Creek in Minnesota showed that all species were limited by a single resource, nitrogen (Tilman & Wedin 1991; Wedin & Tilman 1993). Braakhekke (1980) found no support for the hypothesis in experiments with plants grown in nutrient solutions, but has more recently revived the hypothesis (Braakhekke & Hooftman 1999).

In a study of species growing in British calcareous grassland, Mahdi *et al.* (1989) sought differences between eight species on six niche dimensions by measuring their phenology, and soil depth, pH, and levels of available nitrogen, phosphorus and potassium where each plant grew. An ordination technique was used to find the greatest possible separation of the eight species using the variables measured (Austin 1985). The first two axes of this ordination accounted for nearly 80% of the variance among species, but there was substantial niche overlap (Fig. 8.11). As a test of the resource ratio hypothesis, nitrogen:phosphorus ratios were compared for the eight species, but there were no significant differences among species.

There are always at least two interpretations with which to explain a failure to find niche separation among coexisting species: (i) the wrong niche axes were examined; or (ii) there is no niche separation. Because there is no a priori basis on which to choose correct niche axes, it is impossible to decide between the two, and hence the more general hypothesis that niche separation is required for coexistence cannot be falsified (Silvertown & Law 1987). Should we give up the search for niche separation in plants? The results of some recent studies suggest not.

Niche separation among species requires that they all specialize, but why be a specialist? The answer is that environmental conditions, reinforced by interspecific competition, require it and that **trade-offs** usually prevent a species specialized for one environment beating the specialized competition in another. The role of interspecific competition in forcing specialization upon competing species is well illustrated by the results of a classic experiment performed by Ellenberg (1953). He created a water table gradient in a large concrete tank filled with soil and sowed a number of meadow grasses evenly across the gradient in monoculture

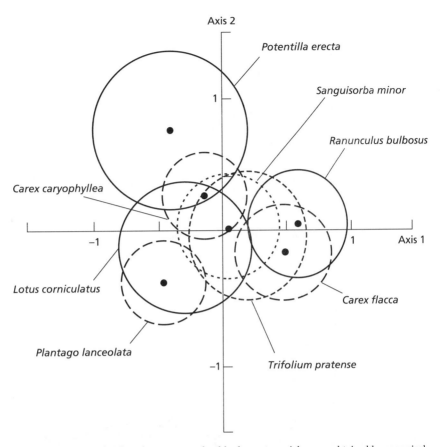

Fig. 8.11 Position of eight calcareous grassland herbs on two niche axes obtained by canonical variate analysis. 95% confidence limits are shown by circles. Axis 1 is influenced mainly by phenology, axis 2 by pH and phenology (from Mahdi *et al.* 1989).

and mixture treatments. When growing on their own, all species did best at the wet end of the gradient (Fig. 8.12a), but when growing in competition with one another, species shifted their optima and segregated along the gradient (Fig. 8.12b). Mean pairwise niche overlap among the species was 94% in monoculture, but 72% when grown together, a highly significant reduction caused by interspecific competition (Silvertown *et al.* 1999). Field data collected on the distribution of more than 50 meadow species in relation to the hydrological conditions where they were found growing within two sites in England support the idea that these species segregate in a niche space defined by two axes, one axis representing tolerance of water-logging and the other a tolerance of drought (Silvertown *et al.* 1999). Plotting the mean values of these two tolerances for each species against each other clearly shows the trade-off that leads to niche separation along hydrological gradients in this community (Fig. 8.13).

It has long been suggested that the guild of gap-colonizing trees found in tropical rainforests might partition the resource represented by tree-fall gaps

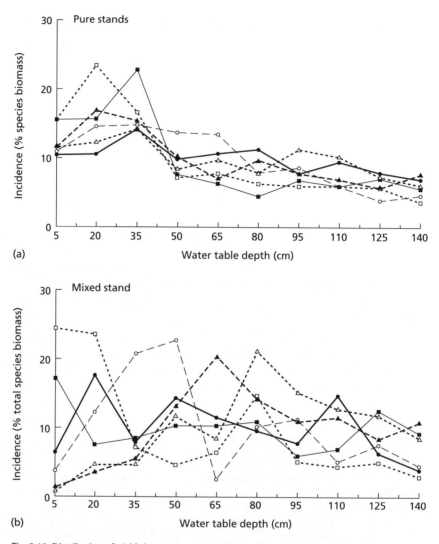

Fig. 8.12 Distribution of yields by six grass species (each shown by its own line) grown on a water table gradient in (a) six monocultures and (b) a mixture of all six species (data from Ellenberg 1953).

according to the different light microenvironments that occur within a gap (Ricklefs 1977) and/or according to size of gap (Denslow 1980). Field tests of this hypothesis, including permanent plot studies at La Selva in Costa Rica (Lieberman *et al.* 1995) and in the 50 ha plot at BCI (Welden *et al.* 1991), did not at first support the idea (Denslow 1987), but more recent studies are beginning to do so. Kobe (1999) followed the survival and growth over a year of seedlings of four species transplanted into field sites across a range of light conditions spanning < 1% to 85% of full sun. Radial growth increased with light intensity in all four species, although the pattern of this increase varied between them. Survival

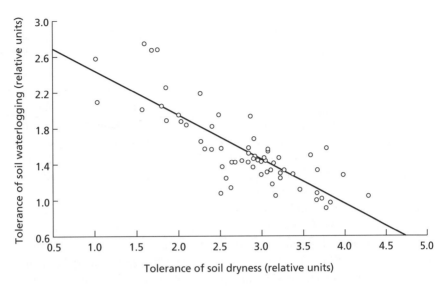

Fig. 8.13 Trade-off between tolerance of soil waterlogging and soil dryness for 64 herbaceous species growing in a wet meadow in Somerset, England (after Silvertown *et al.* 1999).

over one year increased with light intensity for all species between 1% and 20% full sun, but, above 20%, survival decreased in one species, remained constant in another and continued to improve in the remaining two species. While neither survival nor growth on their own indicated that species partitioned the light gradient, when the separate functions describing how survival and growth in each species varied along the light gradient were combined, they showed that each of the four species would dominate a particular segment of the resource axis (Fig. 8.14). This segregation results from a trade-off between growth under high light conditions and survival under low light of the kind Kobe *et al.* (1995) found in temperate forest (Fig. 8.8). Three caveats concerning the interpretation of this study are: (i) that growth and survival were followed for only one year; (ii) that only four species were studied; and (iii) that the clear segregation shown in Fig. 8.14 is partly dependent upon an assumption that only small numbers (2–10) of each species are present. Limited dispersal could ensure this and, as explained in the next section, may also play another role in coexistence.

A comparative study by Davies *et al.* (1998) of the distribution of 11 gap-colonizing tree species in the genus *Macaranga* at Lambir Hills in Sarawak demonstrates that niche separation is possible among a larger guild than Kobe (1999) studied in Costa Rica. A staggering 1175 tree species were enumerated in the 52 ha study plot in Sarawak and it was not uncommon to find 5–8 species of *Macaranga* in a single large gap. The 11 *Macaranga* species studied by Davies *et al.* (1998) segregated the light gradient, with considerable overlap, from species concentrated in high light environments, at one end, to others present mainly in low light. Within the light-demanding group species differed in their frequency of establishment on different types of microsite.

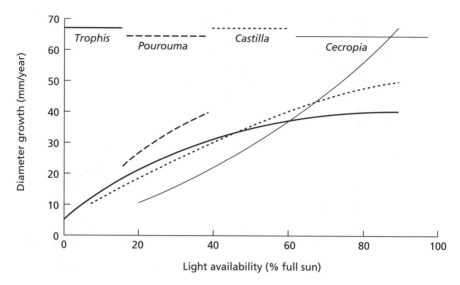

Fig. 8.14 Growth in stem diameter for saplings of four tropical tree species in relation to light availability over ranges where at least one sapling was expected to survive (from Kobe 1999). The range of light availability at which each species dominates is indicated at the top of the figure.

8.4.3 Spatial segregation

As a consequence of clonal growth, limited seed dispersal distances and, indeed, interspecific competition itself, most plant species show a clumped distribution in space (i.e. intraspecific **aggregation**). Aggregation reduces the degree of contact between species (i.e. causes interspecific **segregation**), and hence also the opportunity for interspecific competition to occur. A **spatial refuge** from competitors is provided by the fact that only the individuals on the edge of a monospecific clump compete with other species. Those inside a clump compete with each other. Consequently, the more clumped a species' distribution grows, the less intense interspecific competition will become and the more intense intraspecific competition will become. If this process proceeds to the point where intraspecific competition is stronger than interspecific competition for all competitors, the conditions for coexistence could be met without niche separation (Shmida & Ellner 1984). Pacala (1997) suggests a simple field exercise to see how clumping within a species can result in segregation between species: Randomly select a number of individuals of one species as sampling points and then count the number of individuals of a second species occurring within a fixed radius of these points. Then repeat the exercise, but sample around randomly located points. The two species are spatially segregated if the first estimate is less than the second. The smaller the ratio (first count)/(second count), the greater will be the reduction in the strength of competition between the species, to the point where a ratio of 0/100 would mean the species never encounter one another. But, note that such a result does not mean that interspecific competition has not occurred in the past, because competition during establishment could be the cause of segregation observed among established plants: in effect, competitors repel each other. Nor does a very low ratio necessarily mean that the species do not interact if you have used a sampling radius that is smaller than the spread of roots or canopy.

Ives (1995) constructed a neighbourhood model of interspecific competition in which two processes contributed to the nonrandom distribution of individuals: limited dispersal, which causes intraspecific aggregation, and spatial heterogeneity in the quality of the habitat which causes species that specialize on different resources (i.e. which have different niches) to segregate. Ives used his model to determine under what conditions it would be possible for two species with similar niches to coexist purely on the basis of nonrandom distribution. The answer to this question is framed in terms of an inequality between two ratios: the ratio of inter/intraspecific competition when species occur within the same neighbourhood, and the ratio of intra/interspecific concentration of competition among neighbourhoods. If the value of the first of these ratios is less than one, then intraspecific competition is stronger than interspecific competition and the two species should be able to coexist within neighbourhoods. But what if the ratio is greater than one and they cannot coexist within neighbourhoods? The second ratio effectively measures how often species meet each other, given their relative distributions across different neighbourhoods. If species tend to be concentrated in their spatial distribution (from whatever cause, possibly including some degree of niche separation), then intraspecific competition will tend to be concentrated too. Coexistence is possible if,

within neighbourhoods: among neighbourhoods:

$$\frac{\text{interspecific competition}}{\text{intraspecific competition}} < \frac{\text{concentration of intraspecific competition}}{\text{concentration of interspecific competition}}$$

This means that, with sufficient intraspecific aggregation, competing species can coexist, even when interspecific competition within neighbourhoods is greater than intraspecific competition. Note that if individuals of both species are distributed randomly, then the right-hand ratio becomes unity and the inequality reduces to the mean field result that coexistence is possible if intraspecific competition > interspecific competition.

Very few studies have measured intraspecific and interspecific competition in such a way that their relative strengths can be compared in the field and the role of spatial distribution evaluated. Rees et al. (1996) censused a guild of tiny sand dune annual plants growing at Holkham in Norfolk, England, by counting numbers of four species once a year in one thousand 10 cm × 10 cm permanent quadrats. They fitted data from annual censuses of the four species to competition models of the form given in Eqn 5.9 (page 126), modified to incorporate interspecific as well as intraspecific effects of local density upon population size in individual quadrats. Intraspecific density dependence affected all four species, reducing mean densities within quadrats by up to two thirds from one year to the next, compared to what would be expected without density dependence. By extreme contrast, no significant effects of interspecific competition were detected at all, and so coexistence between the species was certainly to be expected. The interesting question is, *why* was interspecific competition of such negligible effect? Two answers seem likely; first, there was strong intraspecific aggregation and hence marked spatial segregation between species and, second, the strongest competitors were also the rarest species so their numerical impact upon weaker species was low. Rare species have limited recruitment and, as we shall see in the next section, a negative correlation between competitive ability and recruitment can promote coexistence.

Two other studies of competitive interactions using spatial models fitted to field data found rather different results to those of Rees *et al.* (1996). Law *et al.* (1997) estimated competition coefficients in a community of four perennial grass species in a montane grassland in the Czech Republic and found intraspecific competition coefficients to be smaller than interspecific coefficients in many instances. The interspecific coefficients were largely in qualitative agreement with the results of manipulative experiments performed on the same system (Herben *et al.* 1997). Because the competition coefficients were calculated by Law *et al.* (1997) in a manner that incorporated the spatial distribution of competitors, the results imply that spatial segregation was an insufficient force in this community to prevent competitive exclusion. Freeman and Emlen (1995; Emlen *et al.* 1989) used regression models to determine the influence of neighbours upon the seed production of target plants belonging to six species of grasses and shrubs in a relatively species-poor, arid saltbush grassland in Utah that was grazed by livestock. Intraspecific competitive effects were much stronger than interspecific ones in four of the species, and in the two exceptions there was evidence of Allee effects (a positive correlation between performance and density). Paradoxically, intraspecific aggregation may well have been related to the strength of both the negative and the positive intraspecific effects, causing strong intraspecifc competition in some species and arising from nurse-plant effects in others.

8.4.4 Recruitment limitation

Spatial structure can promote coexistence in quite another way from spatial segregation, although it also involves reducing the frequency of encounters between competing species. As we have seen in preceding Chapters, many plant species rely upon vegetation gaps and sites of disturbance for recruitment and this, combined with a stage-specific refuge, provides species that are weak competitors but good dispersers with an opportunity to recruit. When recruitment is limited by dispersal, the meeting of competitors is a probabilistic process and, under the right conditions, there will always be some populations of each species protected from competition because competitors are missing or they arrive at the site too late to matter (Atkinson & Shorrocks 1981; Rosewell *et al.* 1990). The crucial condition for **recruitment limitation** to work as a coexistence mechanism is that there must be a negative correlation (or life history trade-off; see Chapters 9 and 10) between species' competitive ability and their colonizing ability (Skellam 1951): good competitors must be poor colonizers and *vice versa*. Given this condition, an almost unlimited number of species may coexist (Tilman 1994).

Competition/colonization trade-offs have been found among sand dune annuals (Rees 1995; Turnbull *et al.* 1999), savannah grasses (Tilman 1994), prairie forbs (Platt & Weiss 1985; see Fig. 7.7) and forest trees (Pacala *et al.* 1996). Strong circumstantial evidence suggests that recruitment limitation may be important in tropical forests too. Hubbell and Foster (1986b) established a 50-ha study plot at BCI to address the question of coexistence in the forest there. Censuses (and maps) were made of all trees and shrubs greater than 1 cm d.b.h. in 1982 and again in 1985 (and later in 1990, 1995 and 2000). The first (1982) census found 235 895 individuals belonging to 306 species (lianes were not counted). Nearly 17% of all stems belonged to a single shrub species *Hybanthus prunifolius*, but 21 species were represented by single individuals (Hubbell & Foster 1990a). Twelve hundred light-gaps were monitored in the plot between 1985 and 1995 and the species colonizing each one were recorded. Allowing for variation in gap area, individual species were found on average in only 1.4%–6.2% of gaps, indicating that most species were absent from most potential colonization sites (Hubbell *et al.* 1999). Two hundred seed traps were placed in the 50 ha plot, and over 10 years of continuous monitoring, any given trap was on average only reached by the seeds of 12% of the species in the plot. A seedling census found the most common species in only 14% of quadrats. This evidence suggests strongly that most species are recruitment limited and that where they do recruit, the majority of their interspecific competitors are generally absent. This is *not* to say that neighbourhood competition within and between species is not present or strong in light gaps – on the contrary, stem densities in gaps are high and competition is fierce – but rather that which species 'wins' the local lottery in a gap is determined more by which species colonize it than by which species are globally the best competitors.

8.4.5 Natural enemies

Resource-based theories of competition and coexistence assume that populations increase until they are limited by resource availability; but what if populations are limited below this carrying capacity by natural enemies (Section 8.3.3)? If **predators** or **pathogens** operate in a density- or frequency-dependent manner, culling a disproportionately large share from the competitive dominants in a community, they will delay or even prevent competitive exclusion of more minor species. **Grazing** is usually essential to the maintenance of species diversity in species-rich mesic grasslands, almost certainly because of the control that grazing animals exert over grasses which dominate and exclude other species when grazers are removed (Watt 1974). This effect of grazing is seen in fertile conditions, but may be reversed in sites with nutrient-poor soil (Proulx & Mazumder 1998). Grazing may reduce species diversity when primary production is low, because diversity in these conditions is determined more by the number of species that can tolerate herbivory than by interspecific competition.

In tropical rainforests, fungal pathogens, seed predators and other herbivores are thought to be responsible for the widespread occurrence of density dependence in tree populations. Pathogens and many of the herbivores are specialists, attacking only one or a few species of plants (Barone 1998; Coley & Barone 1996). The headquarters of such animals are in the adults of their foodplants, and most of the progeny of these adults appear near them where they are most vulnerable to the herbivores feeding on their parents (Chapter 1). **Janzen** (1970) and **Connell** (1971) first suggested the hypothesis (which now bears their names) that heavy predation on juveniles found in the proximity of adults results in density-dependent mortality among rainforest trees and keeps species at densities so low that they cannot oust each other (Fig. 8.15).

Wright (1983) found at BCI that seeds of the *Scheelea* palm were attacked by bruchid beetles most severely near parent trees and had to be 100 m away to escape. There are examples of such mortality patterns in tropical trees caused by pathogenic fungi (Augspurger 1983, 1984; Kitajima & Augspurger 1989; Gilbert *et al.* 1994) as well as herbivores (Clark & Clark 1984), although density dependence is rarely strong enough to remove aggregation entirely. Because predation reduces but does not eliminate aggregation, and because of the weaknesses of correlative studies of density dependence mentioned in Chapter 4, density dependence cannot

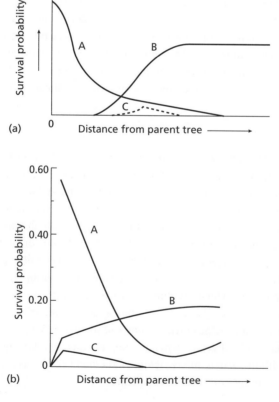

Fig. 8.15 (a) The Janzen–Connell model. Curve A is the distribution of seeds, curve B is the probability of seeds and seedlings escaping predation as a function of the distance from the parent, and curve C, which is the product of curves A and B, is the distribution of surviving plants (after Janzen 1970). (b) Actual curves of seed distribution A, the probability of seedling establishment B, and the distribution of surviving juveniles C, for the tropical dry forest tree *Cecropia peltata* (from Fleming & Williams 1990).

always be detected simply from the spatial relationships of conspecific juveniles and adults. Studies in Australian tropical rainforest (Connell *et al.* 1984) and at BCI (Hubbell 1979) that have been based on these spatial relationships alone have found little evidence of density dependence. It is necessary to measure birth and death rates, as has been done in the 50 ha plot at BCI.

During the three-year interval between the first two censuses at BCI, saplings of some but not all of the commoner species survived and/or grew significantly better if they were beneath a tree of a different species than if they were beneath a conspecific adult (Hubbell *et al.* 1990; Condit *et al.* 1992). The same pattern significantly affected the mortality of rare species as a group, but not their growth (Hubbell & Foster 1990b). Population regulation in the very commonest tree species did seem to be strong enough to account for their existing densities in the BCI plot, but the evidence that density dependence was present or powerful enough in other species, including the very abundant shrub *H. prunifolius*, was still lacking (Hubbell *et al.* 1990).

This picture changed dramatically when data for survival and recruitment occurring over the eight-year interval 1982–90 became available and when more refined methods were used to analyse them. Wills *et al.* (1997) reasoned that, if escape from pests were important, the fate of young trees would not simply be determined by their distance from the nearest adult conspecific, but by the total biomass of trees of the same species nearby (measured as basal area). They therefore analysed recruitment and mortality rates for different species quadrat-by-quadrat, rather than tree-by-tree, for quadrats at a range of scales from 10 m × 10 m up to 100 m × 100 m. Especially at the smaller scales of 10 m and 20 m, Wills *et al.* (1997) found significant negative effects of basal area of conspecifics on recruitment in 67 of the 84 most common species, and on intrinsic rate of increase (r, or $b - d$) in 54 out of 84 species. The cause of these density-dependent effects is not known in most cases, but pest pressure has been demonstrated in some (Kitajima & Augspurger 1989; Gilbert *et al.* 1994), and seems very likely in the others. The critical stage at which density dependence acts in the 50 ha plot appears to be between seed fall and seedling recruitment. Harms *et al.* (2000) compared the densities of seeds of 53 woody species collected in 200 traps with the densities of seedlings of these species growing near the traps and found strong density dependence acting against local concentrations of conspecific seeds. Species diversity (measured by the Shannon Index, H) changed from $H = 2.82$ at the seed stage to $H = 3.80$ at the seedling stage. The latter value is nearly as high as the species diversity of all the stems > 1 cm d.b.h. sampled in the 50 ha plot ($H = 3.93$), indicating the importance of density-dependence during recruitment to the maintenance of diversity.

8.4.6 The storage effect

Fluctuations in recruitment can promote coexistence if the good years for one species are the bad years for others. Such a difference is one aspect of what Grubb

(1977) has called the **regeneration niche**. Because plants become less vulnerable to competition once they are established, in a good recruitment year a species can establish a cohort of juveniles that will be much less sensitive to its competitors when these are recruited later. This can be thought of as a **temporal refuge** from competitors. In effect, the recruits of good years are stored over bad years for the species. For the **storage effect** to work and to promote coexistence, competitors must have overlapping generations and competition between adults must be weak, providing a stage-specific refuge (Warner & Chesson 1985). The first condition applies to many perennials and to annuals with a seed pool. The second condition may be more difficult to satisfy, unless spatial refuges from competition also exist.

Many of the broadleaf forests of North America are dominated by oaks. Whittaker (1969) noticed that although different oak species are dominant in different forests, the two most abundant oak species usually belong to two different subgenera: one to the white oaks and one to the black. Mohler (1990) put this observation to the test using data from 14 forest stands from all over the coterminous USA (i.e. excluding Hawaii and Alaska) and found it to be true in 12 of them. Whittaker had suggested that the pattern could reflect niche differentiation between the two subgenera and Mohler (1990) also examined this idea. Oaks, like many trees, vary a good deal in the size of seed crop from year to year and Mohler found that seed crops of species in the same subgenus were consistently better correlated with each other than were the crops of species in different subgenera. This is probably because black oaks usually require two years to mature their acorns whereas white oaks need only one. The asynchrony in recruitment that this difference generates may create a storage effect that helps to explain the coexistence of white oaks with black. Because white and black oaks differ in other ways too, other explanations of their coexistence are of course possible.

8.4.7 Density-independent mortality

Most types of vegetation, and particularly forests, are prone to periodic disturbance by fire or wind that kills adults and creates opportunities for recruitment (Pickett & White 1985). This mortality is density-independent, but if the frequency and intensity of these episodes of mortality are right and rates of population increase are low, this kind of community disturbance can delay competitive exclusion among similar species almost indefinitely (Fig. 8.16) (Huston 1979).

High rates of disturbance will eliminate those species with populations unable to recover quickly, and low rates of disturbance will allow interspecific competition to take its toll. This idea is known as the **intermediate disturbance hypothesis** (Connell 1978), and is in accord with patterns of diversity in several plant communities. The highest diversities of plants in pine-wiregrass savannah occur in annually burned sites (Walker & Peet 1983), and local disturbances by frost-heave increases diversity in alpine and subalpine communities (Fox 1981; del Moral 1983). In herbaceous plant communities studied by Grime (1979) in England, he found a 'humped-back' relationship between species richness and standing crop

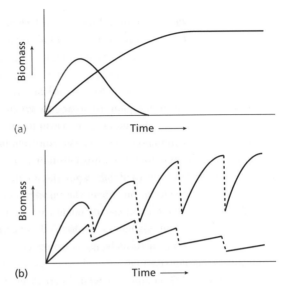

Fig. 8.16 The outcome of competition between two species based upon a model similar to the one described in Section 7.2. (a) without disturbance (b) when there is a disturbance producing a periodic, density-independent population reduction (after Huston 1979).

biomass which also supports the idea that coexistence is favoured between extremes of disturbance (which removes biomass) and competition. This same pattern has subsequently been found by many other workers (Stevens & Carson 1999).

8.4.8 Conclusion

A problem that once seemed to have no solution now has a plethora. In the simple, mean-field world of Lotka–Volterra models the theory stated that plants must occupy different niches to coexist, and yet there did not seem to be enough niches to go round in most kinds of natural vegetation. Advances in theory and empirical studies have modified this view dramatically. First, by revealing new niche dimensions. We now know that plants can respond to very subtle differences in soil moisture conditions that permit them to divide niche space more finely than was once appreciated. Second, new models that incorporate space and temporal variation have revealed the importance of refuges from competition of various kinds. The Janzen–Connell hypothesis is particularly well supported in tropical forest. All of the theories reviewed here, including niche separation, have at least two things in common: (i) each depends upon trade-offs that force species to specialize and prevent any single species from dominating all regions of the habitat; and (ii) none is likely to be the only mechanism operating in nature. The big questions that remain are: Which mechanisms operate where, and what is their relative importance locally and globally?

8.5 Summary

The types of interaction between plants may be classified according to whether R_0 each interacting population is increased (+), decreased (−), or unaffected (0) by the interaction. Five types of pairwise interaction are thus defined: **competition** (−−); **parasitism** (+−); **mutualism** (++); **commensalism** (+0); and **amensalism** (−0). The classification may be further refined to include the effect of populations on an **intermediary** in the interaction such as a nutrient, which may reveal the mechanism of the interaction. This approach shows how natural enemies may generate **apparent competition** and **apparent mutualism**, for example.

The dynamics of two competing populations may be represented in a **joint-abundance diagram**. By using simple equations that quantify the numerical effect of competitors on each other in a mixture, in terms of **competition coefficients** and the **carrying capacities** of the environment for each population when they are growing alone, the conditions that permit stable coexistence of competing species may be defined. This leads us to the **competitive exclusion principle** which states that ecologically equivalent competing species cannot coexist. Significant ecological differences between species, or **niche separation**, permit stable coexistence.

Two-species **competition experiments** of various designs may be used to explore the parameter space of a joint abundance diagram. The **partial additive** design consists of mixtures in which the density of one species is fixed and the other varies. In the **replacement design** the total density of the mixture is fixed and the proportions of the two species vary. In **additive series** and **complete additive** designs densities and proportions of both species are systematically varied. The additive design has been used to determine the effect of the number of neighbours on a target plant, and shows that the effect of each additional neighbour tends to diminish rapidly as their number rises. Additive designs are also often used to compare the relative effects of root and shoot competition. The replacement design may be used to calculate the **relative yield total** (RYT) of a mixture, which under restricted circumstances may indicate the existence of niche separation between competitors. **Yield advantage** in mixtures can be assessed by the **Relative Resource Total** (RRT) or by the **Land Equivalent Ratio** (LER).

Additive series and complete additive designs can be used to determine the location of the **zero isoclines** in a joint abundance diagram. They may also be used to calculate the parameters of the **reciprocal yield model,** which allows competition coefficients to be calculated and the effects of intraspecific and interspecific competition to be clearly separated. In general, competition experiments demonstrate that competition coefficients are sensitive to growing conditions and density and that the outcome is contingent upon these. Competition in the field usually involves simultaneous interactions between many more than two species. **Removal experiments** show that interspecific competition is the rule rather than the exception in the field and that relationships are usually **transitive** and **size-dependent**. **Individual-based neighbourhood models** can be built from regression models of how neighbours influence each others' survival, growth and fecundity. These models may make different predictions from those of **mean field models** that ignore the existence of spatial variation. Competition between plants in the field invariably takes place in the presence of predators (including grazers), pathogens and symbiotic microorganisms that may influence or even wholly reverse the outcome. **Mycorrhizal fungi** can have a variety of effects upon competitors and may either benefit or disadvantage dominants.

Some of the richest plant communities contain far more species than simple theory based upon **Lotka–Volterra models** of interspecific competition would predict can stably coexist. A variety of refinements and alternatives to the simple theory suggest how **coxistence** is possible. Some common elements of these theories are a **stage-specific refuge** from competition, spatial **aggregation, lottery competition,** and **trade-offs** of various kinds that enforce specialization. The Lotka–Volterra model requires species to **partition resources** by occupying distinct **niches.** There is some evidence that this mechanism is important in meadow grasslands. **Spatial refuges** from competition caused by intraspecific aggregation are a possible aid to coexistence in many communities. The mechanism of **recruitment limitation** depends upon a trade-off between dispersal and competitive ability. In the **Janzen–Connell hypothesis** natural enemies cause density-dependent recruitment that prevents species monopolizing local areas. In the **storage effect** asynchronous recruitment combined with a stage-specific refuge from competition may permit competitors to coexist. The **intermediate disturbance hypothesis** predicts that coexistence is promoted through the periodic destruction of dominant species by distubance.

8.6 Further reading

Grace, J. B. & Tilman, D. (1990) *Perspectives in Plant Competition*. Academic Press, London.
Tokeshi, M. (1999) *Species Coexistence*. Blackwell Science, Oxford.

8.7 Questions

1 Describe the range of types of plant–plant interaction and the role played by other organisms in deciding the nature and outcome of these.

2 Discuss the different approaches available for the study of interspecific plant competition in the glasshouse and the field.

3 Write a simple computer model of interspecific competition based upon a version of Eqn 8.6 for two species. Make up some values for starting densities, S_{mi} and competition coefficients and plot on a graph how the two species behave over 10 or more generations when these parameters are varied one at a time between runs.

4 How do the predictions of simple Lotka–Volterra models of interspecific competition fail to explain the composition of most types of natural vegetation?

5 In what ways can the incorporation of space into competition models explain the coexistence of weaker competitors with stronger ones?

9 The evolution of plant life history: breeding systems

9.1 Introduction

In Chapters 9 and 10 we explore the evolution of life history traits whose ecological and genetic importance has been discussed earlier in the book (Table 9.1). In the present chapter we see how some of the kinds of breeding systems that were outlined in Chapter 2 can evolve. Three fundamentally important issues in plant populations are dealt with: (i) sex vs. apomixis; (ii) why species that reproduce sexually go to the trouble of mating with other individuals (outcrossing), instead of self-fertilizing; and (iii) why outcrossing plants have often evolved two sexes, instead of remaining hermaphroditic. Even though hermaphroditism is the predominant sexual state in angiosperms, separate sexes have evolved many times, in many different plant families. In thinking about these questions, it is important to understand that evolutionary transitions between different breeding systems have been repeated events in plant evolution, and so plants offer some of the best opportunities to test theories about breeding system evolution.

In addition to being important for understanding the genetic structure of plant populations (see Chapter 3), plant reproductive systems are very interesting in their own right. Few people are aware of their diversity and evolutionary interest. Yet one can see much of the diversity in any garden, or among the plants seen on a walk in the country, or even among city weeds. Most people are astonished to learn that plants have evolved separate sexes (from an initially hermaphroditic condition) completely independently of animals, and that they have done this hundreds of times, in different taxonomic groups of flowering plants. They have even evolved sex chromosome systems (usually male XY, female XX, just like the human or fruit fly system) at least a dozen times. Most of the evolutionary changes that have happened in plant breeding systems have occurred in both directions, i.e. reversals are not uncommon. In this chapter, we illustrate some of the major concepts that help us understand these changes and counter-changes.

9.1.1 The importance of theoretical models

This chapter will also illustrate theoretical modelling of particular evolutionary processes, such as the evolution of plant selfing rates, or of unisexuality from

Table 9.1 Overview of the contents and organization of life history topics dealt with in Chapters 9 and 10.

Topic	Trait and alternate states		Section
Breeding systems:			9
	Sex	Apomixis	9.2
	Outcrossing	Self-fertilization	9.3
	Self-incompatibility	Self-compatibility	9.4
	Dioecy	Hermaphrodistism	9.5
Age at maturity	Precocity	Delayed	10.2
Seed crop variation	Masting	Nonmasting	10.3
Seeds:			10.4
Size	Large	Small	10.4.1
Dispersal	Far	Near	10.4.2.1
	Dormant	Nondormant	10.4.2.2
Growth	Clonal	Nonclonal	10.5
Senescence	Senescent	Immortal	10.6
Life history strategies	Various		10.7

hermaphroditism. Such population genetical analyses are useful for at least two reasons. First, explicit models help us think clearly about the factors that are likely to be important. Second, once we have some clear ideas, we can design empirical tests to obtain evidence about the actual importance of factors that we think might be involved, in the natural populations in which the plants live.

Table 9.2 shows some of the most important factors influencing the evolution of plant breeding systems. The evolution of **sex** depends strongly on one important process, the transmission of genes to progeny, but the populations' ecological circumstances (such as the presence and density of potential mates, and occurrence of pollinator visits) nevertheless often determine whether apomixis will be favoured, or whether sexuality will be maintained (Table 9.2). The evolution of **selfing rates** can be understood in terms of *two* major genetic factors. The transmission of genes is, of course, still important, but the relative survival of inbred vs. outbred progeny is also critical. Again, ecological circumstances will, of course, interact with these two factors to determine whether selfing or outcrossing will be favoured, or whether a mixed mating system, with some selfing and some outcrossing, will evolve (Table 9.2). The evolution of **separate sexes** introduces a *third* important concept: the allocation of resources between male and female functions (Table 9.2). Before we explain these three major topics in detail, however, it is worth thinking briefly why theory is so important to our understanding of breeding systems.

An important part of making theoretical models in science is deciding what should be included, and what may be safely ignored. Should biologically relevant events, such as pollinator behaviour, or the modular structure of plants, be incorporated? Is it necessary to include an explicit genetic model, or can we work

Table 9.2 Major factors influencing the evolution of plant breeding systems

Breeding system	Transmission of genes	Relative survival of inbred versus outbred progeny	Allocation of resources	Mate reproductive availability
Apomixis	✓			✓
Self-fertilization	✓	✓		✓
Separate sexes (dioecy)	✓	✓	✓	✓

just in terms of **ecological factors**, and merely consider which types of situation are most likely to lead to evolutionary change or maintenance of some breeding system. For an understanding of the evolution of sexual systems, it is often essential to include in our models the genetic control of the sex types, because the genetic details assumed in models can change the results, as will be illustrated below. A very important point that will be illustrated by the theories outlined here is that the breeding system does not evolve in order to control levels of genetic diversity (although, as explained in Section 3.4, diversity patterns are strongly affected by breeding systems). Unlike what was thought in the past (Stebbins 1950), individual advantages and disadvantages seem to drive these evolutionary changes (Maynard Smith 1978; Holsinger, 2000).

9.1.2 Invasion of populations by new phenotypes: evolutionary stable strategies

It is important to understand that, even though breeding systems often have strong effects on population characteristics such as homozygosity, mean fitness, and even on populations' levels of genetic variability and evolutionary potential (as explained in Chapter 3), these are consequences of the changes *within* the populations, and *not* the reasons why their breeding systems evolved. Evolutionary changes in breeding systems occur when a population with a given breeding system is invaded by a new type, i.e. the new type increases in frequency and may take over the population, or become a substantial proportion of it. A general approach for studying many different evolutionary questions is therefore to imagine an initial population with a particular set of characteristics (its phenotype), and then to imagine a different phenotype being introduced at a low frequency into the population (as would happen if the new type arose by a mutation). For example, we might ask whether females will invade a population of hermaphrodites. This kind of problem can be reduced to establishing when the new type will be expected to increase in frequency (i.e. will have a higher fitness, as opposed to having low fitness and being eliminated from the population).

Finding conditions for populations to be invaded, vs. resisting invasion, offers a simple but rigorous way to study evolution of phenotypes. In tests for evolutionary stability vs. invasibility of plant breeding systems, fitness values are therefore assigned to phenotypes, and such models are called **phenotypic selection models**. The fitnesses must, of course, take into account success in reproducing, and also any effects on survival caused by reproductive behaviours. For instance, a plant that makes a very large fruit may be able to make only a few of them; this is the principle involved in thinning fruit crops, and in growing giant pumpkins, when only one fruit is allowed to develop, and is often called a **size–number trade-off** (Fig. 9.1). Trade-offs between growth and reproduction will be an important part of life history evolution theories discussed in Chapter 10.

A useful kind of phenotypic selection analysis is to ask when a population will resist invasion by a new phenotype, i.e. to find the phenotype that is 'best' in the sense that other variants will be eliminated from the population. This is called the **evolutionarily stable strategy** or **ESS** (see Charnov 1982; Maynard Smith 1978). It has been used widely in modelling the evolution of breeding systems. By focusing on the simple situation of a low-frequency variant in a population, we can analyse situations with complex phenotypic differences, including differences in self-fertilization rates, phenotypic differences involving trade-offs, and situations in which the fitnesses of different phenotypes depend on their frequencies. Later in this chapter, we shall see examples of this approach, and other phenotypic selection analyses, and ESS analysis will also be used in Chapter 10 when we deal with life histories.

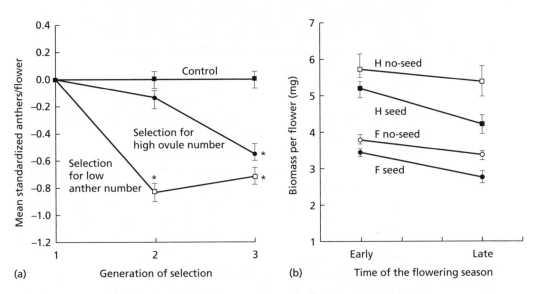

Fig. 9.1 Examples of trade-offs. (a) Negative response of anther number in plants of *Spergularia marina* selected for high ovule number. Note that the response was similar to that obtained when low anther number was itself selected (Mazer *et al.* 1999). (b) Negative response of biomass per flower in response to seed production in females and hermaphrodites of the gynodioecious mallow, *Sidalcea oregana* ssp. *spicata*. Plants were either allowed to form fruit, or prevented from doing so, and the flowers were measured either early or late in the flowering season. Hermaphrodites (H) had larger flowers than females (F) and were more strongly affected by the treatment (from Ashman 1992).

9.2 Evolution of sex

It seems logical to start with the evolution of sex, but there is a difficulty: we still do not have a simple explanation for why sexual reproduction occurs, or (to put it differently) why most organisms do not reproduce asexually. We therefore deal with this question briefly. An excellent review is given in *The Evolution of Sex* by Maynard Smith (1978), and the topic is also dealt with by Stearns and Hoekstra (2000) and by Holsinger (2000). An important point that Maynard Smith explains clearly is that, to account for the maintenance of sexual reproduction in a species with males and females, we must explain what prevents females from producing all-female families, which would give them the advantage of transmitting more genes to the progeny generation. Instead of devoting half their egg production to males, a female that reproduced asexually, via all-female progenies, would be represented in the offspring generation by twice as many copies of herself as others in the population. There is thus a twofold 'cost' of refraining from asexual reproduction (see also Holsinger 2000). Similarly, a hermaphroditic plant could throw up an apomictic mutant that makes ovules instead of ovules and pollen. Such new types would save the twofold **cost of males**, and should rapidly invade populations, even if plenty of potential mates are available (see also Futuyma 1999). This has clearly usually not happened. This simple theoretical insight poses a challenge: to explain why sexual reproduction is not more frequently lost.

Many hypotheses have been suggested (Barton & Charlesworth 1998). To mention just a few possible advantages of sexual reproduction, sexual organisms have the ability to produce genetically varied offspring, which may be critical to the ability to evolve sufficiently speedily to survive attacks by fast-evolving pathogens or other kinds of environmental changes that require populations to be able to incorporate advantageous mutations. Alternatively, sexually reproducing individuals may have greater ability to rid themselves of disadvantageous mutations, in both the short and the long term. Attempts to explain the maintenance of sexual reproduction therefore often involve modelling these processes, in order to ask whether they are strong enough to outweigh the advantage of asexuality. Similarly, empirical work on species in which both sexual and asexual organisms coexist (such as *Antennaria parvifolia*), to ask what disadvantages asexuals actually seem to suffer (Bierzychudek 1987b). For example, in *Ranunculus auricomus*, apomictic cytotypes abort a significantly greater proportion of ovules than sexual diploids (Izmailow 1996).

A different approach that may offer hope of shedding light on what factors are important in the maintenance of sexual reproduction is to find out, by comparisons between species, which kinds of circumstances are usually associated with loss of sexual reproduction (Bierzychudek 1987a). This approach has produced several promising hypotheses, but none that convincingly explains all observations (Holsinger 2000). It is not even clear whether the critical disadvantage of becoming asexual is a short-term one that we could hope to measure in field studies (Bierzychudek 1987a), or longer term effects that cause asexual lineages to have short evolutionary life spans, though there is some evidence that this is true (see Quattro *et al.* 1991), and it is, of course, quite possible that different forces are at work in different circumstances. This is a major gap in our knowledge and understanding of breeding system evolution. No brief review can do justice to this complex field, and readers are advised to read one of the references just cited.

9.3 Selection pressures on the selfing rate

9.3.1 The transmission advantage of selfing

Before discussing models for the evolution of selfing rates, we should first emphasize that self-fertilization has many advantages (see Table 9.2). It spares organisms from the need to seek mates, or to expend resources attracting and courting them (or attracting pollinators, in the case of plants), and this may be particularly important in low-density populations where selfing offers **reproductive assurance**. Selfing also allows locally adapted genotypes to persist, and outcrossing between too distant populations sometimes leads to low progeny fitness (**outbreeding depression**; see Waser & Price 1989). These effects can favour selfing in some special situations such as that of populations adapted to the polluted earth of lead mines (Antonovics 1968; but see also Chapter 3).

The advantages just mentioned are easy to understand intuitively, and can be considered as ecological advantages of selfing. There is another advantage of selfing that is not immediately evident, and requires us to think about the transmission of genes to the next generation. Similarly to asexuality, selfing has a strong advantage resulting from the high relatedness between parents and offspring produced by selfing. This **cost of outcrossing** arises because a maternal plant that reproduces by selfing, and can also sire offspring by pollinating other plants, provides both the alleles at any locus in the selfed progeny, in addition to those in the seeds sired (see Holsinger 2000 for a detailed discussion). Their total contribution per reproductive individual is therefore three alleles in the next generation. Outcrossing individuals with the same total reproductive success (in terms of seeds matured and seeds sired on other plants) will, however, transmit alleles to only two offspring on average, one via seeds fertilized by unrelated pollen donors, and one via pollen that fertilizes another plant (Fig. 9.2a). A genotype that reproduces by selfing thus has a considerable (50%) **transmission advantage** (Table 9.2) if it is introduced at low frequency into an outcrossing population, even if it has exactly the same output of seeds and pollen as the rest of the population (Fisher 1941), and this can outweigh a large disadvantage in terms of other characteristics, for instance viability (Fig. 9.2b).

With all these advantages, it is thus surprising that organisms are not all self-fertilizing hermaphrodites, and the maintenance of outcrossing, like the maintenance of sexual reproduction, is puzzling. Like asexual reproduction, self-fertilization may also lead to reduced long-term population survival. Now that phylogenetic relationships between different species can be estimated using the

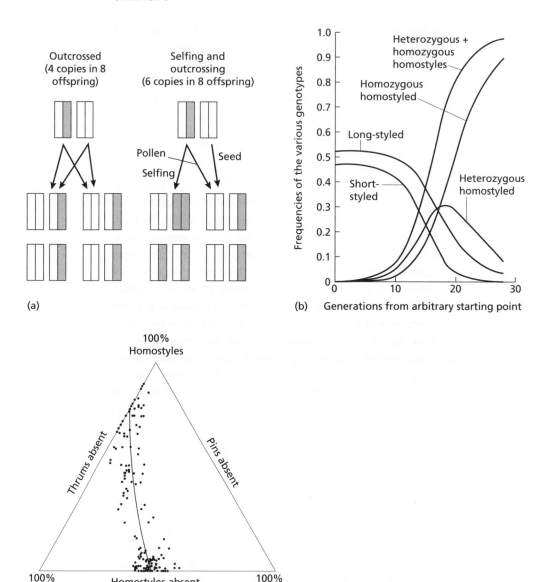

(a) Diagrammatic illustration of the transmission advantage of self-fertilization. (b) Results
of a model calculation showing the spread of a completely selfing homostyle morph, assuming that
homostyles have the same survival rate as the heterostyled morphs (from Crosby 1949). (c)
Distribution of morph frequencies in British populations of *Primula vulgaris* where homostyle
plants have been found (from Crosby 1949). The frequencies show a good fit to the theoretical
prediction (solid line), which is based on assuming that homostyles have a survival rate 45% lower
than that of the heterostyled morphs. Even with this disadvantage, the transmission advantage of
self-fertilization is strong enough that homostyles replace the other morphs (see Chapter 2 for
detailed description of heterostyled plants; 'thrum' = short-styled, and 'pin' = long-styled).

Fig. 9.2

extent of DNA sequence differences between them, evidence is slowly accumulating that selfing plant taxa may tend to have shorter evolutionary life spans than related outcrossing ones (Schoen *et al.* 1997). Here, however, we will focus on the short-term consequences of selfing, because (unlike for the evolution of asexual reproduction) some of the major disadvantages of selfing are clearly understood.

9.3.2 Inbreeding depression and the evolution of selfing rates

What can select for outcrossing against so many advantages of selfing? An important characteristic of different reproductive behaviours is whether it causes inbreeding depression (low survival or fertility of inbred progeny; see Table 9.1 and Section 2.8.2). It therefore seems likely that natural selection should favour mechanisms to ensure that progeny are produced by outcrossing (Darwin, 1877). Additional factors may be important in the evolution of separate sexes, where one might argue that allocation of resources can be optimized by specializing in one sex function (Darwin 1877; Charnov *et al.* 1976; Charlesworth & Morgan 1991). We will consider the evolution of separate sexes later, and incorporate such factors into our hypotheses. But when we consider outcrossing systems such as self-incompatibility systems without any morphological differences, inbreeding depression seems likely to be more important.

The 50% transmission advantage of selfing explained above suggests that outcrossing will be maintained only if progeny of selfing have fitnesses less than half that of those from outcrossing (i.e. inbreeding depression exceeds 0.5; Lloyd 1979b). This threshold 50% inbreeding depression level can be demonstrated if we consider the simplest possible theoretical model, which assumes that plants with different selfing rates do not differ in any other way. For instance, identical pollen output is assumed, regardless of the selfing rate. This may not be biologically plausible. Plants often evolve selfing by reducing flower size, so that the anther–stigma distance is smaller, and pollen output is also frequently low (see Chapter 2). A significant proportion of the pollen might therefore be used up in selfing, lowering selfing individuals' ability to cross-fertilize other plants. Alternatively, if pollinators carry pollen from flower to flower on the same plant, this loss of pollen that might otherwise have been used in outcrossing can be total. Reduced pollen export (**pollen discounting**) of selfing forms will clearly reduce the selfer's advantage compared with a phenotype that is well able to fertilize other plants (see Holsinger 2000). Thus, a lower inbreeding depression threshold can block the evolution of selfing.

This very simple view of inbreeding depression as a fixed quantity for a population, leaves out important details that influence invasion of populations by mutants with higher or lower selfing rates, and determine what will happen afterwards. If inbreeding depression is caused mainly by a few lethal and major mutations, a selfing form can rapidly eliminate these mutations. Once a selfing line has begun to invade a population, its load of mutations will be reduced by natural selection (purging; see Chapter 2), which removes any hindrance to its further increase in the population. Such a selfing line would have high fitness, and so could spread in a population, even if the initial inbreeding depression was high (Lande & Schemske 1985). A selfing form that can invade will therefore be expected to take over entirely. If, however, the inbreeding depression was the consequence of many mutations with small effects on fitness, purging would be unlikely and selfing forms would be eliminated. The genetic details of a population's inbreeding depression, and not merely its overall magnitude, therefore determine whether or not selfing phenotypes can invade.

9.3.3 Causes of evolutionary changes in selfing rates

Despite the probable importance of automatic selection for selfing, and of inbreeding depression, in the evolution and maintenance of outbreeding systems, these factors alone cannot explain evolutionary changes in breeding systems. The automatic selection for selfing is an inevitable consequence of Mendelian transmission, and the inbreeding depression of a population that has established an outcrossing system is likely to stabilize (because it depends mainly on the mutation rate to deleterious alleles; see Chapter 2), and so is unlikely to trigger a change in the breeding system. Thus, to understand evolutionary changes in breeding systems, we must focus on those ecological factors that may alter a population's circumstances and promote invasion by genotypes with different breeding systems.

An important possibility, first suggested by (Darwin 1876), is that an advantage selfing has is reproductive assurance under conditions of low pollinator service or low density (or absence) of mates after long-distance dispersal (Baker 1955; Schoen et al. 1997; Pannell & Barrett 1998). In this hypothesis, the ecological change causing the evolutionary breakdown of outcrossing systems is pollination failure. Critical tests of this mechanism in the field, using comparisons between related selfing and outcrossing populations, are, however, few (e.g. Piper et al. 1986; Barrett & Shore 1987), and we are still in a poor position to gauge the general importance of reproductive assurance in mating-system evolution. In a study of 167 populations of the tristylous plant *Eichhornia paniculata* in NE Brazil, Husband and Barrett (1993) found self-pollinating variants in 86% of populations that had lost one of the three morphs, and 50% of populations that had lost two morphs, compared with only 9% of trimorphic populations. Because morph loss is a clear indicator of small population size, this supports the hypothesis that selfing variants tend to evolve in small populations, probably because of a shortage of compatible mates. We shall discuss breakdown of tristyly in more detail in Section 9.4.2.

Later in this chapter, we shall discuss mechanisms that plants have evolved to avoid selfing (outcrossing systems, including self-incompatibility and separation of the sexes). As reviewed in Section 2.7.3, some self-compatible plants nevertheless largely outcross, because of temporal or spatial separation of the male and female flower parts. Some species have flowers with separate sexes. Monoecious species may differ in selfing rates depending on the proportion of male and female flowers. The genetic basis of such characteristics is often not simple (they are often quantitative traits, like the flower size and other examples in Chapter 2). The evolution of outcrossing in such cases can therefore be treated simply as the evolution of the selfing rate, as in the previous sections of this chapter, without including any genetic details. It can be predicted that outcrossing will be maintained only if inbreeding depression exceeds one half; inbreeding depression less than this value will lead to loss of outcrossing, and selfing will evolve.

9.3.4 Intermediate selfing rates

Why, then, do intermediate selfing rates exist in plant populations (see Chapter 2)? It turns out that the existence of intermediate selfing rates is not difficult to explain, once we extend the limits of the rather oversimplified theories outlined above. In the real world, different factors may act simultaneously, so that it is difficult to disentangle them. Different factors are probably important in different plants, so that results from one species do not carry over to others. Here we have no space to do more than list some of the main ideas.

If the disadvantage of selfing increases with successive generations of inbreeding, evolutionarily stable intermediate selfing rates can evolve (Maynard Smith 1978; Lloyd 1980a). Even a small number of loci with overdominance (see Section 3.5.5) can produce this situation (Charlesworth & Charlesworth 1990). If loci with overdominance are important in natural populations, then the puzzle of intermediate selfing rates could therefore be immediately resolved. As already mentioned, however, there is debate over whether or not overdominance often occurs.

Local adaptation can also lead to a stable balance between selection for selfing and outcrossing (Schoen & Lloyd 1984; Holsinger 1986), although in certain situations it can favour selfing, to maintain locally successful genotypes (see Antonovics 1968). When populations are patchily distributed, as will usually be the case when there is local adaptation, **biparental inbreeding** (when crosses between different plants involve close relatives) will often occur within localities. In such situations, the transmission advantage of selfing relative to crossing is reduced, and this can lead to the maintenance of intermediate levels of selfing (Uyenoyama 1986).

Another biologically plausible possibility is that different selfing rates also imply other differences that help to maintain different phenotypes in populations. It is quite plausible that **trade-offs** between male and female reproduction occur, for example if individuals with high male function have lower levels of female functioning. Fertility is then frequency-dependent, i.e. the fertility of a given phenotype depends on the frequencies of the different phenotypes in the population, and it can be shown that intermediate selfing rates will invade populations with more extreme values (Charlesworth & Charlesworth 1978). A similar result arises if pollen competes for ovules to fertilize, and if greater pollen output produces higher outcross fertilization rates (Holsinger 1991).

Given all these hypotheses for the existence of mixed mating systems, it should be possible to devise tests of their assumptions and predictions. At present, however, we have little basis for rejecting any of them, and there are currently few good data for testing between the different hypotheses (reviewed in Jarne & Charlesworth 1993).

9.4 Self-incompatibility

The models outlined above assume that selfing rates can evolve, but leave out the details of how levels of self-fertilization are controlled. Such simplified models are helpful in developing an understanding of the selective forces acting on selfing rates, but they do not fully represent the biological situations in which plant breeding systems actually evolve. When biologically known means of preventing selfing, such as self-incompatibility (see Chapter 2), are put into the models, even higher inbreeding depression levels than the value of 0.5 discussed above are necessary to prevent the evolution of selfing; thus, a value of 50% represents a minimum for the evolution of outcrossing.

9.4.1 Homomorphic self-incompatibility

Imagine a population with no self-incompatibility, in which individuals can potentially self-fertilize (we can write their genotype as (S_0/S_0)). If a new self-incompatibility allele (S_1, say) arises with sporophytic expression in pollen, none of the pollen of the plant carrying this allele (whose genotype will be S_0/S_1) will be able to fertilize this plant's ovules (i.e. it prevents self-fertilization), but pollination of other plants will not be affected, because they do not have the S_1 allele. Thus, this kind of self-incompatibility allele can invade populations provided that inbreeding depression exceeds 0.5, as explained above (Charlesworth 1988; Uyenoyama 1989). Now imagine a newly arisen self-incompatibility allele with gametophytic action, i.e. pollen grains with the S_1 allele will be unable to fertilize ovules of any plant that carries the same allele. S_1 pollen again cannot fertilize the plant that produces it, but the pollen carrying the other allele at the locus (S_0) will be able to do so, and both alleles will be equally able to fertilize other individuals in the population. The new 'self-incompatibility' allele therefore does not succeed in preventing selfing (although it may reduce selfing to some extent, because it halves the amount of compatible pollen), but only prevents its own transmission to selfed progeny. It is thus a 'suicide allele' and suffers an even greater cost than other kinds of genes that cause outcrossing. In general, very high inbreeding depression is needed for gametophytic self-incompatibility to evolve and to be maintained in the presence of self-fertile mutant forms (Charlesworth & Charlesworth 1979; Uyenoyama 1988).

Once self-incompatibility has evolved, any new S-allele that arises by mutation from an existing one has an advantage because, while it is still rare, pollen carrying it will not encounter stigmas that have the same allele. There is thus frequency-dependent selection (see Section 3.5.6) favouring each new incompatibility-type allele, which will rise in frequency until equilibrium is reached with equal allele frequencies. The equilibrium number of alleles will thus be high, depending on population sizes (because alleles will be lost by chance in finite populations) and mutation rates to new specificities (because frequent mutation increases allele numbers; see Wright 1939). This explains why populations of self-incompatible plants have very large numbers of such alleles, in many cases studied as high as 50, or even more (e.g. in the poppy *Papaver rhoeas*; see O'Donnell & Lawrence 1984).

9.4.2 Heterostyly and the breakdown of self-incompatibility

For an understanding of the distribution and patterns of plant breeding systems, it is just as important to understand how and when outbreeding systems break down as to know how they evolved. Although we do not yet fully understand how heterostyly evolved (Barrett 1992b), plants with this breeding system have been very important in studies of their breakdown and reversion to selfing. In a pioneering study, Crosby (1949) studied primrose (*Primula vulgaris*) populations that contain homostyle selfing forms (with the female parts of the long-styled and the male parts of the short-styled morphs; see Fig. 2.8b). Crosby showed how their rapid spread in initially heterostyled populations will occur by Fisher's transmission advantage (see Section 9.3.1), and his computer calculations of genotype frequencies as the selfing form increases from generation to generation agree very well with observed frequencies (Fig. 9.2c). At high frequencies of the selfing morph, however, some disadvantage of homozygotes for the allele producing this morph seems to prevent it taking over the entire population. The homostyle primroses probably enjoy high reproductive assurance, because their anthers are close to the stigmas of their flowers, and their pollen is compatible with their female reproductive tissues. Although they might suffer from pollen discounting, too little pollen is deposited on their own stigmas to noticeably reduce their ability to pollinate pin plants' stigmas (Piper *et al.* 1986).

A different way heterostyly breaks down is by chance loss of morphs in tristylous populations (see Section 3.2.2). At equilibrium, a tristylous population is expected to contain equal numbers of each morph. Because of the genetic basis of tristyly, such equilibrium populations have a low frequency of the allele necessary for the short-styled morph to be present (see Barrett *et al.* 1989). In a small local population, this allele can by chance fail to be transmitted to any progeny, resulting in its loss from the population. We can therefore predict that the short-styled morph will sometimes be absent from populations of tristylous species (Heuch 1980), i.e. that some populations of such species will have all three morphs, but others just the long- and mid-styled types. This is exactly what is observed (Fig. 9.3). Once the short-styled morph is lost, the populations then often evolve in response to their new floral morphology,

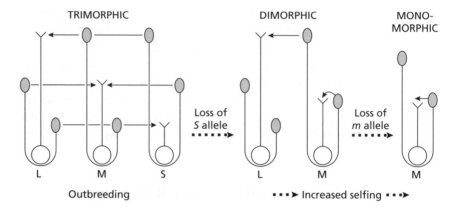

Fig. 9.3 Breakdown of tristyly. Floral morphs in tristylous *Eichhornia paniculata* and in populations that have lost morphs and become distylous or monomorphic; in distylous populations, the mid-styled morph may evolve (as shown) so that its lower anther whorl is higher than normal, and is close to the stigma, so that this morph has a higher selfing rate than the long-styled morph (from Glover & Barrett 1986).

and sets of anthers move into new positions that correspond with the positions of the stigmas in the morphs they are most likely to pollinate successfully (Fig. 9.3).

These studies of heterostyly illustrate the evolutionary flux and inconstancy of plant breeding systems (see Stebbins 1957). Outcrossing systems evolve, but are not necessarily maintained forever; they repeatedly break down to selfing again, or change to different outcrossing systems. Distyly is even thought to have evolved into dioecy, with the long-styled morph becoming female and the short-styled morph male (Casper & Charnov 1982).

9.5 Evolution of separate sexes

9.5.1 Advantages and disadvantages of separate sexes: inbreeding depression and resource allocation

Another important way in which plants ensure outbreeding is to be male *or* female, which of course completely prevents self-fertilization. We will next discuss the evolution of dioecy. Theories for the evolution of dioecy focus on the conditions for invasion of populations by **unisexual** types, so we usually combine thinking about dioecy and gyno- and andro-dioecy (see Chapter 2 and Fig. 9.7 for definitions of these terms). An obvious advantage for females might be that other plants must pollinate them, so that their offspring are not likely to suffer inbreeding depression (**inbreeding avoidance**).

Inbreeding avoidance is not the only advantage for females and males. Unisexuals may also differ from hermaphrodites in the amounts of resources they expend on reproduction. Females may, for example, save resources that

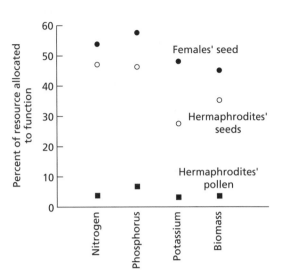

Fig. 9.4 Resource allocations in females and hermaphrodites of the gynodioecious mallow, *Sidalcea oregana* ssp. *spicata*, measured as percentages of four quantities (biomass and amounts of three important elements). Hermaphrodites' percentage allocation to pollen plus seeds was similar to females' percentage allocation to seeds. This suggests that the amounts of limiting resources are similar for both sexes, but that hermaphrodites' allocation to pollen reduces resources available for seed production (data of Ashman 1994).

hermaphrodites devote to anther and pollen production. An important question is therefore how **investment of reproductive resources** affects male and female fertility (Darwin, 1877). It seems reasonable to think that reproductive resources are finite, some being allocated to female functions, and the remainder left over for male functions, e.g. anthers. Thus, an increase in female function should **trade-off** as reduced male function in a similar way to trade-offs between reproduction and growth (Fig. 9.1). For instance, pollen production in a gynodioecious mallow, *Sidalcea oregana*, appears to lower fruit production (Fig. 9.4, Ashman 1994).

Bringing resource allocation into our models of the evolution of plant reproduction connects them to the plants' ecological environments, and may help to solve a puzzle about the environmental correlates of dioecy. Dioecy is often found in arid or other harsh conditions (e.g. Barrett 1992a). This is the opposite of what one would expect if harsh conditions often lead to low density that limits females' ability to be pollinated. It might, however, make sense if in these conditions resources become limiting and plants are unable to sustain both sex functions (as suggested by Darwin, 1877). Females might then have an advantage in such environments. This has been observed in *Hebe strictissima* (Delph 1990). Harsh conditions may also intensify inbreeding depression, if fitness differences are greatest in poor environments, as is sometimes found (Fig. 9.5).

9.5.2 The evolutionary stability of dioecy: when will hermaphrodites invade dioecious populations?

To understand why some, but not all, species are dioecious, we need to understand not only how they evolve in the first place, but also what maintains dioecy. Dioecious populations often include a few hermaphrodites; why do these

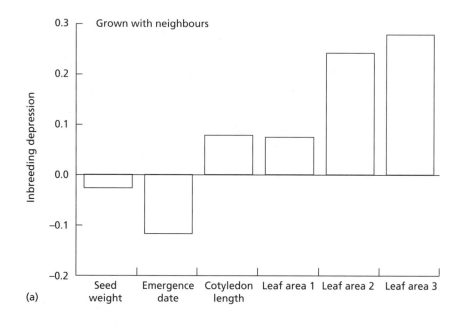

(a)

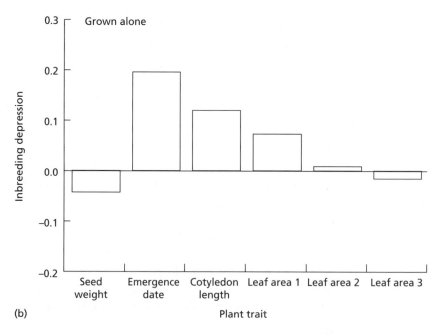

(b)

Fig. 9.5 Inbreeding depression for different traits in plants of *Hydrophyllum appendiculatum* grown either with neighbours of the same species, or alone (from Wolfe 1993). Leaf area was measured at three stages (1–3).

not take over? If we consider outcrossing hermaphrodites, there is no advantage to unisexuals resulting from inbreeding avoidance, and this simplifies the thinking. It is intuitively obvious that a dioecious population will resist invasion by outcrossing hermaphrodites or cosexuals, if these have lower reproductive fitness than unisexual individuals with just male or female function (Charnov *et al.* 1976). We therefore need to know how realized male or female fertility depend on the plants' allocations to different reproductive functions. Such relationships are

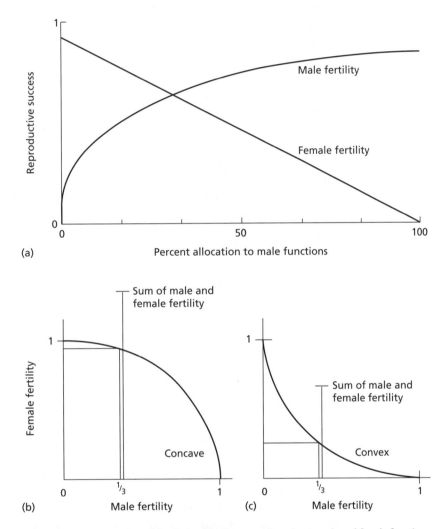

(a)

(b)

(c)

Fig. 9.6 (a) An example of possible relationships between allocation to male and female functions, and reproductive success through those functions and (b & c) gain curves resulting from such relationships. (b) When the gain curve is concave with respect to the axes, the sum of male and female fertility values is always highest when both male and female fertility are intermediate. The lines in the figure show, as an example, that if allocation to male function is such that male fertility is 1/3 of the maximum possible when all resources are devoted to male functions, then the sum of male and female fertility values is higher than it would be if all resources were allocated to one or other of the other sex function alone. (c) A curve that is convex with respect to the axes illustrates the case where allocation to male function is again 1/3 of the maximum possible, but results in low male fertility. Here, unisexuality can give a higher fitness than an intermediate allocation to both sex functions.

called **gain curves** (see Fig. 9.6 and Charnov *et al.* 1976). In an outcrossing population, **reproductive fitness** is the sum of the reproductive output via seeds produced and by seeds sired. A 'convex' relationship between male and female fertility means that either extreme of maleness or femaleness confers higher fertility than a combination of both sex functions (Fig. 9.6b). This evidently favours unisexuals, because they have the highest reproductive fitness. Conversely, a population of outcrossing cosexuals cannot be invaded by either male and female forms if the relationship between male and female fertility is concave, i.e. cosexuals have higher reproductive fitness (Fig. 9.6c)

Although gain curves are hard to measure, we can guess how they will behave. Increased female function will probably trade-off against male function, as just discussed. Clearly we also expect the fertility benefits from high allocation to either male or female functions to level off. Male fertility, for example, must level off because the numbers of available ovules, and of pollinators, are finite. There is thus no advantage in producing unlimited amounts of pollen (although in wind-pollinated plants the relationship between pollen output and numbers of ovules fertilized might be close to linear, and of course such plants often do produce vast quantities of pollen). Seed production probably increases in linear proportion to allocation to female functions, but female fertility would probably still level off because of limited seed dispersal, leading to competition between seeds of individual plants (see Section 10.4). The curves might thus be as shown in Fig. 9.6.

To give an example of this approach, imagine a plant whose success in both sex functions depends strongly on a large inflorescence. Without the inflorescence pollinators would not be attracted (in which case both male and female functions would both be useless), and hence cosexuality is expected to be the most advantageous strategy, i.e. a population of males or females would be invaded by a cosexual. This may be the reason why so many plants are cosexual, rather than having separate males and females. If this is the reason, we might expect other sessile organisms to also be cosexual, and this has indeed been found to be true (Charnov *et al.* 1976).

Of course, this oversimplifies biological reality. Reproduction is not instantaneous, but takes time in the plant's life cycle. If we equate female functions with seed production, this ignores additional resources that become available to developing fruits and seeds after pollination has occurred. It also ignores the possibility that different flowers may be in different stages simultaneously: some will be maturing seeds when others are still at the pollination stage. Finally, fitness via male or female function usually requires investment in structures such as nectar or petals to attract pollinators, which is also ignored (Lloyd 1987a). Such biological realism can be incorporated, but the simpler version captures the essentials, and can help predict when cosexes will resist invasion by males or females. It is much harder to include a non-annual life history, in which allocation to reproduction might lower the probability of future reproduction (Zhang & Wang 1994).

9.5.3 Pathways to dioecy from hermaphroditism

An even more interesting question than the maintenance of dioecy is its evolution. Because most angiosperms are hermaphrodites, we know that dioecy evolved many times among these taxa (see Chapter 2). The approach outlined in the previous section can help us tell when we should expect females or males to invade cosexual populations. However, it ignores the role of inbreeding avoidance, which we have already mentioned as an important factor. It also leaves out genetic details that are an essential part of the evolutionary change from an initially monomorphic hermaphroditic or monoecious population to one with males and females. The evolution of two sexes requires at least two genetic changes, one (male-sterility) creating females and the other (female-sterility) producing males (Fig. 9.7). It is thus impossible to have 'a mutation to dioecy' though the reverse change could occur (see Fig. 9.3).

The first step in the evolution of dioecy (invasion of cosexual populations by either females or males) must be followed by alteration of the remaining cosexuals to become complementarily unisexual (the gynodioecy and androdioecy pathways in Fig. 9.7a). Alternatively, invasion might occur by cosexual forms evolving a bias towards one sex or the other. These might be termed 'partial sterility' morphs. We will first discuss models with male sterility mutations of major effect as the first step, as gynodioecy appears to be a common step in the evolution of dioecy and there is evidence for the involvement of such muta- tions in the evolution of gynodioecy (e.g. Kohn 1988; Weller & Sakai 1990; Mayer & Charlesworth 1992). One must bear in mind, however, that unisexual types could arise through a longer series of genetic changes, each giving only partial sterility. For instance, dioecy can evolve from monoecy, in graded stages (Fig. 9.7b) (Lloyd 1980b; Renner & Ricklefs 1995; Weiblen *et al.* 2000). Dioecy may also evolve from distyly (see Chapter 2); one morph could become specialized as male and the other as female, presumably by gradual stages (Darwin 1877; Lloyd 1979a).

The models we shall describe are oversimplified, and biologically realistic situations may introduce several complications. We shall mostly ignore the possi- bility that females may receive less pollen than needed for fertilization of all their ovules, even though this pollen limitation could become severe when female frequencies increase in a population. Thus, females may reach lower frequencies than those predicted by our simple models. This seems especially likely in small populations and populations at low densities.

9.5.4 Gynodioecy: when will females invade hermaphrodite populations?

Invasion of cosexual populations by females often seems to be the first step towards dioecy. Male sterility (femaleness) in plants can be caused by mutations in nuclear genes or in mitochondrial loci that are maternally transmitted (called cytoplasmic male-sterility or CMS factors; Lewis 1941, Lloyd 1975).

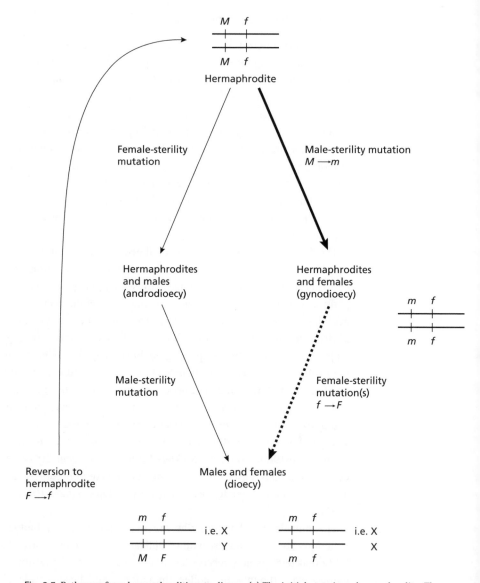

Fig. 9.7 Pathways from hermaphroditism to dioecy. (a) The initial state is an hermaphrodite. The right-hand side shows a recessive male-sterility mutation $M \rightarrow m$ producing females. If these invade and become established in the population, a gynodioecious population results. In this population a mutation ($f \rightarrow F$) that suppresses female functions, wholly or partially, may later happen. As explained in the text, this may be advantageous in the presence of females, and so the hermaphrodites in the gynodioecious population may evolve towards a more and more male phenotype. The end result will be females with the m allele and the original f allele, and males with M and F heterozygous with m and f. If the two genes are linked on the same chromosome, and recombine rarely, this situation resembles an XX female XY male sex determination system. The left-hand side of the figure shows the reverse possibility (female-suppressing mutation first, then male-sterility mutation, evolving from hermaphroditism to dioecy via androdioecy). In addition, the possibility of reversion from dioecy to hermaphroditism is shown. This requires only the loss of the female-suppressing gene of males.

9.5.4.1 *Cytoplasmic male sterility (CMS)*

As one would expect intuitively, a male sterility mutation inherited maternally (exclusively via ovules) will invade a cosexual population if the females have any increase in female fertility at all (Lewis 1941). This probably explains why male sterility in natural plant populations often involves CMS factors, which are now known to be inherited in the mitochondrial genome (Levings *et al.* 1980; Schnable & Wise 1998). The advantage of the male sterility mutation is independent of the frequency of females, so a population that had been invaded by a male sterility mutation would end up all female (and go extinct), but for the fact that the pollen supply diminishes when females reach high frequencies. Females will therefore stabilize at an intermediate frequency. Males cannot, of course, spread in the same way (because increased male fertility gives no advantage to cytoplasmic factors that are transmitted only via ovules).

However, simple cytoplasmic male-sterility is probably rare in natural plant populations. Many species that have cytoplasmic male sterility also have **restorer alleles** at nuclear loci; these turn individuals with sterility cytoplasms into functional hermaphrodites (Table 9.3). This is the situation in male sterile maize, and in gynodioecious populations of the perennial herbs ribwort plantain *Plantago lanceolata* (Damme 1983), wild thyme *Thymus vulgaris* (Belhassen *et al.* 1991) and bladder campion *Silene vulgaris* (Charlesworth & Laporte 1998).

9.5.4.2 *Nuclear male-sterility*

It is harder for nuclear male-sterility mutations to spread so readily in cosexual populations, because nuclear genes acquire fitness through male as well as female functions. The loss of male functions is a strong disadvantage that can be outweighed only by a correspondingly strong increase in female performance. Nuclear male-sterility mutations cannot therefore invade outcrossing cosexual populations unless they cause female fertility to increase to at least twice that of hermaphrodites, a much larger increase than for a cytoplasmic sterility factor. The increase in female fertility required is, however, smaller if the cosexual is partly self-fertilizing (Lloyd 1975), because females can gain additional fitness benefits from inbreeding avoidance, which is advantageous if there is inbreeding depression (Section 2.8.2).

The condition for invasion can be found by comparing the fitness of unisexual females with that of a cosexual. For partially self-fertilizing cosexes, we calculate

Table 9.3 Genetic basis of gynodioecy in a plant with cytoplasmic male-sterility (S cytoplasmic type) and a recessive restorer of male fertility (*rf*). N is a non-sterility cytoplasmic type.

Cytoplasmic type	Nuclear genotype		
	Rf/Rf	*Rf/rf*	*rf/rf*
N	Hermaphrodite	Hermaphrodite	Hermaphrodite
S	Hermaphrodite	Hermaphrodite	Female

fitness by counting up the numbers of gametes contributed to the next generation (Haldane 1937; Lloyd 1975; Charlesworth & Charlesworth 1978), taking into account contributions via both female and male reproduction (i.e. via both ovules and pollen). This ensures that the total progeny produced via pollen equals the total produced via ovules, so that the individuals in the population as a whole each have one male and one female parent, as they should, and that male reproduction is limited by the available ovules, because competition occurs between all pollen producing individuals.

The number of progeny produced by a cosexual (its reproductive fitness) depends on its female and male fertility values, f_{cosex} and m_{cosex}, the cosexual's selfing rate, S, and the inbreeding depression (δ). This ignores several biologically important effects. For instance, all ovules are assumed to be fertilized, whereas in reality pollination may limit female fertility (see Section 9.5.3). Also we assume that self-fertilization occurs at a fixed frequency S per ovule, independent of pollinator service or other variables. Extensions to deal with other situations are possible but the simple version illustrates the type of fitness measure needed.

To ask whether a female mutant will invade a cosexual population (i.e. whether rare unisexuals will have higher fitness than the initial cosexual), we compare the fitness of a phenotype without any male function ($m_f = 0$ for females) with that of the cosexual. This yields a simple inequality involving the females' increase in fertility, relative to that of the original cosexuals. Writing k for this increase in female fertility, we find the condition:

$$k = \frac{f_f - f_{cosex}}{f_{cosex}} > 1 - 2S\delta. \tag{9.1}$$

If there were no inbreeding depression ($\delta = 0$), females could thus invade provided that their fertility was doubled ($k > 1$). Dioecy rarely evolves from self-incompatible cosexes, however, even in families such as the Asteraceae (daisy family) in which self-incompatibility is common, and this implies that inbreeding avoidance is an important factor in the evolution of dioecy (Charlesworth 1985).

On the other hand, if there were no increase in fertility ($k = 0$), Eqn 9.1 shows that females could invade only if inbreeding depression were so severe that the product of the selfing rate and inbreeding depression exceeds 1/2. It seems unlikely that inbreeding depression would be as intense as this (for $S = 0.7$, for instance, δ would have to exceed 0.7). Experiments on inbreeding depression in the gynodioecious species *Hebe subalpina*, for example, yielded a value of about $\delta = 0.45$ (Delph & Lloyd 1996). It is more plausible biologically that both advantages to females are present simultaneously, i.e. both the quantity and quality of seeds are higher. A population of cosexuals with selfing rate S and inbreeding depression 50% would require an increase in fertility of $1 - S$. This shows that, even if inbreeding depression is rather intense, a large increase in female fertility is usually required. Clearly, there is a corresponding condition for invasion of a population by males. We will return to this below.

9.5.4.3 *Testing the theories for the advantage of females*

Sexually polymorphic populations, such as populations with gynodioecy or androdioecy, are ideal for testing possible selective factors affecting the evolution of breeding system variants (Baker 1963). If a male-sterility allele can invade a population, it will reach equilibrium (because a population cannot become all female), and the population will be gynodioecious. Clearly the frequency of females predicted by theory will depend on the female fertilities of the cosexuals and females, the selfing rate of cosexuals and the inbreeding depression experienced by progeny of selfing:

$$\text{Equilibrium female frequency} = \frac{k + 2S\delta - 1}{2(k + S\delta)}. \tag{9.2}$$

Few suitable populations exist for testing this theory, both because nuclear male-sterility polymorphisms are certainly not as common as cytoplasmic male sterility (which also often have restorer genes so their genetics is complex, as explained in Section 9.5.4.1), and also because gynodioecious populations probably often quickly evolve into dioecious ones (as discussed later in this chapter). A few suitable populations have, however, been studied, and the facts fit the models well. In the gourd *Cucurbita foetidissima*, females produce 1.5 times as many seeds as the monoecious cosexes in the same populations, and females' seeds survive at much higher rates, probably because of inbreeding depression in the progeny of cosexuals, which are often produced by selfing. In the populations studied, these two relevant quantities were high enough to satisfy the equation for the presence of females at the observed frequency of 32% (Kohn 1988).

A more indirect test is to compare female frequencies in different populations. For example, in a set of species of New Zealand Umbelliferae female frequencies increased with the difference in female fertility between females and cosexuals (k), just as expected from the model (Webb 1979), and the same was found for different populations of the New Zealand shrub *Hebe strictissima* (Delph 1990). In Hawaiian *Bidens* species, higher female frequencies were found with increasing selfing rates of cosexuals (S), again as expected (Sun & Ganders 1986); the data are consistent with a k-value of 1 (females having double the female fertility of cosexuals), and an inbreeding depression of about 0.7 (Fig. 9.8).

9.5.5 Androdioecy

Now consider mutations producing males. The increase in male fertility, K, which gives males higher fitness than cosexuals is as follows:

$$1 + K > \frac{2(1 - S\delta)}{1 - S}. \tag{9.3}$$

Just like females, males can thus invade only if their fertility is at least doubled ($K > 1$) compared with an outcrossing initial cosexual form. Invasion of males into partially selfing populations, however, requires even higher fertility

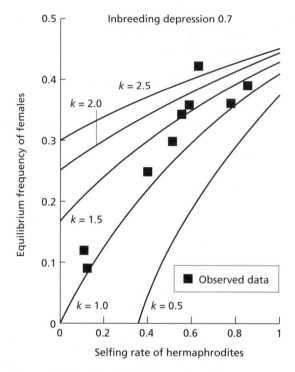

Fig. 9.8 Observed (squares) and predicted (lines) frequencies of females in gynodioecious populations. The predicted frequencies are data from eight Hawaiian *Bidens* populations whose selfing rates (*S*) were estimated from data on isozymes (Sun & Ganders 1986). The predicted frequencies are equilibrium frequencies using Eqn 9.3. Because neither the magnitude of the inbreeding depression nor the females' increase in fertility (*k* in the equation) are known, the predicted frequencies were calculated for a range of *k*-values, and the inbreeding depression (*δ* in the equation) is assumed to be 0.7. It can be seen that the observed data (black squares) are consistent with $k \approx 1$, for this value of *δ*. If *δ* were 0.5, female frequencies should never reach as high as the observed 40% unless we assume $k > 2$, but in that case they are always high (above 20%), which is inconsistent with the observed frequencies of about 10% in some populations.

differences than invasion of a population by females. This is because, with partial selfing, ovules are less available to pollen from other individuals; it is thus harder to increase reproductive fitness by increasing pollen output. An important conclusion from these results is that male sterility is more likely to evolve than female sterility.

The pathway to dioecy via androdioecy (Fig. 9.7) therefore seems implausible. A few androdioecious plant populations have, nevertheless, been discovered. Most have turned out not to be truly androdioecious, but instead to be functionally dioecious populations in which the females are morphologically hermaphroditic, and it is thought that this happens when pollinators are likely to perceive the absence of pollen and discriminate against female plants (Charlesworth 1984). Both pollinators and botanists have been deceived by this **floral mimicry**. A few well-authenticated cases of functional androdioecy have recently been found in wind-pollinated species (Fritsch & Rieseberg 1992; Pannell 1997b). One of these, *Datisca glomerata*, provides an interesting test case for the theory above. There is enough information to show that the conditions are satisfied for polymorphism of males and hermaphrodites to be stably maintained (Fritsch & Rieseberg 1992). Interestingly, these few cases of functional androdioecy may have arisen by breakdown of dioecy, because the most closely related species are dioecious. Most likely a dioecious population was invaded by modified females with reversion to partial male fertility; the resultant cosexuals, with low male fertility, could replace the females in environments where female fertility is low (perhaps due to low density) because they can self-fertilize (Pannell 1997a). If the cosexuals are really modified females this could explain why they have much lower male fertility than the males (which, as explained above, is the necessary condition for the population to be polymorphic for cosexuals and males).

9.5.6 Sex allocation theory in hermaphrodites and other cosexes

So far, we have assumed arbitrarily that unisexuals have increased fertility (values of k or K greater than 1 in the equations above). Why might unisexual females or males have higher fertility than the cosexuals from which they arose, and how big a fertility increase is likely? If fertility stems from resources allocated, it is natural to think that a unisexual mutant that ceases to allocate resources to one sex function would automatically increase its remaining sex function. The theory of reproductive resource allocation predicts how much of a fixed **reproductive resource pool** will be allocated to male and to female functions, respectively, and therefore how much unisexuals could gain in one sex function by abolishing the other. Allocation theory is thus intimately connected with models of the evolution of unisexuality, and can help to bring explicit ecological features of the plants being considered into our reasoning. Because of their importance in the theories dealing with invasion of populations by other sex morphs, the shapes of gain curves have been used to predict what biological situations are likely to promote the evolution of gynodioecy and dioecy (Charnov *et al.* 1976; Thomson & Brunet 1990). A diminishing returns relationship often implies stability of the cosexual state, while a convex one leads to the potential for unisexuals to invade (Charnov *et al.* 1976). This is a useful summary, but we should not forget that it ignores the role of inbreeding avoidance in making female unisexuality somewhat more likely.

First let us find expected allocations of cosexuals, assuming some relationships between resources allocated to male and female functions and fertility (the f_{cosex} and m_{cosex} that come into the general fitness measure mentioned above). The gain curves for male and female functions (see Section 9.5.2), determine the allocation pattern that is expected to evolve. In order to find the expected proportional allocation of reproductive resources to male functions, we search for a value that cannot be invaded by any other slightly different value or, in other words, for the evolutionarily stable strategy or ESS (see Section 9.1.1). The allocation to female functions is obviously just the remainder of the total reproductive resources.

Allocation models usually assume simple exponential gain curves of the form:

$$\text{fertility} = \text{resource invested}^c, \tag{9.4}$$

where c is a parameter that determines whether the curve accelerates ($c > 1$) or decelerates (meaning that there are diminishing returns to further investment in the same function; see Fig. 9.6). Exponential gain curves lead to a particularly simple rule (provided that at least one of the gain curves shows diminishing returns, which is biologically highly plausible; see Section 9.5.2): the allocations to different functions will be proportional to the exponents of the respective gain curves (Lloyd 1984). If M is the fraction of reproductive resources allocated to male functions (and $1 - M$ to female functions), and if we assume the exponential functions $m_{\text{cosex}} = M^b$ for male and $f_{\text{cosex}} = (1 - M)^c$ for female reproduction, the prediction for completely outcrossing cosexuals is:

$$\frac{M}{1 - M} = \frac{b}{c}. \tag{9.5}$$

This makes intuitive sense. Allocation will be pushed highest for functions whose gain curves have high exponents. This result also has an interesting implication: ESS allocation values to the two sex functions do not necessarily equal 1/2 as one might have expected. Thus, sex allocations in cosexual species are different from sex ratios in dioecious ones (where 1:1 ratios, at least among seeds of plants, or embryos of animals, are usual; see Maynard Smith 1978).

For populations with partial selfing with a selfing rate of S, the right-hand side of the above expression is multiplied by

$$\frac{1 - S}{1 + S(1 - 2\delta)}. \tag{9.6}$$

If $S > 0$, this is less than 1, showing that selfing leads to lower male, and higher female, allocations of reproductive resources (Charlesworth & Charlesworth 1981). This conclusion fits with a large amount of data showing that measures of relative allocation to male functions decline with levels of inbreeding (see Section 2.8.3).

9.5.6.1 *Sex allocation theory and the evolution of separate sexes*

By including allocation patterns in our thinking, we can now calculate how much fertility plants can gain by re-allocating their entire reproductive resources to female functions ($M = 0$). We then see that it is only when cosexuals have large fitness gains from investing in *male* functions (high b, and thus high M), that females can gain a high fertility benefit from re-allocating to female functions, and be able to invade (Seger & Eckhart 1996).

Once unisexual females have established in a population, the availability of females should lead to evolution of increased pollen output of the cosexual morph, i.e. a more male-biased allocation of reproductive resources. Full or partial female steriles can therefore invade gynodioecious populations more readily than in the absence of females. It turns out that increased maleness is selectively advantageous only if male fertility is increased more than female fertility decreases; in other words, some reallocation of resources from female to male fertility is necessary. Thus, while a role for inbreeding avoidance is strongly suggested, dioecy cannot evolve solely to avoid inbreeding, and re-allocation of resources is also important (Charlesworth & Charlesworth 1978).

Our conclusions about evolution from gynodioecy to dioecy are only partial, however. They tell us only that the direction of selection on the cosexual morph is towards greater maleness; even when this selection pressure exists, however, no evolution will occur unless a suitable mutation arises in the population, to respond to this selection. Even if modifiers arise that make the cosexual more male in function, they will not necessarily invade the population. Our assumption of trade-offs implies that such modifiers will also reduce female fertility (see Fig. 9.1), i.e. they will cause partial female-sterility. Because progeny carrying both male sterility and female sterility alleles would be neuter, there is counter-selection against the spread of such factors (Table 9.4). Consequently a modifier producing a more male form can often invade only if it is linked to the initial male sterility locus. There is also selection for tighter linkage between the male-sterility locus and modifier loci, to reduce the frequency of progeny carrying both male sterility and female sterility alleles (Charlesworth & Charlesworth 1978). The evolution of dioecy will therefore probably lead to a cluster of linked loci in one chromosomal region, and this may start the evolution of a sex chromosome system (see Charlesworth 1991).

Including some genetic details into our models also shows that evolution from gynodioecy to complete dioecy is not inevitable. Populations may often become subdioecious, with cosexes as well as females and males. An example of a population like this is the spindle tree (*Euonymus europeaus*), which has females plus a range of forms from fully male to hermaphrodite. These cosexuals have long been known to be strikingly variable in fruit and seed output (Darwin 1877). This may provide material for testing whether genotypes with greater male function tend to have lowered female function, as required by the trade-off assumption in allocation models (which has not yet been extensively tested). There is no evidence of this kind yet, but it seems very likely, because if there were no cost to increased male function there seems to be no selective reason why gynodioecy should evolve into dioecy; instead, cosexuals could simply increase their male function while remaining cosexual.

Table 9.4 Effects of the same modifier gene change ($f \rightarrow F$) on cosexuals and females. When females are present (as a result of a male-sterility allele, m, at a different locus), the modifier's effect on cosexuals is good, because it increases male fertility, but the effect on females is bad because it decreases female fertility.

	Genotype at male-sterility locus	
Genotype at modifier locus	M/m or M/M	m/m
f/f	Hermaphrodite	Female
F/f or F/F	Male	Neuter

9.5.7 Testing the theory

We have already mentioned several pieces of evidence that support the models, particularly their assumptions, and give evidence on why cosexual populations sometimes become dioecious. However, it must be realized that it is not easy to subject models of evolutionary events to rigorous testing, and we cannot predict when dioecy will and will not evolve.

A possible source of illumination about the forces that promote the evolution of sexual dimorphism is evidence of associations with particular circumstances, or particular features of the species that undergo these evolutionary transitions. An example is the observed association between dioecy and animal dispersal of fruits. This might suggest that the number of visits to plants by fruit dispersers is an accelerating function of fruit numbers, so that investment in fruiting yields accelerating gains, giving a convex gain curve (Bawa 1980). However, data on fruit removal by animals do not support accelerating gains, in the few well-studied species (Denslow 1987). It is particularly difficult to be sure which factor or factors are causally connected with the evolution of dioecy, because factors are often correlated with one another; for instance shrubs are often understorey species with wind pollination and fleshy fruits, both characters that are associated with dioecy (Muenchow 1987). Co-occurrence of a particular characteristic with dioecy does not necessarily imply that its presence promotes the evolution of dioecy; it may merely show that both dioecy and the characteristic tend to evolve in similar situations.

The best tests currently available are therefore the detailed ones, testing the models' assumptions, which we have already outlined. In the future, better comparative tests should be possible. Trustworthy phylogenetic trees of groups of species within which dioecy has evolved should allow us to tell how many times dioecy has evolved from hermaphrodite progenitors. For instance, in the genus *Silene* alone, dioecy has evolved at least twice (Desfeux *et al.* 1996), and the same is true in the genus *Schiedea* in the relatively short amount of evolutionary time since the Hawaiian archipelago was formed (Weller *et al.* 1995). Phylogenies would also enable us to estimate the phenotypic and ecological characteristics of the progenitors of dioecious species (from the characteristics of the closest relatives). Our current knowledge is confined to the characteristics of present-day dioecious plants, but it should become possible to ask whether the ancestors most likely had an outcrossing breeding system, or other factors that may be important, such as living in a harsh environment.

9.6 Summary

Three fundamentally important questions concerning the evolution of breeding systems of plant populations are: first, why do most plants reproduce sexually rather than by apomixis; second, why do species that reproduce sexually outcross, instead of self-fertilizing; and, finally, why have outcrossing plants often evolved two sexes, instead of remaining hermaphroditic. These questions are studied by theoretical modelling of particular evolutionary processes, such as the evolution of plant selfing rates, or of unisexuality from hermaphroditism. Such models help clarify the most important factors, which can then be tested empirically. Models of breeding system evolution often use the **evolutionarily stable strategy** or ESS method. This asks when a population will resist invasion by a new phenotype, and can be used to find the phenotype that is 'best' in the sense that other variants will be eliminated from the population. The **evolution of sex** depends strongly on the transmission of genes to progeny, and ecological circumstances (such as the presence and density of potential mates, and occurrence of pollinator visits) often determine whether apomixis will be favoured, or whether sexuality will be maintained. A major gap in our knowledge and understanding of breeding system evolution is that the reason for the maintenance of sexual reproduction is not yet fully understood.

The evolution of **selfing rates** can be understood in terms of both the transmission of genes (**cost of outcrossing**) and also the relative survival of inbred vs. outbred progeny (the magnitude of **inbreeding depression**). A 'cost of outcrossing' arises because a plant that can reproduce by selfing, and also sire offspring by pollinating others, contributes more alleles to the next generation than genotypes that only outcross. However, if the progeny produced by selfing have low survival or fertility, i.e. if inbreeding depression is severe enough, or if self-fertilization significantly reduces the amount of pollen remaining to fertilize ovules of other plants (**pollen discounting**), outcrossing can be maintained. Ecological circumstances will, of course, interact with these two factors to determine whether selfing or outcrossing will be favoured, or whether a mixed mating system, with some selfing and some outcrossing, will evolve. Simple theoretical results have been derived that quantify the level of inbreeding depression and pollen discounting that will prevent selfing forms from successfully invading an outcrossing population, and will maintain self-incompatibility systems in populations. Intermediate selfing rates can also be maintained under some models.

The **evolution of separate sexes** depends, in addition, on the allocation of resources between male and female functions. In the simplest models, cosexual populations may be invaded by females, which can benefit in two ways, by producing outcrossed progeny, and by devoting all their reproductive resources to seed and fruits, instead of having to produce pollen. The theory is well worked out for both cytoplasmic and nuclear male sterility, and has been tested with data from real gynodioecious populations. Gynodioecy may sometimes evolve into dioecy, by re-allocation of the hermaphrodites' reproductive resources towards male functions, thus converting them into males.

9.7 Further reading

Futuyma, D. J. (1999) *Evolutionary Biology*, 3rd edn. Sinauer, Sunderland, MA.

Holsinger, K. E. (2000) Reproductive systems and evolution in vascular plants. *Proceedings of the National Academy of Sciences of the USA*, 97, 7037–7042.

9.8 Questions

1 Why do self-incompatibility loci have high numbers of alleles?

2 What is pseudogamy and what does it tell us about the evolution of apomixis?

3 What is gynodioecy? What factors may be expected to lead to the evolution of gynodioecy in a plant species?

4 What is a trade-off? Give an example that is important in plant mating systems.

5 Explain why allocation of reproductive resources should be higher in self-fertilizing than outcrossing plant species.

10 The evolution of plant life history: reproduction, growth, senescence and death

10.1 Introduction

In this chapter we look at the evolution of a diverse set of life history traits starting with age at reproductive maturity and progressing through variation in seed crops, seed size, dispersal and dormancy, clonal growth, ageing and death (Table 9.1). As in Chapter 9, we compare the relative fitnesses of genotypes with differing life history characteristics and analyse how ecological circumstances direct evolutionary change. After discussing plant life history traits individually, we finally examine the idea that suites of traits evolve in correlated fashion to form a limited number of recognizable life history strategies.

If fitness can be measured by the lifetime sum of age-specific survival and fecundity ($\Sigma l_x m_x$), then it must clearly be influenced by how survival and fecundity change with age; but what are the relative effects on fitness of changes in l_x or m_x at different ages? The answer to this question provides the ground rules for the evolution of life histories. For a population at equilibrium, the sensitivity of fitness to small changes in log survival (log p_x) for an individual aged x is simply what remains of lifetime reproductive success (Charlesworth 1994):

$$\sum_{i=x+1}^{i=\infty} (l_i m_i). \tag{10.1}$$

This expression shows how the **force of selection** acting on survival, measured by the sensitivity of fitness, changes with age. It has profound implications because it means that small changes to survival rate late in life, when Eqn 10.1 has a low value, may have very little effect on fitness compared to similar changes much earlier in life (see Fig. 1.3). We shall see in Section 10.5 how this can explain the evolution of ageing. Correspondingly, the sensitivity of fitness to small changes in fecundity (m_x) in a population at equilibrium is simply l_x. Thus, because l_x decreases with each passing year, small changes in fecundity will have larger effects on fitness the earlier in life they occur. Other things being equal, this should favour the evolution of the earliest possible age of reproductive maturity and a concentration of reproduction at this age, ultimately causing semelparity. To understand why plants delay reproduction at all (Section 10.2) and why semelparity is the exception rather than the rule (Section 10.6) we need to consider how the costs of reproduction and mortality from various sources shape life history.

10.2 Reproductive maturity

Reproduction utilizes resources and meristems accumulated during prior periods of growth. This is why bigger individuals produce more seeds than smaller ones (Chapter 4). Because reproduction depends upon prior growth (and of course survival), trade-offs between reproduction and growth/survival also lead to trade-offs between reproduction now and reproduction later. Law (1979) found that in annual meadow grass *Poa annua*, high rates of reproduction early in life lead to lower rates of reproduction and smaller plant size thereafter, and that this trade-off had a genetic basis. In short-lived plants, such trade-offs often arise from the fact that a limited number of meristems must be shared between reproduction and growth (Chapter 1). In the annual *Polygonum arenastrum* Geber (1990) found that the rate of production of new meristems declined as soon as flowering began, and that this was necessarily followed by reduced fecundity, growth, and longevity. There were strong negative correlations with a genetic basis between age-specific growth and fecundity, reflecting genotypic differences between plants in the age at which they began flowering.

If reproduction incurred no cost to future survival or fecundity, seed production should commence immediately at birth because the sooner it starts the greater will be the lifetime sum of $l_x m_x$, and hence fitness. In reality, age at first reproduction varies greatly between species and tends to be positively correlated with lifespan (Fig. 10.1), suggesting that the two traits evolve together because early reproduction negatively affects subsequent survival. The optimum age at which an individual should begin reproduction depends upon how reproduction at a particular age (x) would affect later survival and reproduction. The cost of reproduction in a plant reproducing at young age may slow its growth and increase the risk of death, because small plants are more vulnerable (Chapter 4). Thus, when reproduction bears a cost, delaying its onset may increase $l_x m_x$. Conversely, delaying maturity may sometimes expose plants to an increased mortality risk that prematurely curtails reproduction, causing selection for an earlier age of first reproduction. For example, Kadereit and Briggs (1985) found that groundsel *Senecio vulgaris* sampled from the flower-beds of the Cambridge Botanic Garden developed more quickly and flowered significantly earlier than plants of the same species sampled from habitats where gardeners do not rogue out this weed. An excellent example of how the optimum timing of the start of reproduction may be determined by the balance between the risks and benefits of delay is provided by a study of the annual woodland herb *Impatiens pallida* in Illinois. Schemske (1978, 1984) observed that a population in the interior of a wood flowered significantly earlier than a population only 64 m away at the woodland edge. During six years of observation, the interior population was destroyed annually in July by a chrysomelid beetle, but the edge population

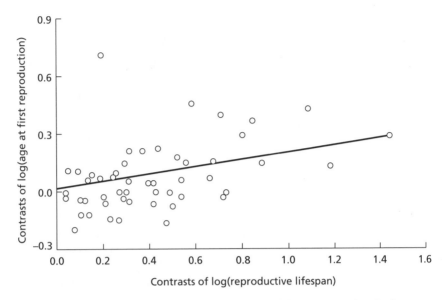

Fig. 10.1 The relationship between age at first reproduction (α) and the mean length of the reproductive lifespan (i.e. total lifespan $- \alpha$) for 63 species of herbs and trees. ($r = 0.58$, $P < 0.0001$). The variables are plotted as phylogenetically independent contrasts (see Section 10.4.1.2 for further explanation. From Silvertown *et al.* 2001).

escaped beetle attack. Plants in the woodland interior that delayed flowering produced drastically fewer seeds, because late-flowering plants were eaten. Plants transplanted reciprocally between woodland interior and woodland edge had lower fitness than natives. Plants from the edge did less well in the interior than natives because they were eaten before they flowered, while interior plants did less well at the edge than natives because they flowered for a shorter period of time and consequently produced fewer seeds.

Quantitatively, the optimum age of reproductive maturity is reached when no further increase in $l_x m_x$ can be obtained by any further delay. The simplest case is for semelparous species because these reproduce only once, so reproduction should occur at the age when $l_x m_x$ is at a maximum. In fact, under several life history models a delay in reproduction is also favoured for iteroparous species until $l_x m_x$ reaches a peak (Bell 1980; Charlesworth 1980b). The optimal age at first reproduction depends upon the fitness measure appropriate for the situation in question (R_0 in the cases discussed here) rather than on semelparity or iteroparity (Roff 1992). This allows general conclusions to be drawn from semelparous plants, in which the evolution of age at maturity is better studied than in iteroparous ones. Kachi and Hirose (1985) tested theoretical predictions of age at maturity in the semelparous perennial *Oenothera glazioviana* in a sand dune system in Japan (Fig. 10.2). With survival diminishing and fecundity increasing with age, a maximum value of $l_x m_x$ was reached at 4 years of age, in reasonable agreement with the observed mean age of reproductive maturity in the population, which was

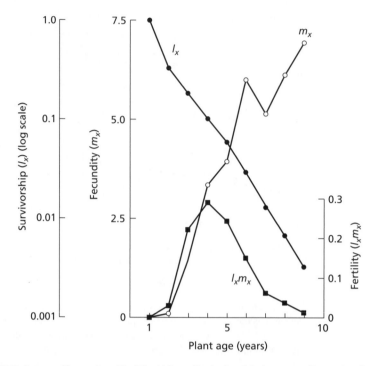

Fig. 10.2 Age-specific survivorship (l_x) and fecundity (m_x) and their age-specific product $l_x m_x$ in a population of the semelparous perennial *Oenothera glazioviana* (from Kachi & Hirose 1985).

4.7 yr. The minor discrepancy between the predicted and observed mean ages of reproduction may be a consequence of the fact that in *O. glazioviana*, as in other semelparous perennials (e.g. Gross 1981), there is a size threshold for reproduction, and fecundity and survival are more closely correlated with size than with age. Size and age are imperfectly correlated, so age-based models contain an inbuilt source of error, even though they are still useful because of their simplicity. The principles that determine the optimum *age* of reproductive maturity also apply to determining the optimum *size* at which reproduction should begin in populations where size rather than age determines survival and fecundity (Takada & Caswell 1997).

It is important to remember that there is usually variation in age at maturity between years and among individuals in a population. Indeed, in semelparous species like *O. glazioviana*, variation between individuals is used by investigators to determine how m_x varies with age. There is a norm of reaction for age at maturity (Lotz 1990; Berrigan & Koella 1994; Kudo *et al.* 1996), but in several populations where it has been investigated there are heritable differences between individuals too (Wesselingh & de Jong 1995; Wesselingh & Klinkhamer 1996; Van Dijk *et al.* 1997). In plant species that span a broad geographical range there is commonly a positive correlation between the latitude of a population and the age at which plants reach reproductive maturity. Studies by Bøcher (1949; Bøcher &

Larsen 1958) of the iteroparous herbs *Prunella vulgaris* and *Holcus lanatus* in Europe were among the first to show that northern populations flowered later than southern ones, and that the difference had a genetic basis. In addition to reaching a size threshold, semelparous perennials in more northerly populations are also typically prevented from flowering just before winter by a vernalization requirement that is broken by chilling. In wild beet *Beta vulgaris* ssp. *maritima* the vernalization requirement is cancelled by a dominant allele that is found at high frequency in the Mediterranean region, but which is absent from populations in northern Europe (Van Dijk *et al.* 1997).

Heritable variation in size thresholds for reproduction in semelparous perennials sometimes correlates with soil nutrient status as well as with latitude, and both kinds of variation can be understood in terms of the effect on fitness of demographic differences between populations (see Fig. 2.2a). In North American populations of wild carrot *Daucus carota* (Lacey 1986, 1988) and *Verbascum thapsus* (Reinartz 1984), there was variation within and between populations in the length of the prereproductive period, which varied from 1 year in the south to 3 years in the north. In *D. carota*, first-year survival in North Carolina populations (36°N) was double that in Ottawa (45°N), but survival in the second year showed the opposite trend and was 10 times greater in Ottawa than in North Carolina. This result supports the hypothesis that differences in age-specific survival explain the diferences in age at maturity between wild carrot populations at different lattitudes. High mortality in the second year in North Carolina was thought to be caused by two fungal pathogens that are a particular hazard to cultivated carrots in SE USA. Why this mortality factor should be age-specific is not clear.

Because flowering in *Daucus carota* occurs when plants reach a threshold size, latitudinal differences between *Daucus* populations in the age at maturity could result either from latitudinal differences in the growth rate, which determines how quickly the reproductive threshold is reached, or from differences in the threshold size itself. In carrot there was some evidence for genetic differences in growth rate (Lacey 1986), but in *V. thapsus* Reinartz (1984) found that the threshold size at which rosettes flowered was significantly larger in Canadian populations than in populations in southern USA, and that this difference was maintained in a common garden.

Wesselingh *et al.* (1993, 1997) found that the threshold size for flowering in the semelparous perennial *Cynoglossum officinale* was significantly greater at a sand dune site in Norfolk, England, than at a site at similar latitude on the coast of Holland. Two factors appeared to account for this difference. First, larvae of a root-boring weevil preferentially attacked and killed larger overwintering rosettes in the Dutch population but were absent from the Norfolk population, thus creating a difference in size- and age-specific mortality between the populations that should select for earlier reproduction in the former population compared with the latter. Second, relative growth rates were three times greater in Norfolk than in Holland, favouring reproduction at larger size in the Norfolk population. We can see why a higher growth rate should select for a larger threshold size, using a simple reformulation of the rule that the optimal age for reproduction in a semelparous perennial occurs when $l_x m_x$ reaches its maximum value. By this rule, reproduction should be delayed in year x if $l_{x+1} m_{x+1} > l_x m_x$. Rearranging the terms we get $m_{x+1}/m_x > l_x/l_{x+1}$. Remembering that $l_{x+1}/l_x = p_x$ (Chapter 1), the condition for delaying reproduction is:

$$m_{x+1}/m_x > 1/p_x. \tag{10.2}$$

This expression means that a plant aged x should delay reproduction if the relative increase in the number of seeds it can produce by waiting a year compensates for the mortality risk of doing so (de Jong et al. 1987). Plant growth rate enters implicitly into Eqn 10.2 because fecundity depends upon plant size and so the ratio m_{x+1}/m_x depends upon growth rate. Thus, at a particular value of p_x, a higher growth rate should select for delaying reproduction until a larger size is reached.

10.3 Mast variation in seed crop size

In iteroparous species total crop size per plant often fluctuates from year to year and the magnitude of this variation can be very different in different species. Among trees it is extreme in beech (*Fagus* spp.), oak (*Quercus* spp.), and pine (*Pinus* spp.) in the temperate zone and in dipterocarp trees in SE Asia. These species produce vast crops of seed, called **mast**, in some years but few or no seeds in the intervening periods between mast years. Typically, seed production by different individuals in such populations is synchronized, although some individuals may skip some mast years (Koenig et al. 1994; Crawley & Long 1995; Herrera 1998). In conifers, mast synchrony may occur between species over hundreds of kilometres (Koenig & Knops 1998). Mast years tend to be correlated with climatic variables; for instance, the European beech *Fagus sylvatica* may mast in the year following a hot summer but rarely in the year following a cold one. Size of acorn crops correlated with weather conditions in four of five species studied by Koenig et al. (1996) in California. In SE Asia, mast years in dipterocarp trees appear to be triggered by slight climatic changes associated with El Niño, an oceanic event that occurs many thousands of miles away in the Pacific, but which has global climatic repercussions (Ashton et al. 1988; Curran et al. 1999).

Why do some species miss opportunities to reproduce, and then produce large crops in mast years? The null hypothesis to explain this phenomenon must be that climatic conditions simply suit seed production better in some years than in others and that barren periods result because trees take time to recover from the effort of reproduction. This idea, known as the **resource-matching hypothesis**, would certainly seem to explain why fruit trees such as pear and apple only bear large fruit crops on a biennial cycle, but it cannot explain why other species differ so greatly from each other in seed crop variability. An alternative hypothesis, prompted by the observation that masting seems to waste opportunities for reproduction (Waller 1979), is that these lost opportunities are in some way compensated because the habit increases the fitness of individual trees. One argument is that synchronized flowering improves the chance of cross pollination in wind-pollinated species (Smith *et al.* 1990). There is some evidence for the **wind pollination hypothesis** in trees such as *F. sylvatica* and *Pinus sylvestris* in which there is a smaller percentage of unfilled seeds in mast years than in years when crops are small (Sarvas 1968; Nilsson & Wästljung 1987). The hypothesis cannot apply to dipterocarps because they are insect-pollinated and it has proved inadequate in some wind-pollinated species too (Sork 1993). Kelly and Sullivan (1997) evaluated the effect of crop size upon fertilization success in the endemic New Zealand bunch grass *Chionochloa pallens* and found that the beneficial effect of large flower crops was only slight, although significant.

One thing all large trees, and indeed all plants, have in common is predation on their seeds by animals. This is the basis of a more general explanation for the possible advantage of the masting habit. The argument is that seed predators consume a large proportion of small seed crops but that they cannot consume a plant's entire crop in a mast year. Hence, the probability of a seed escaping predation is greatest when crops are large. This is known as the **predator satiation hypothesis**. It predicts that there should be a negative relationship between the probability of a seed being eaten and the size of the current seed crop in a masting population. This has been found in many forest trees (Silvertown 1980b; Sork 1993; Kelly 1994) (Fig. 10.3), among which the masting habit appears to be most pronounced in those populations where seed predation is strongest (Silvertown 1980b). The closest thing to a natural experimental test of the predator satiation hypothesis comes from a comparison of the seeding behaviour of two populations of the tropical tree *Hymenaea coubaril*. This tree occurs both on the island of

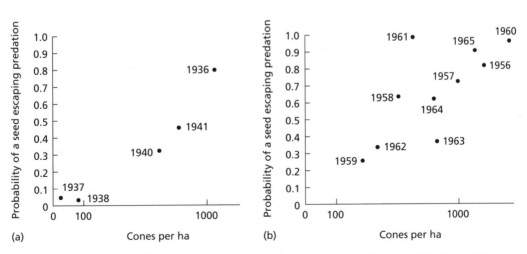

Fig. 10.3 The relationship between annual cone crop size and the probability of a seed of ponderosa pine escaping predation by (a) chalcid wasps, or (b) abert squirrels. Study (a) was carried out in California by Fowells & Schubert (1956), study (b) in Arizona by Larson & Schubert (1970).

Puerto Rico, where one of its major insect seed predators is absent, and on mainland Costa Rica, where these predators are present. The mainland tree population shows the masting habit but the island one does not. Other morphological features of *H. coubaril* fruit that help deter predators in Costa Rica are also absent from populations in Puerto Rico (Janzen 1975b).

It follows from the argument that masting prevents animals from consuming an entire tree crop and that trees with fleshy fruits and animal-dispersed seeds should not mast but should produce fruit regularly. The seeds in such fruits generally pass through the gut of the dispersal agent intact, so that the only effect of masting in a fleshy-fruited species would be to prevent efficient seed dispersal. In a sample of North American trees, Silvertown (1980b) found that most species with nonfleshy dispersal units masted to some degree, while most of those with fleshy dispersal units did not; Herrera *et al.* (1998) found a similar trend for a more widely drawn sample of woody species. The exceptions in the latter dataset involved fleshy-fruited species whose reproductive structures were heavily preyed upon by specialized predators, serving as a reminder that predation and dispersal operate simultaneously on the fitness of plants.

The **frequency** of large crops, and **synchrony** among individuals, are potentially important parts of the predator satiation mechanism. It would be disastrous for a plant if large crops were produced regularly, because predators might show a **numerical response**, simply building up their numbers from one year to the next on succeeding bumper crops. If predators are capable of tracking seed crop size by a numerical response, then predation on seeds should vary negatively with the magnitude of the difference between seed crops in successive years, not just the size of crop. In the New Zealand grass *Chionochloa pallens*, large seed crops reduced seed predation by vastly more than they improved fertilization, but only when large crops followed small ones, indicating the existence of a strong numerical response (Kelly & Sullivan 1997). Surprisingly, numerical responses have not been found in trees (Silvertown 1980b; Nilsson & Wästljung 1987).

Reproductive structures are usually prey to a whole guild of insect pests (Turgeon *et al.* 1994) and there appears to be at least one case where pre-emptive competition between two seed-feeding insects reduces the numerical response of the more damaging species, to the benefit of the tree. Harris and Chung (1998) found that the pecan nut casebearer *Acrobasis nuxvorella*, by consuming pecan nuts (*Carya illinoensis*) in years when crops were small, removed the food source for the pecan weevil *Curculio caryae*, which feeds later in the season. *Acrobasis nuxvorella* caused little damage itself in mast years and, by simulating seed losses to *C. caryae* in mast years with and without *A. nuxvorella* present in between, Harris and Chung (1998) estimated that *A. nuxvorella* increased the lifetime seed production of an average pecan tree by 71 116 nuts, at a cost of only 210 nuts consumed. Whether similar interactions among seed predators nullify numerical responses in other tree species deserves investigation.

The selective advantage of seed crop synchrony is the aspect of masting that has been least investigated in the field, although its evolution has been explored in several theoretical models. An initial level of spatial synchrony among individuals can be produced by plants' resource matching a variable environment in a mechanism known as the **Moran effect**. The evolution of greater synchrony depends upon predator searching behaviour. Predators that are specialized and aggregate their searching are most likely to select for synchronous seed production among plants (Ims 1990; Lalonde and Roitberg 1992). In situations where the mechanism of the wind pollination hypothesis is important, Isagi *et al.* (1997) have shown that the extra resource drain on trees in years when seed crop and fertilization success are both high can induce synchronous seed production among individuals.

Seeds are extremely important food resources for animals in many habitats, with the consequence that masting causes significant coupled fluctuations in populations of rodents, birds, insects and even bears. In New England, mast years for oak species are followed by increases in populations of mice and local densities of deer that are both hosts to a tick that vectors the causative agent of Lyme disease, *Borellia burgdorferi*. The risk of this disease to the human population of the area can be predicted from the occurrence of mast years (Jones *et al.* 1998).

10.4 Seeds

In this section we shall look at the evolution of seed size, seed dispersal and seed dormancy. All of these traits are largely under maternal control, which means that their evolution is driven by selection on the reproductive success of the maternal parent. Seed traits that maximize maternal reproductive success are quite often different from those that maximize the fitness of individual seeds themselves, but mother calls the shots.

10.4.1 Seed size

The principle of allocation dictates that a plant can package its reproductive effort into a few large seeds or many small ones, obeying a **size–number trade-off** (Section 9.1.2). An example of this trade-off was observed in the perennial thistle, *Cirsium arvense*, by Lalonde and Roitberg (1989) when seed number was experimentally altered by controlling pollination. Fully pollinated flower heads produced more, smaller seeds than heads receiving limited amounts of pollen. Similar trade-offs have been observed as a result of natural variation in pollination success in *Ipomopsis aggregata* (Wolf *et al.* 1986), *Primula vulgaris* (Boyd *et al.* 1990) and several other species. In this section we consider two basic questions about the balance between seed size and number: (i) how and why does seed size vary within populations?, and (ii) how and why do species differ in seed size?

10.4.1.1 *Variation within species*

The size of a seed is especially closely correlated with its seedling's success in competitive situations and when germinating in shade. For example, Black (1958) planted large and small seeds of subterranean clover *Trifolium subterraneum* in a mixture and found that the percentage share of light interception by seedlings from small seeds was reduced virtually to zero after 80 days (Fig. 10.4). Wulff (1986b) compared the growth of plants from large and small seeds of another legume, *Desmodium paniculatum*, when raised in monoculture with their perform-ance when competing in a mixture. Pure stands grown from small seeds produced the same yield per plant as pure stands grown from large seeds, but in mixtures the seed production of plants from small seeds was less than half that of plants from large ones. In an experiment with the wild oat *Avena fatua*, whose seeds varied fivefold in weight, large seeds produced more than twice the number of offspring compared with small ones, when growing in competition with barley (Peters 1985). However, bigger doesn't always mean better (Janzen 1969). Moegenburg (1996) found that seed-feeding bruchid beetles placed in choice chambers were 10 times as likely to lay an egg on 8 mm diameter seeds of cabbage palm *Sabal palmetto* than on seeds half the size. Bruchids are very widespread pre-dispersal seed predators in the neotropics (Janzen 1980) and similar results have been obtained in several other studies.

A model following Smith and Fretwell (1974) predicts, from the perspective of the mother, that the optimum size offspring should be when there is a size/number trade-off. If offspring survival (s) is a function of offspring size (m) (Fig. 10.5a), then the optimum seed size corresponds to the highest value of s/m. This value can be determined by the point at which a line through the origin lies tangentially to the curve of offspring size vs. offspring survival (Fig. 10.5a) and indicates where increasing investment by the parent brings diminishing returns in survival of offspring (Lloyd 1987b). The curve shown in Fig. 10.5a assumes that small seeds will be inviable or have low fitness and that, beyond a certain size, offspring survival is size-independent. Seeds much bigger than the optimum waste maternal resources that could be put to better use by provisioning more seeds, and are therefore a fitness

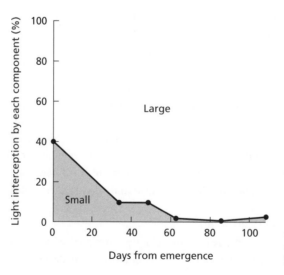

Fig. 10.4 The percentage of light intercepted by plants from large and small seeds of *Trifolium subterraneum* when grown in mixture with each other (from Black 1958).

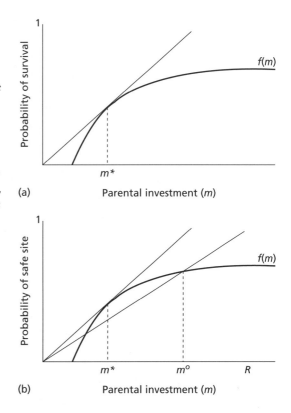

Fig. 10.5 Two models for the evolution of seed size based upon parental investment per seed and the size-dependent fate of offspring. (a) The Smith–Fretwell model predicts an optimum seed size at the point (m^*) where the diagonal line, which represents constant returns in offspring survival for constant increments in seed size, meets the size/survival (s) curve at a tangent. (b) The Geritz ESS model is based on a size/germination (g) function, with seedling survival dependent upon relative seed size in safe sites. A range of seed sizes (m^*–m^o) is predicted between a lower limit g/m and an upper limit where the function is intersected by a diagonal from the origin to the point $g = 1$, m = total maternal resources (after Haig 1996).

cost to the parent. It may be to the future advantage of the individual *embryo* in a seed to grow as large as possible, but the maternal parent controls the flow of resources to its seeds, such that fitness costs to the parent place an upper limit on the 'profitability' of large seeds (Casper 1990). The heritability of seed size in wild plants tends to be low and such heritable variation as does exist is largely between maternal individuals, rather than embryo or paternal plant genotypes (e.g. Schaal 1980; Primack & Antonovics 1981; Waller 1982; Mazer 1987; Roach 1987; Wolfe 1995; Mojonnier 1998).

According to the Smith–Fretwell model, the optimum size is unaffected by the resources available to a plant, so that when these vary, for example with plant size, plants should respond by changing seed number rather than seed size. This model may explain why the size of some seeds is so constant. Seeds of the carob tree *(Ceratonia siliqua)*, for example, have been used to define a unit of weight (the carat) for trading in gold. But clearly, not all species are as reliable as this and anyone who chose the seeds of *Trifolium subterraneum* as a unit of measurement would find they had a standard that could vary 17-fold in weight! (Black 1959). Indeed, phenotypic variation in seed size is significant in many species. In a survey of seed size variation in 39 North American species, Michaels *et al.* (1988) found significant variation in seed weight within populations of 37 of them, with most of the variance arising from differences between seeds on the same plant. Other studies of individual species have found much the same thing, with most of the observed variation related to position on the plant (e.g. Waller 1982; Stanton 1984; Wulff 1986a), to plant size (e.g. Hendrix 1984), or to season (e.g. Cavers & Steele 1984; Wolfe 1995). Why is there such phenotypic plasticity when the Smith–Fretwell model predicts that there should be a single optimum size? One possible answer is that **constraints** sometimes prevent a plant from determining seed mass (McGinley *et al.* 1990) or seed number, such as when pollination success is poor, so that in the circumstances seed size cannot be kept uniform. However, an alternative kind of evolutionary model offers another answer.

In the Smith–Fretwell model the success of a seed of a particular size is independent of the frequency with which seeds of different size occur in the population. This is not realistic if success is determined by competition among seedlings (as opposed to competition between seedlings and established vegetation) because it is *relative size* that determines the outcome in such situations (Chapter 4): hence, the average success of a seed depends upon the frequency with which it encounters smaller competitors. Geritz (1995) constructed a game-theory model in which seeds are distributed at random among 'safe sites', so that each site may receive none, a few or many according to a Poisson distribution. Some density-independent mortality is assumed before competition occurs between the seeds reaching each site, with the largest seed among those present winning the contest. In this situation, a parent that concentrated all its resources in one massive seed would be sure to win the contest wherever that seed landed, but it would be at a numerical disadvantage to parents producing many small seeds, because these would colonize a larger number of safe sites. The evolutionarily stable strategy (ESS), or the strategy to beat all others, in this situation is to produce a range of seed sizes between a lower size limit set by g/m, where g is the probability that a seed in a safe site will germinate, and an upper size limit set by the number of seeds required to replace the parent with one surviving offspring (Fig. 10.5b). This model predicts that a single seed size is not evolutionarily stable. The ESS distribution of seed sizes within the prescribed limits may be produced by each parent producing an ESS mix of seed sizes, or by a mix of parents in ESS proportions each producing seeds of one size. The former situation would correspond to observed patterns of seed size variation within plants, and the latter situation to communities in which species with different mean seed sizes typically coexist (Rees & Westoby 1997).

The Geritz (1995, 1998) model makes some predictions about how variation in seed size should evolve under a variety of ecological scenarios. The most important difference between the predictions of the Geritz and the Smith–Fretwell models – that no single seed size is optimal – depends upon the strength of competitive asymmetry between large seeds and small ones (Geritz *et al.* 1999). If this asymmetry is low, then the advantage of larger seeds over smaller ones is small, frequency dependence becomes unimportant, and there is a single optimum size as in the Smith–Fretwell model. In the Geritz model, size variability is also predicted to be less when density-independent seedling mortality is greater, ultimately leading to a single optimal seed size when mortality is high enough (Geritz *et al.*

1999). Unlike the Smith–Fretwell model, the Geritz (1995) model predicts that plants with more resources, for example because they are large, should have larger seeds and a greater range of seed sizes than plants with fewer resources, as has indeed been found in a number of populations (Venable 1992).

Seed size variation within plants is usually continuous, but there are some cases where it is markedly discontinuous. The annual *Amphicarpum purshii* produces a fairly constant number of large seeds early in life, and increasing numbers of small ones later as the plant grows larger, if resources and time allow (Cheplick & Quinn 1982; Cheplick 1983). The large seeds are produced from subterranean flowers and the small ones from aerial ones. Although this syndrome (**amphicarpy**) is rare, it has evolved independently in at least nine different families (Cheplick 1987). Amphicarpic species appear to be playing a sophisticated, mixed strategy in which, very unusually for plants, seed *number* is strictly limited for one seed type: the large, buried ones. These seeds come from cleistogamous flowers and are thus full-sibs, so it is easy to see why a parent would not benefit from unlimited numbers of its identical offspring competing for a single safe site. Aerial seeds, on the other hand, are dispersed more broadly and thus the more that can be produced, the more safe sites can potentially be colonized. We shall return to this subject in the discussion of the evolution of dispersal in Section 10.4.2.

10.4.1.2 Variation between species

An enormous range in seed size occurs between species, from the tiny seeds weighing 10^{-6} g produced by the orchid *Goodyera repens* to the monstrous seed of the double coconut palm *(Lodoicea maldivica)* which weighs over 10^4 g (18–27 kg) (Harper *et al.* 1970). Even within regional floras, seed size typically ranges across five or more orders of magnitude (Hammond & Brown 1995; Leishman *et al.* 1995). Considerable study has gone into understanding the correlates of this variation and into partitioning it into three major components resulting from phylogenetic, life history and habitat differences between species.

All species are ultimately related to one another through common ancestors during their evolutionary history, but how much of the similarity between related species results from their common phylogenetic origins? Studies of seed size variation that have taken phylogeny into account (or taxonomy as its surrogate) have found this to be overwhelmingly the strongest of the three components (Kelly & Purvis 1993; Kelly 1995; Lord *et al.* 1995; Mazer 1989, 1990). For example, not just *Goodyera repens*, but all species in the family Orchidaceae have tiny seeds (although by no means all palms have large ones). Lord *et al.* (1995) analysed seed size variation in six temperate floras and found that in each of them around 90% of the variance could be accounted for by species' taxonomic membership (e.g. to which genus, family or higher taxonomic group they belonged), and that taxonomic differences accounted for nearly as much of the variance (86%) in seed mass between regions. One might think that, with so much of the variance in seed size accounted for by phylogeny, there would be no scope left for covariance between seed size and life history or habitat, but this would be a misinterpretation. Although it simplifies matters to think of phylogeny as a source of seed size variation separate from life history and habitat, it tends to be correlated with these variables and cannot be meaningfully separated from them.

Consider two alternative interpretations of why a particular phylogenetic lineage might conserve a trait such as seed size. Under one intepretation, little evolutionary change occurs in the trait because there is a **phylogenetic constraint** that prevents it. Phylogenetic constraint is a notoriously slippery concept (Antonovics & van Tienderen 1991), but its essential feature is that *adaptive* change is influenced by a species' evolutionary history. Examples of phylogenetic constraint tend to be speculative because it is so difficult to demonstrate why something has *not* happened, especially when evolution is able to produce something as improbable as a 27-kg coconut! Hodgson and Mackey (1986) have suggested that the number of ovules per carpel, which tends to be a conservative trait in plant families, constrains seed size evolution. In keeping with this, seed size is larger in species with one ovule/carpel than in those with more than one in the floras of the Sheffield region in Britain (Hodgson & Mackey 1986) and also in the tropical woody flora of Guyana (Casper *et al.* 1992). Janson (1992) analysed the frequency of evolutionary transitions between different seed dispersal syndromes in Neotropical genera and found highly nonrandom patterns suggesting the existence of constraints.

An alternative interpretation for the lack of change in a trait is that the lineage in question possesses a suite of traits, one of which might be seed size, that fit it very well to the particular ecological niche it occupies and in which it can speciate, hence producing many species with the characteristic seed size (say). This scenario is known as **phylogenetic niche conservatism** (PNC) and its important feature is that lack of change results from the trait having high adaptive value in a particular niche (Harvey & Pagel 1991). Minute seed size has certainly not hampered speciation in the orchid family, which is estimated to number between 20 000 and 25 000 species and is probably still speciating (Dressler 1990). In the orchids minute seed size is possible only because orchid seedlings are mycotrophic, living parasitically upon a fungal symbiont in the early stages, so there is no doubt in this case that the characteristic seed size is part of a highly successful syndrome of interdependent traits.

How can one distinguish between phylogenetic constraint and PNC in a dataset, and why does it matter? It matters because the existence of PNC in seed size evolution could reconcile the finding that the trait is strongly correlated with phylogeny with the seemingly contradictory evidence that it is also strongly correlated with habitat. Lord *et al.* (1995) suggest that PNC is indicated when there are suites of traits correlated with phylogeny, but the trouble here is that one cannot tell by this fact alone which of the suite of traits is important to fitness and which is not. Statistically independent variation between seed size and other traits is necessary to make such a distinction. One way of obtaining this is to use the method of **phylogenetically independent contrasts** (PIC). At its simplest, this approach compares pairs of related species that differ in just one trait, say habitat, and asks whether there is a consistent difference within pairs in another trait such as seed size. For example Salisbury (1942, 1974; Thompson & Hodkinson 1998) compared seed size between pairs of congeneric species from the British flora, where one species lived in shaded (closed) and the other in open habitats. In a significant excess of pairs (23/29) the shaded habitat species had a higher seed weight than its open-habitat congener (Fig. 10.6). Because species in the same genus share a more recent common ancestor than do species in different genera, differences (contrasts) in seed size between one pair of congeners must have evolved independently of contrasts within other genera. Each such contrast is therefore phylogenetically and statistically independent of other cases and can be used as a replicate in a test of association between, for example, seed size and habitat. In more sophisticated analyses PICs need not be confined to congeneric comparisons, so long as the contrasts are drawn between taxa that branch from a common point (node) in the phylogeny (Harvey & Pagel 1991).

The pervasive influence of phylogeny on seed size variation between species means that correlations between this trait, life history and habitat that treat each species as an independent datum (i.e. 'cross-species' correlations) must be interpreted with caution. However, because PNC may have been important in the evolution of seed size, cross-species correlations should not be ignored (Westoby et al. 1997). Cross-species correlations show that the mean weight of seeds increases progressively through herb, shrub and tree species in the floras of Britain,

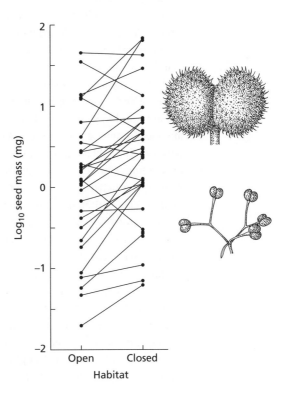

Fig. 10.6 Comparisons of seed size between congeneric pairs of closed-(shaded) and open-habitat perennial species compiled by Salisbury (1942, 1974). Each pair is joined by a line. Twenty-three out of 29 comparisons show the seed weight of the closed-habitat species to be greater than that of the open-habitat species, giving a highly significant overall trend ($P = 0.0004$; Thompson & Hodkinson 1998). Two-seeded fruits of a closed-habitat (*Galium aparine*, upper) and an open-habitat (*G. parisiense*, lower) pair of congenes are shown (from Salisbury 1942).

California, tropical forests, and on a worldwide scale (Levin & Kerster 1974; Rockwood 1985; Metcalfe & Grubb 1995). Seed weight increased significantly with plant height and growth form (herb → woody) in regional samples of the Australian, North American and British floras analysed by Leishman *et al.* (1995). In tropical forest species from Peru, Foster and Janson (1985) found that taller plants had significantly larger seeds than shorter ones of the same life form. Strangely, tropical species seems to have larger seeds than temperate relatives with the same growth form (Lord *et al.* 1997). Among herbs in California, the mean seed weight of perennials is significantly greater than that of annuals (Baker 1972), and a similar correlation with lifespan also occurs in Britain when PICs are used to remove the positive effect of shady habitats upon seed size (Hodkinson *et al.* 1998). Rees (1997) used PICs to analyse seed weight variation in the Sheffield flora and found independent, positive relationships with plant height and adult longevity, but only among species that were not animal dispersed. In total, these studies constitute formidably strong evidence that seed size correlates with other life history variables and suggests that, as the Geritz (1995) ESS model predicts, better resourced, larger plants have evolved larger seeds.

Growth form, life history and habitat are of course inextricably interrelated, and the fact that tall species live in habitats with tall competitors must also favour their evolving larger seeds. Tropical forest trees that are able to regenerate in small gaps or in shade have significantly larger seeds than pioneer trees and those which need large light gaps (Foster & Janson 1985; Hammond & Brown 1995). Nine out of 13 PICs supported this trend in secondary tropical forest in Singapore (Metcalfe & Grubb 1995), but in Australian tropical rainforest Grubb and Metcalfe (1996) found that the same trend could only be found with PICs drawn within families, and did not occur for PICs within genera. Grubb and Metcalfe (1996) suggested this was evidence that the more recent evolution of seed size in their species had been driven by factors other than shade tolerance. Among North American trees, PIC analysis has shown that shade tolerators have larger seeds than shade intolerant species among the angiosperms, but not among the conifers (Hewitt 1998). Seed weight in the Californian flora appears to be more strongly related to the risk of seedling mortality resulting from drought than from shade, and a positive relationship between seed weight and the dryness of the habitat occurs among herb species within the same genus, for whole herb communities and among Californian trees. California has a large number of introduced species that fit the same patterns of seed weight and environmental conditions shown by native species (Baker 1972).

Experimental studies comparing species with different seed sizes suggest that larger seeds confer a number of benefits upon seedlings including greater height (Leishman & Westoby 1994a) and better survival in shade (Grime & Jeffrey 1965) (Fig. 10.7), better mineral nutrition (Milberg & Lamont 1997; Milberg *et al.* 1998), less nutritional dependence upon mycorrhizal symbionts (Allsopp & Stock 1995), and greater tolerance of drought (Leishman & Westoby 1994b) and herbivory (Armstrong & Westoby 1993; Mack 1998).

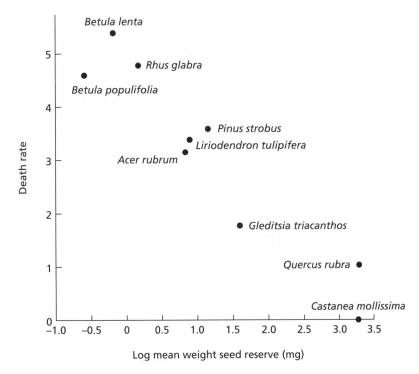

Fig. 10.7 The relationship between death rate in shade conditions and log mean seed weight in nine tree species (from Grime & Jeffrey 1965).

10.4.2 Dispersal in space and time

Seed dispersal, defined in the broad sense as any mechanism which causes a seedling to experience a different microenvironment from that of its mother, occurs in two forms whose evolution is closely linked: dispersal in space and dispersal in time. Dispersal in time can be achieved by seed dormancy. When the environment varies in space or time, there are four distinct ways in which either kind of broad-sense dispersal can increase maternal fitness, by: (i) **risk reduction** in a variable environment (Cohen 1966; Venable & Brown 1988); (ii) **escape** from the negative consequences of high density (Chapter 4); (iii) reduction of **sib competition** (Nilsson *et al.* 1994; Ellner 1986); and (iv) **prediction** of favourable times for germination or directing offspring to predictably favourable germination sites (Cohen 1967; Venable & Lawlor 1980; Venable & Brown 1993).

10.4.2.1 *Seed dispersal in space*
Significant fractions of most seed crops are dispersed to a distance from the mother plant, often aided by wind or by animals attracted by the fruit that seeds come packed in (Chapter 3). Most dispersed seeds perish, but a simple theoretical model shows that despite this, quite high levels of dispersal should be favoured by

selection to avoid sib competition, even when there is no temporal or spatial heterogeneity and the other three mechanisms favouring broad-sense dispersal mentioned above do not operate. In a landmark paper, Hamilton and May (1977) used the ESS approach to compare the success of two hypothetical dispersal genotypes inhabiting permanent microsites that they vacate by dying at the end of the year. One genotype keeps all offspring at home and has a high probability of replacing itself, while the other genotype disperses a fraction of its offspring to other sites where they compete with the residents' offspring to occupy new sites. The dispersers will always be replaced by new dispersers (because non-dispersers, by definition, can't get to their microsites) and they will also capture some fraction of the non-disperser's microsites. Non-dispersers suffer more severely than dispersers from competition among sibs. The dispersing type will therefore increase at the expense of the non-disperser and ultimately replace it. Dispersing your offspring is therefore the evolutionarily stable strategy, but what fraction of offspring should be dispersed? Hamilton and May (1977) found the answer to this question to be a surprisingly high proportion. If the survival of dispersing offspring is p, then the fraction v^* that the ESS type should disperse is:

$$v* = 1/(2 - p). \tag{10.3}$$

v^* increases with the survival chances of dispersed offspring, but even when these are very low Eqn 10.3 tells us that a mother should disperse half her offspring. Amphicarpic species and others such as the annual *Hypochoeris glabra* (Baker & O'Dowd 1982) that disperse an increasing fraction of their seeds as fecundity rises, support the notion that sib-competition favours dispersal.

The Hamilton–May model assumes that: all adults are annuals, vacating their microsites each year; each site holds only one individual; and offspring are genetically identical to their parents. There are many variants of this model that modify its original assumptions, but Eqn 10.3 is a robust result in many of them (Johnson & Gaines 1990). Introducing spatial and temporal heterogeneity into ESS dispersal models alters the value of v^* (Levin *et al.* 1984). Temporal variability tends to increase v^* because dispersal reduces risk. By contrast, spatial heterogeneity tends to reduce v^* because, when each microsite is allowed to contain many individuals and sites vary in carrying capacity, individuals will tend to become concentrated in high-quality patches and dispersal will therefore carry offspring to microsites of poorer quality than those occupied by their mothers (Johnson & Gaines 1990). In metapopulations, v^* is expected to change with population age (Olivieri *et al.* 1995). New populations are likely to be founded by the farthest dispersed genotypes, but once patches are colonized, selection within populations will tend to reduce v^*. Decreases in dispersability with population age have been observed in thistles *Carduus* spp. (Olivieri & Gouyon 1985), red maple *Acer rubrum* (Peroni 1994), *Pinus contorta* (Cwynar & MacDonald 1987) and in wall lettuce *Lactuca muralis* colonizing small islands off the coast of British Columbia, Canada (Cody & Overton 1996).

10.4.2.2   *Seed dormancy: dispersal in time*

Substantial spatial dispersal is an ESS even in uniform, unvarying environments but, by contrast, temporal dispersal is *not*, because it delays reproduction and should thus be selected against (Section 10.1). In fact, seed dormancy of the kind which spreads the germination of a plant's seeds over more than one season, or **germination heteromorphism**, is common and is often correlated with variation in seed size within a plant (Hendrix 1984; Silvertown 1984). Theoretical models predict that germination heteromorphism should be favoured when there is variation in offspring success between seasons, because it reduces risk by allowing a mother to spread the germination of her offspring (or hedge her bets) over several years (Cohen 1966; Venable & Brown 1988; Rees 1994). Venable and Pake (1999) found that, among annuals in the Sonoran Desert of SW USA, smaller-seeded species, which are more vulnerable to drought than larger-seeded ones, had a higher variance in net reproductive rate (R_0) measured over 10 years, and that species experiencing higher variance held a larger fraction of their seeds dormant in the soil. The **risk-reduction hypothesis** for germination heteromorphism can be tested by calculating the average fitness of genotypes producing a monomorphic clutch with that of genotypes producing a heteromorphic clutch over a run of years. The appropriate kind of average to use in this calculation is the *geometric mean* fitness because the long-term success of a genotype is determined by multiplying together its success each year (Venable 1985). This point is fundamental to adaptive explanations of germination heteromorphism and other risk-reducing strategies.

Seeds of the annual grass *Bromus tectorum* growing in Washington State display germination heteromorphism and germinate over a prolonged period between autumn and spring. Mack and Pyke (1983) found at a dry study site that which cohorts did best depended on the year. Taking this variation into account over a run of years, genotypes with heteromorphic seeds turn out to have higher geometric mean fitness (R_0) than would monomorphic producers at this site (Fig. 10.8) (Silvertown 1988). Germination heteromorphism correlated with differences in morphology between seeds in the same seed head are very common in the

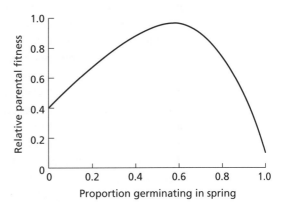

Fig. 10.8 The expected geometric mean fitness of plants producing seed crops with different proportions of spring-germinating seeds, estimated from annual variability in the fitness of different cohorts of *Bromus tectorum*. Genotypes producing a mixture of seed germination morphs are favoured (from Silvertown 1988).

Asteraceae (Venable & Lawlor 1980). Venable *et al.* (1987) studied the annual *Heterosperma pinnatum* in Mexico and found that seeds (more properly called 'achenes' in the Asteraceae) from the centre of flowering heads germinated with the early spring rains. These seedlings risked being caught by a later drought, but survivors could reach large size. Achenes from the margin of the fruiting head were slow to germinate and consequently did not risk drought, but produced smaller plants that themselves yielded fewer seeds. Germination heteromorphism is a heritable trait that varies between sites in both *B. tectorum* (Meyer & Allen 1999a,b) and *H. pinnatum* (Venable & Búrquez 1989). This is consistent with a character whose fitness is affected by the climate, because selection of variable direction will maintain genetic variation within and between populations (Chapter 3).

According to ESS theory, not only seed dormancy but also spatial dispersal, larger seed size (Venable & Brown 1988) and greater adult longevity (Rees 1994) can reduce risk in variable environments. Because the evolution of any one of these risk-reducing traits would diminish the fitness advantage of the others, it is expected that the four traits should be negatively correlated. Using PICs on a sample of British herbs, Rees (1993) confirmed the existence of negative correlations between seed dormancy and adult longevity, seed size and seed dispersal. As we have already seen (Section 10.4.1.2), seed size tends to be positively correlated with adult longevity and size for other reasons.

If spatial dispersal is limited, competition among sibs falling around a mother plant can reduce maternal fitness (Cheplick 1992) and favour seed dormancy as a means of ameliorating this (Nilsson *et al.* 1994). This **sib competition hypothesis** is supported by many cases where individual plants produce poorly dispersed seeds with high dormancy and better dispersed seeds with less dormancy (Venable & Lawlor 1980; Silvertown 1985; Cheplick 1996), and also by some other species where the dormant fraction increases with the size of the seed crop, and hence with increased risk of sib-competition (Philippi 1993b; Zammit & Zedler 1990).

10.5 Clonal growth

Because of their modular construction, it is not fanciful to argue that clonal growth comes naturally to plants. A tree adds modules to its branches and so extends vertically, a clonal plant adds modules at its base and extends horizontally. A few clonal shrubs and trees do both. The difference between these growth habits is mostly one of scale and it has been said that a clonal herb is just a tree lying on its side (Harper 1981) (Fig. 1.7a), although of course this omits the essential difference that the branches of a clonal herb are ramets capable of an independent existence. Clonal growth is evolutionarily ancient and widespread among plant taxa (Mogie & Hutchings 1990). Salisbury (1942) estimated that more than two-thirds of the commonest perennials in the British flora show pronounced clonal growth. Even greater proportions of woodland and aquatic herbs show the habit (van Groenendael *et al.* 1997).

10.5.1   Costs and benefits of clonal growth

The advantages of clonal growth are many. It permits rapid increase in the size of genets, mobility, the capture of space and resources, storage, and vegetative reproduction without the costs of sex or the hazards of recruitment from seed. Given these substantial advantages, and the predisposition of plants to clonal growth, the conundrum is why a significant number of plants, including most trees and all conifers, are *not* clonal, rather than the other way about. Like all other features of life history, clonal growth has costs and some of these can be understood in terms of their relationship to three other life history traits that we have already considered: sexual reproduction, reproductive allocation and dispersal.

Clonal growth can be thought of as a form of *asexual* reproduction that does not involve seeds. The genetic uniformity of clonal offspring can make them particularly susceptible to disease. When elm disease appeared in the Netherlands, 95% of the elms belonged to one susceptible clone, and some cities lost 70% of their elms (Burdon & Shattock 1980).

Trade-offs between allocation to sexual reproduction and clonal growth occur. Sutherland and Vickery (1988) found a trade-off between allocation to sexual reproduction and clonal growth in five closely related North American species of *Mimulus*. All five species produced rooted stolons or rhizomes as well as flowers that were chiefly pollinated by hummingbirds. Across the five species there was a perfect negative correlation between the number of closely spaced ramets per unit of plant weight and all measures of investment in sexual reproduction, such as nectar volume, fruit set and seed production. Comparing how different grass species colonized experimental gaps Bullock *et al.* (1995) found a negative correlation between species' colonization by seed and their clonal ingrowth from the gap margin.

Selection to reduce the cost of clonal growth to sexual reproduction may explain an interesting phenomenon seen in rosettes in the genus *Hieracium*, which produce stolons only after they have flowered (Bishop *et al.* 1978). In some cases clonal growth may postpone rather than replace sexual reproduction, ultimately increasing lifetime fecundity in species such as the semelparous bamboos because it increases the final size of the genet. To the extent that the reproductive costs of clonal growth delay flowering, clonal growth must increase reproductive success if its costs are to be repaid. We should therefore expect clonal growth to be absent where selection favours early reproduction, for instance in highly disturbed habitats where the risk of adult mortality is high (Section 10.2). Disruptive selection caused by differences in the degree of disturbance in different habitats may be responsible for polymorphisms in degree of clonality in several grasses and sedges (Mogie & Hutchings 1990; Skalova *et al.* 1997).

The daughter ramets produced by terrestrial clonal species tend to fill the space around their parents and thus potentially incur competition from sexually produced sibs. Local crowding is a cost of clonal growth that will favour the better dispersed propagules produced by sexual reproduction. Interestingly, the same does not apply to aquatic clonal plants, which produce a variety of types of vegetative propagules that are dispersed long distances by water. Asexual populations of these species are frequent, perhaps because sexual organs are not required for the production of propagules for dispersal (Grace 1995).

10.5.2 Clone morphology and physiological integration

Depending upon the length of connections (or **spacers**) and the frequency of branching, clonal growth can produce linear arrangements, networks or clumped distributions of ramets (Fig. 1.7). There is a continuum of growth form from the solid advancing front of ramets, called a **phalanx**, to the production of widely spaced ramets that infiltrate the surrounding vegetation in **guerilla** mode, sometimes with short-lived connections (Lovett Doust 1981). In the flora of central Europe guerilla species tend to occur in moist, fertile habitats where light availability is low, by contrast with phalanx species which tend to occur in the opposite kinds of conditions (van Groenendael *et al.* 1997). However, these patterns have many exceptions. For example, among eastern North American forest herbs the clone regularly fragments in *Medeola virginiana*, but in *Aralia nudicaulis* the connections between ramets are permanent (Cook 1983; Edwards 1984; Hutchings & Bradbury 1986). The explanation for the variation in clonal morphology probably lies in the benefits and costs associated with two functional aspects of the clonal growth habit: the degree to which ramets are physiologically dependent upon one another, and the degree to which clonal growth helps plants forage for resources.

In most clonal species in which the rhizome, stolon or other connection remains intact, the genet behaves as a physiologically integrated unit. Radio-tracer studies, in which CO_2 labelled with the isotope ^{14}C has been fed to leaves, have shown for most species investigated that young ramets are supported by a flow of carbohydrates from older ones. Depending on the species, these carbon flows generally cease as young ramets mature, but they may be re-established if a ramet is shaded or defoliated (Marshall 1990). Physiological integration appears to be the rule for clonal plants in nutrient-poor environments (Jónsdóttir & Watson 1997). For example, the arctic herbs *Carex bigelowii* and *Lycopodium antoninum*, and the sand dune plant *Carex arenaria*, possess the guerilla structure, but clones are physiologically integrated throughout life. Only the youngest tillers or shoots are photosynthetically active, and these export carbohydrate to an extensive rhizome system that returns nutrients captured from a wide area (Carlsson *et al.* 1990; Jónsdóttir & Callaghan 1990; D'Hertefeldt & Jónsdóttir 1999).

Translocation of carbohydrates, water and nutrients is also commonplace between mature ramets in many clonal species (Marshall 1990). This will benefit the clone as a whole if a transfer of resources between ramets increases the contribution to genet fitness by the recipient more than it decreases the contribution of the donor (Caraco & Kelly 1991). It is easy to see how this situation arises between mother and daughter ramets, but it can also occur between mature ramets in heterogeneous environments if there is a **division of labour** between them caused by each having superior access to a different resource, say nitrogen and light. Although greenhouse and garden experiments have shown the potential for this in many clonal species (Hutchings & Wijesinghe 1997), evidence from field experiments is mainly lacking. Slade and Hutchings (1987a,b) grew connected ramets of the woodland herb *Glechoma hederacea* in individual pots that differed in nutrient supply and light environment. Older ramets in good conditions supported the growth of younger ones in poor environments, without suffering any detectable cost compared to controls, but the relationship was not reversible when younger ramets were in the better environment and older ones in the poorer. Physiological integration in this species appears to be constrained by the direction assimilates can travel (Price *et al.* 1992). Comparable experiments have been carried out with the beach strawberry *Fragaria chiloensis*, which lives in patchy dune habitats in California where nitrogen is available but light is poor beneath shrubs, in contrast to areas of bare sand where nitrogen is very low but light is high. Bidirectional transfers of water, nitrogen and carbohydrate occur in this species and connections were capable of keeping two dune ramets alive in conditions where both would have died if disconnected (Alpert & Mooney 1986). Division of labour, and the advantage to a genet of sharing resources between ramets, should break down in a homogeneous environment. In confirmation of this, Alpert (1999) found that resource transfer between ramets, of *F. chiloensis* sampled from an homogenous grassland habitat was weaker than between ramets belonging to genotypes from patchy dune habitat.

10.5.3   Foraging by clonal plants

Foraging is a manifestation of phenotypic plasticity (or behaviour) that increases resource capture in an environment that is spatially heterogeneous for the resource. This definition includes behaviours with two distinct functions: (i) a **searching function** triggered in resource-poor patches and which improves the probability of encountering a new resource patch and (ii) a **holding function** triggered in resource-rich patches and which improves their utilization once found. Foraging is not, of course, confined to clonal species and the shoots and roots of many, if not all, plants respond in some way to the availability of light or below-ground resources. Among clonal species, both the searching and holding functions may potentially be filled by changes **in spacer length** affecting whether ramets are locally concentrated or dispersed from the current location, and the holding function by an increase in ramet size or number in rich patches through root and/or shoot **proliferation and branching**.

Models of foraging suggest that plants growing in resource-rich microsites should consolidate their hold over the site, while those in poor sites should place ramets elsewhere (Sutherland & Stillman 1988; Oborny 1994; Cain *et al.* 1996). In experiments with *F. chiloensis*, Alpert (1991) found that nitrogen allocation appeared to follow the predicted pattern. The nitrogen a ramet received from its own roots promoted its own growth and the production of new stolons, but when nitrogen was acquired from another ramet, this went only into stolon production, not growth (Alpert 1991). Thus, ramets growing in resource-poor patches did not make any further investment in the spot they occupied, but instead allocated resources to stolons that had a chance of rooting in a more nutrient-rich site.

Local proliferation of ramets in response to resource-rich patches of light and nutrients has been observed in most clonal plants investigated (de Kroon & Hutchings 1995). When growing in nutrient-rich conditions, *Glechoma hederacea* branched profusely and produced many ramets on short internodes, but in poor conditions it branched less and produced fewer ramets on longer stolons (Slade & Hutchings 1987b). Rhizomes of couch grass *Elymus repens* may proliferate in response to patches of soil rich in nitrogen (Mortimer 1984) and selectively invade bare ground (Kleijn & van Groenendael 1999). In other grasses, root proliferation in nitrogen-rich patches increases competitive ability for nitrogen (Robinson *et al.* 1999; Hodge *et al.* 1999). A lengthening of spacers occurs in response to low light in many stoloniferous species, but not in rhizomatous ones (in which the spacers are buried) or in most cases (except *Glechoma hederacea*) in response to local nutrients (de Kroon & Hutchings 1995). Models of clonal plant foraging parameterized with experimental data suggest that observed changes in spacer length may play a greater role in locating new resource patches than in concentrating ramets within occupied patches (Cain *et al.* 1996). Because the scale of natural patchiness affects forgaging success, precisely how effective plastic changes in spacer length might be in locating new patches cannot be determined until clonal plants are studied in naturally patchy environments, rather than in the artificial arenas in which nearly all experiments on this topic have been conducted to date.

10.6 Senescence and death

It has been said that only two things are unavoidable in life: death and taxes. This cannot have been written by a botanist since some plants escape both, although cheating the grim reaper is a rarity. The 11 000–year-old clones of the creosote bush *Larrea tridentata* mentioned in Chapter 1 are effectively immortal, and bristlecone pines *Pinus aristata* in Nevada that are over 4500 years old demonstrate that immense longevity is not confined to clonal species. However, these species are unusual and most plants have finite lifespans because plants, like most animals, senesce. **Senescence** (or ageing) is defined as a decline in physiological state with age, and manifests itself in life tables through an increase in mortality rate (q_x) with age. This occurs abruptly in annuals and semelparous perennials, but more gradually in iteroparous species. Silvertown *et al.* (2001) analysed the life tables of 64 iteroparous herbs and woody plants and found senescent increases in q_x in over half the species; senescence could not be ruled out for the remainder because changes in q_x are difficult to detect among the small number of individuals that remain when a cohort reaches an advanced age.

Because at least some plants appear to be capable of near-immortality, why do most senesce and die? The explanation is to be found in Eqn 10.1 (page 271), which describes how the force of selection on survival varies with age *x*. As already discussed in Section 10.1, fitness is much less sensitive to changes in survival late in life than to changes that occur earlier. There are several ways in which this can result in the evolution of senescence (Rose 1991), but the familiar trade-off between early reproduction and later survival is perhaps the most universal. When selection favours early or concentrated reproduction, the trade-off with later survival has limited effects upon fitness. Thus, senescence and death can be the unavoidable consequences of maximizing early fitness, given the survival costs of reproduction. Annual and semelparous perennial life histories are the clearest cases of this and are discussed in detail in Sections 10.6.1 and 10.6.2 below.

But, what of those rare cases of extreme longevity and clonal immortality? Are these plants a challenge to the theory of the evolution of senescence? Extreme longevity in nonclonal species such as the bristlecone pine is probably the result of selection for survival at the expense of reproduction in an environment where opportunities for seedling recruitment are very, very rare. These trees probably age, but do so very, very slowly. Long-lived clones are a different case, and here the fact that the genet splits into ramets, so that vegetative reproduction contributes substantially to genet fitness, may mean that the force of natural selection does not decline with age in a manner that permits senescence to evolve. Clonal reproduction does not automatically exempt genets from the evolution of senescence, but it does create the possibility of escaping senescence, particularly if sexual reproduction is limited (Orive 1995; Pedersen 1995; Gardner & Mangel 1997).

10.6.1 Annuals

Annuals are commonest in arid areas and in habitats disturbed by human activity. They constitute about 25% of species in the flora of the Sonoran desert, USA, but there are practically no annuals in the native flora of New Zealand (Harper 1977). Annuals from Europe have displaced perennial grasslands along the western seaboard of the USA, replacing them with monocultures of cheatgrass *Bromus tectorum* in the north-west (Mack 1981) and species-rich annual grasslands in California. Most annuals have evolved from iteroparous perennial ancestors, although there are a few known cases of the reverse (Barrett *et al.* 1997). A simple model can give us a feel for situations in which an annual genotype can have superior fitness to an iteroparous perennial one (Schaffer & Gadgil 1975). An annual genotype will increase at a rate λ_a, given by the product of survival from seed to flowering adult (c) and mean seed production per individual (m_a):

$$\lambda_a = cm_a. \tag{10.4}$$

The equivalent expression for an iteroparous perennial with prereproductive survival c and adult survival p is:

$$\lambda_p = cm_p + p. \tag{10.5}$$

An annual will then have superior fitness to a perennial when:

$$m_a > m_p + (p/c). \tag{10.6}$$

The advantage of the annual life history depends upon the ratio adult/juvenile survival. If adult and juvenile survival are equivalent, then the value of $p/c = 1$, and this means that an annual can match the fitness of a perennial by producing just one more seed than the perennial does (Cole 1954). This is the key to the success of annuals and explains why they typically allocate more resources to reproduction than do their perennial relatives (Silvertown & Dodd 1997). The model predicts that iteroparous perennials will be favoured in environments where seedling mortality is high, as typically occurs in mesic habitats with a closed cover of vegetation, while annuals are favoured in habitats that have low vegetation cover as a consequence of aridity or disturbance and where seedlings can attain high rates of survival if germination is opportunely timed.

10.6.2 Perennials

Perennial semelparity is a relatively rare kind of life history among plants but it has evolved independently in many different groups, from the humble carrot to the massive Talipot palm *Corypha umbraculifera* which garners its resources for one gargantuan, fatal act of reproduction on a 7-m tall inflorescence laden with seeds the size and shape of large marbles. Why should any plant evolve so bizarre a life history? Theory suggests that the answer lies in how fecundity (m_r) and the costs of reproduction in terms of survival (p_r) and growth (g_r) depend on the proportion of resources devoted to reproduction, that is with **reproductive effort** (RE). Costs of reproduction will ensure that the product of growth and survival $p_r g_r$ will fall as RE rises, while m_r will increase with RE. Fitness (R_0) is maximized at the RE that produces the largest value of the sum $m_r + p_r g_r$ (Schaffer 1974). If the cost and fecundity curves are both convex with respect to the x-axis (like a dish), $m_r + p_r g_r$ is maximized at 0% or 100% RE (Fig. 10.9a,b,c), indicating that semelparity is optimal once reproduction occurs. Curves of fecundity and its costs that are concave with respect to the x-axis (like a hill) indicate that iteroparity is optimal because fitness is maximized at an RE value between zero and 100% (Fig. 10.9d,e,f).

What ecological conditions will result in curves of these shapes? Generally speaking, circumstances which reduce the resource cost per seed (number of seeds

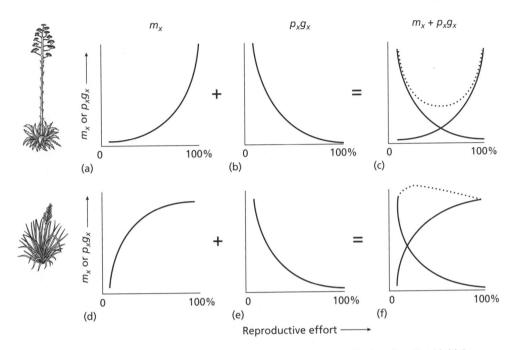

Fig. 10.9 Relationships between fecundity m_x, p_xg_x and reproductive effort that (a)–(c) favour semelparity and (d)–(f) favour iteroparity (see text for more details). A semelparous agave (top) and an iteroparous yucca (bottom) are pictured.

per unit of RE) as the size of a seed crop increases, or which increase the probability that a seed will itself survive to reproduce as the size of the seed or seedling cohort increases, will produce a convex (accelerating) fecundity curve. Seed predators can exercise this kind of effect because, even if a small plant devotes all of its resources to reproduction, it can only produce a relatively small crop of seeds. As we have already seen, small crops generally suffer proportionately more losses to seed predators than large ones, and this creates a situation in which a plant may gain a disproportionate release from predation by delaying reproduction until it is big enough to increase its crop size and swamp its seed predators. This is the explanation that has been suggested for the most spectacular delays of reproductive maturity known, which are found in species of semelparous bamboo, among which a 20-year prereproductive period is common. Before they reproduce, some Indian semelparous bamboos appear to grow at an exponential rate, causing clonal genets to expand at 10% a year until they flower (Gadgil & Prasad 1984). A Japanese species, *Phyllostachys bambusoides,* waits 120 years to flower and die. A few semelparous bamboos also synchronize reproduction within cohorts of the same age (Campbell 1985). This synchrony is probably important because it reduces the risk that predator populations will move from one bamboo population to another as they fruit (Janzen 1976), although Keeley and Bond (1999) question whether predators could select for the very long prereproductive periods observed in many semelparous bamboos and they suggest an alternative hypothesis.

Sib-competition resulting in density-dependent seedling mortality between seeds from the same mother will produce the opposite effect and result in a concave fecundity curve that favours iteroparity. One might expect that the huge crops of seed produced by semelparous perennials would compete with one another as seedlings and that this might alter the shape of the fecundity curve towards iteroparity (Young & Augspurger 1991), but presumably this effect is not as strong as the forces operating in the other direction.

Reproductive efficiency, or economies of scale in seed production, seem to explain semelparity in a number of long-lived semelparous species. The genus *Agave* contains a number of semelparous perennials which grow in the deserts and chaparral of western North America. Some Agave species produce vegetative bulbils and consequently the genet can be considered iteroparous (Arizaga & Ezcurra 1995). On the other hand, individual rosettes of most of these plants are semelparous. A typical species is the century plant (*A. deserti*) which delays reproduction for many years, storing water and carbohydrates in its rosette leaves. When it finally flowers, a rosette of only 60 cm in diameter is capable of producing an inflorescence up to 4 m tall. This feat can only be achieved by a massive translocation of water and assimilates from the rosette into the growing inflorescence, which obtains 70% of its carbon from this source (Tissue & Noble 1990).

Agaves have a similar habitat and morphology to plants in the genus *Yucca,* but yuccas are mostly iteroparous. Rosettes of yuccas and agaves both have stiff xeromorphic leaves and bear a central spike of insect-pollinated flowers. In an attempt to explain the difference in reproductive habit between species in these two genera, Schaffer and Schaffer (1977, 1979) compared the life history of seven *Agave* and five *Yucca* species. One *Yucca (Y. whipplei)* proved to have semelparous rosettes and one of the agaves had iteroparous ones (*A. parviflora*). This demonstrated that differences in reproductive habit between the two genera were not simply a result of different evolutionary descent, but had evolved separately in each genus. Schaffer and Schaffer suggested that semelparity in these plants was favoured by the selective behaviour of pollinating insects, which they showed visited the larger inflorescences disproportionately more often than smaller ones. Larger inflorescences also produced more fruits per flower than smaller ones in the semelparous species of both genera, but this pattern was absent in iteroparous species.

A number of semelparous giant rosette herbs occur in the alpine zone of tropical mountains in South America and Africa. Young (1990) studied populations of a semelparous species *Lobelia telekii* and compared them with populations of an iteroparous species *L. keniensis* on Mount Kenya, in Africa. Both species are pollinated by birds, and the semelparous species produced inflorescences up to two and a half times as long as the iteroparous species. As in the semelparous agaves and yuccas, the relationship between RA and fecundity in semelparous *L. telekii,* but not *L. keniensis,* was convex, and there were more seeds per unit length of inflorescence in large flower spikes than small ones. Strangely enough, seed set in *L. telekii* was apparently *not* limited by pollination, so this was not the cause of the disproportionate seed set on large inflorescences. However, *L. telekii* and *L. keniensis* occupied different habitats, with the semelparous species occurring in drier sites than the iteroparous one. Young concluded that the costs of reproduction were higher in dry conditions, and that this was decisive in favouring semelparity at these sites.

10.7 Life history strategies

In this chapter we have examined plant life history traits one at a time, but as we have progressed through the life cycle from birth to growth and death two recurring themes have become apparent: first, that traits do not evolve independently of one another, and secondly, that age-specific mortality drives the evolution of many of them. In a sense the interaction among traits provides a kind of internal structure to the evolution of life history, impeding certain trait combinations and facilitating others, while mortality is the lever by which the environment tips the scales of natural selection in favour of one trait combination or another. These trait combinations are usually referred to as life history strategies and many different schemes have been devised to classify them and to predict which kinds of strategy should be found in which kinds of habitat (Southwood 1988). A **life history strategy** is a specialized phenotype of correlated traits that has evolved independently in different populations or species exposed to similar selection pressures. Specialization is necessitated by trade-offs, so a good reality-check for any scheme is whether observed life-history trade-offs are necessary and sufficient to predict the existence of the strategies proposed.

One of the first schemes was the *r*- and *K*-strategy theory of MacArthur and Wilson (1967), elaborated by Pianka (1970). This proposed that species vary along a continuum from those with short generation time, high fecundity and good dispersal (*r* strategists) that are adapted to colonize habitats where mortality is density-independent, to species with the opposite traits which are adapted to thrive in habitats where mortality is density-dependent (*K* strategists) (Boyce 1984). The theory is supported by the existence of an expected trade-off between dispersal and competitive ability (Chapter 8), but its fundamental assumption of a trade-off between rapid population growth and resistance to density-dependent mortality does not hold (Stearns 1992).

Grime (1977, 1979) devised a three-strategy scheme intended as a refinement of *r*/*K* theory which classifies plants in the *established phase* of their life history as **Competitors** (similar to *K*-strategists), **Stress-tolerators** that grow slowly but are successful in nutrient-poor habitats, and **Ruderals** (similar to *r*-strategists). The so-called **CSR scheme** has been criticized from a number of points of view (Harper 1982; Grubb 1985, 1998; Loehle 1988), but it is the most developed scheme proposed for plants to date. Independent theoretical support for CSR comes from a spatial model of plant competition for resources in which Bolker and Pacala

(1999) found that there were only three possible resource-aquiring strategies and that these corresponded closely to C, S and R. There is also empirical support for the trade-offs implied by the CSR scheme. Grime *et al.* (1997) used principal components analysis (PCA) to ordinate 53 morphological, chemical and life history traits for mature plants of 45 herb species. The first PCA axis accounted for 22% of the variation among the species for these traits, with traits associated with stress-tolerance at one extreme and those associated with competitive ability in nutrient-rich environments at the other. The second axis of variation was associated with trait differences between monocotyledons and dicotyledons and the third axis, accounting for only 8% of total variation, separated species by life history, with ruderals at one extreme and competitors at the other. These results will have been influenced by the choice of traits and species, which is perhaps why life history and the C–R axis appeared subordinate to the C–S trade-off, and to phylogenetically correlated differences between monocotyledons and dicotyledons.

Trade-offs between rapid growth and lifespan are strong among trees, and support a third strategy scheme: the **fast–slow continuum** hypothesis. This has been successfully applied to life history variation in mammals and other vertebrates in which age at maturity, life span and body size are strongly correlated across species, and life history variation falls along a single axis from small animals that breed fast and die young to large ones that breed slowly and live long (Charnov & Berrigan 1990; Promislow & Harvey 1990). Life history variation among trees in the 50 ha plot at BCI (Chapter 8) follows a similar pattern falling along an axis from trees with shade-intolerant, fast-growing saplings that suffer high mortality at one extreme, to shade-tolerant species with slow-growing saplings with low mortality at the other (Condit *et al.* 1996). Among tropical tree species in Costa Rica, Enquist *et al.* (1999) found a very general, inverse relationship between lifespan and wood density, on the one hand, and growth rate, on the other, that also supports the fast–slow continuum hypothesis. Similar correlations supporting the fast–slow pattern occur in North American temperate trees, particularly among the broadleaved species (angiosperms) (Loehle 1988). Whether anything is to be gained by trying to fit more complex strategy schemes to the life history variation observed in plants remains to be seen. What does seem certain is that plant ecologists will continue to search for them (Westoby 1998; Weiher *et al.* 1999).

In the last two chapters of this book we hope to have demonstrated how the topic of life history evolution brings together the different strands of plant population biology. Textbook authors tend to dislike loose ends, but we freely admit that there are many. Find them and tie them!

10.8 Summary

The evolution of life history can be analysed in terms of how changes in age-specific demographic parameters affect R_0, used as a measure of fitness. The force of selection acting on survival diminishes with age, and thus changes in l_x early in life affect fitness more than changes in old age. This permits the **evolution of senescence**. Correspondingly, the force of selection acting on fecundity is proportional to l_x, favouring early reproduction (other things being equal). **Delayed reproduction** will be favoured if the risks of delay are outweighed by, for example, greater growth and larger size leading to more offspring. Intermittent reproduction, such as is commonly observed in forest trees, will be favoured if concentrating the resources accumulated between bouts of reproduction into a few very large **mast** crops disproportionately increases the number of seeds produced and/or the proportion that escape predation.

Seed traits such as size, dormancy and dispersal, are largely maternally determined and thus selected through their effect on the fitness of the mother. A **size–number trade-off** governs **seed size**, which the **Smith–Fretwell model** predicts should result in a single optimum size of seed for a plant. The **Geritz model** envisages seeds competing for safe sites and predicts some variation in seed size within a plant's crop. In reality, seed size does vary within plants, but the greatest variation is found between species. Interspecific seed size variation is correlated with phylogeny, plant size and habitat. **Seed dormancy** and **seed dispersal** in space may both be considered to be mechanisms of dispersal in the broad sense, causing offspring to experience a different envionment from that of their mother. Four ways in which broad-sense dispersal may increase maternal fitness are: **risk reduction** in a variable environment, **escape** from intraspecific competition or from **sib-competition**, and **directed dispersal** to predictably good sites.

Clonal growth is widespread but not universal in plants. Its disadvantages are apparently few, while its advantages include **division of labour** among **physiologically integrated** ramets, improved resource acquistion through **foraging** and better exploitation of patchily distributed resources. **Senescence** is expected to evolve when the force of natural selection declines with age. Clonal growth may prevent senescence evolving in some, but not all, cases. **Semelparity** involves rapid senescence after reproduction and is predicted to increase fitness when reproductive success scales disproportionately with **reproductive effort**. Life history traits are correlated with one another, leading to the idea that there exist distinct **life history strategies**, or specialized phenotypes that have evolved independently across many genotypes exposed to similar selection pressures. Two such strategy schemes are **CSR** and the **fast–slow continuum** hypothesis.

10.9 Further reading

Baskin, C. C. & Baskin, J. M. (1998) *Seeds*. Academic Press, San Diego, CA.

de Kroon, H. & van Groenendael, J. (Eds) (1997) *The Ecology and Evolution of Clonal Plants*. Backhuys Press, Leiden.

Silvertown, J., Franco, M. & Harper, J. L. (Eds) (1997) *Plant Life Histories: ecology, phylogeny and evolution*. Cambridge University Press, Cambridge.

Stearns, S. C. (1992) *The Evolution of Life Histories*. Oxford University Press, Oxford.

10.10 Questions

1 Herrera *et al.* (1998) have argued that, because annual seed crop variability ranges from slight to extreme across woody species as a whole, and plants do not fall into distinct groups of 'masting' and 'nonmasting' species, the term itself should be abandoned. Consider whether you would expect masting to be a bimodally distributed trait and whether you agree or disagree with the suggestion made by Herrera *et al.* (1998).

2 What plausible ecological factors would alter the shape of the functions relating seed size to survival and germination shown in Fig. 10.5? Would you expect such changes to affect the predictions of the Smith–Fretwell and Geritz ESS models and if so, how?

3 Consider the similarities and differences between spatial dispersal and seed dormancy from an evolutionary viewpoint. What are the reasons for and against regarding both as forms of broad-sense dispersal?

4 Compare the mean fitness of annuals when calculated as the arithmetic mean and the geometric mean of 10 year's annual seed production. Use data from the literature or random numbers to provide data.

5 Estimate some values for seedling and adult survival from one or more published demographic studies (see Chapter 5 for some references) and use them in the Eqn 10.6 to calculate the fecundity an annual genotype would require to match the fitness of a perennial. Consider the kinds of assumptions you have had to make in making this calculation. Are they justified, and how might they be tested?

Appendix: symbols and terms used in this book

In choosing symbols we have tried to be consistent and to follow convention. Inevitably, some symbols have more than one meaning; however, it should be obvious from the context which the correct one is in any particular case.

Symbol	Definition
A	Genetic neighbourhood area
a	The area necessary to achieve w_m
α, β	Competition coefficients
b	(1) Exponent of the yield/density equation (Eqn 4.3), and the competition–density effect (Eqn 4.4). When the law of constant final yield applies, $b = 1$. (2) The exponent whose value determines the shape of the recruitment curve in a model of density-dependent population dynamics (Eqn 5.11). (3) The birth rate.
c	(1) Intercept on the vertical axis of the self-thinning rule (Eqn 4.5) relating w and density. (2) The rate at which vacant sites in a metapopulation are colonized.
d	The death rate.
d_x	Mortality between ages x and $x + 1$ in a life table (e.g. Table 1.1), i.e. $l_x - l_{x+1}$.
δ	A measure of inbreeding depression. $d = 1 - W_i$, where W_i is the relative fitness of inbred progeny.
F_{ST}	$1/(1 + 4 N_e m)$
G	Fraction of a seed population that germinates
G_{ST}	Genetic differentiation among populations $= (H_T - H_S)/H_T$
H	An index of genetic diversity, equivalent to the probability that two randomly chosen alleles will be different.
h^2	Heritability $= V_A/V_P$
H_B	Heritability in the broad sense $= V_g/V_p$
H_N	Heritability in the narrow sense $= V_a/V_p$
H_S	Genetic diversity within a population
H_T	Total genetic diversity at the species level

k	Exponent in the self-thinning rule (Eqn. 4.5) relating yield and density.
K_i	The maximum sustainable density of a species in a mixture. Subscripts denote different species. $K = m^{-1}$.
λ	The annual (or finite) rate of population increase. $\ln \lambda = r$
l_x	Proportion of individuals in a cohort surviving to age x.
m	(1) Migration rate (Eqn 3.9). (2) Reciprocal of the maximum value that density can achieve $(m = K^{-1})$, limited by self-thinning (Eqn 5.12).
m_x	Fecundity of an individual of age x.
N	The number (as distinct from the density) of individuals in a population.
N_e	Effective population size.
N, N_t, N_i, N_x	Population density. Subscripts denote time (t), different species (i), or age (x)
p	(1) An allele frequency. (2) The proportion of sites in a metapopulation that are occupied.
p_x	Survival rate per individual alive at the start of an age interval in a life table (e.g. Table 1.1), i.e. $1 - (d_x/l_x)$
P_t	The number of occupied sites in a metapopulation at time t.
q	An allele frequency.
q_x	Mortality rate per individual alive at the start of an age interval in a life table (e.g. Table 1.1), i.e. d_x/l_x
r	Intrinsic rate of increase
R	Response to selection $= Sh^2$ (see p. 29).
R_0	Net reproductive rate, i.e. $N_{t+\tau}/N_t$.
S	(1) Selfing rate. (2) Select differential.
s	(1) Selection coefficient. (2) Mean seed production per plant in a population.
s_m	Maximum possible seed production by a plant in uncrowded conditions.
T	The total number of sites in a metapopulation $(V_t + P_t)$
t	(1) A subscript denoting the time associated with a parameter, e.g. N_t is the population size at time t. (2) t outcrossing rate.
τ	The length of a generation in years.
V^*	Optimal dispersal fraction (see p. 288).
V_A	Additive genetic variance
V_E	Environmental variance
V_G	Genetic variance
V_P	Phenotypic variance
V_t	The number of vacant sites in a metapopulation at time t.
W	Darwinian fitness (its maximum value is 1).
w	Mean plant weight.

w_m	Maximum possible weight of a plant growing in uncrowded conditions.
x	(1) The rate at which the occupied sites in a metapopulation become extinct. (2) A subscript denoting age for an age-specific variable.
Y	Total plant yield of a population or plot.

References

Aarssen, L. W. & Epp, G. A. (1990) Neighbor manipulations in natural vegetation: a review. *Journal of Vegetation Science*, 1, 13–30.

Aarssen, L. W. & Turkington, R. (1985) Biotic specialization between neighbouring genotypes in Lolium perenne and Trifolium repens from a permanent pasture. *Journal of Ecology*, 73, 605–614.

Abbott, R. J. & Gomes, M. F. (1988) Population genetic structure and outcrossing rate of *Arabidopsis thaliana* (L.) Heynh. *Heredity*, 62, 411–418.

Abrams, P. (1983) The theory of limiting similarity. *Annual Review of Ecology and Systematics*, 14, 359–376.

Ågren, J. (1996) Population size, pollinator limitation, and seed set in the self-incompatible herb *Lythrum salicaria*. *Ecology*, 77, 1779–1790.

Ågren, J. & Zackrisson, O. (1990) Age and size structure of *Pinus Sylvestris* populations on mires in central and northern Sweden. *Journal of Ecology*, 78, 1049–1062.

Ahokas, H. (1996) Unfecund, gigantic mutant of oats (*Avena sativa*) shows fecundity overdominance and difference in DNA methylation properties. *Euphytica*, 92, 21–26.

Allen, E. B. & Allen, M. F. (1990) The mediation of competition by mycorrhizae in successional and patchy environments. *Perspectives on Plant Competition* (eds J. B. Grace & D. Tilman), pp. 367–389. Academic Press, San Diego, CA.

Allsopp, N. & Stock, W. D. (1995) Relationships between seed reserves, seedling growth and mycorrhizal responses in 14 related shrubs (Rosidae) from a low-nutrient environment. *Functional Ecology*, 9, 248–254.

Alpert, P. (1991) Nitrogen sharing among ramets increases clonal growth in *Fragaria chiloensis*. *Ecology*, 72, 69–80.

Alpert, P. (1999) Clonal integration in *Fragaria chiloensis* differs between populations: ramets from grassland are selfish. *Oecologia*, 120, 69–76.

Alpert, P. & Mooney, H. A. (1986) Resource sharing among ramets in the clonal herb, *Fragaria chiloensis*. *Oecologia*, 70, 227–233.

Alvarez-Buylla, E. R., Chaos, A., Pinero, D. & Garay, A. A. (1996) Demographic genetics of a pioneer tropical tree species: Patch dynamics, seed dispersal, and seed banks. *Evolution*, 50, 1155–1166.

Alvarez-Buylla, E. R. & Garay, A. A. (1994) Population genetic structure of *Cecropia obtusifolia*, a tropical pioneer tree species. *Evolution*, 48, 437–453.

Alvarez-Buylla, E. R. & García-Barrios, R. (1991) Seed and forest dynamics: a theoretical framework and an example from the neotropics. *American Naturalist*, 137, 133–154.

Alvarez-Buylla, E. R. & Martinez-Ramos, M. (1990) Seed bank versus seed rain in the regeneration of a tropical pioneer tree. *Oecologia*, 84, 314–325.

Anderson, G. J. & Symon, D. E. (1989) Functional dioecy and andromonoecy in *Solanum*. *Evolution*, 43, 204–219.

Antlfinger, A. (1981) The genetic basis of microdifferentiation in natural and experimental populations of *Borrichia frutescens* in relation to salinity. *Evolution*, 35, 1056–1068.

Antlfinger, A. E., Curtis, W. F. & Solbrig, O. T. (1985) Environmental and genetic determinants of plant size in *Viola sororia*. *Evolution*, 39, 1053–1064.

Antonovics, J. (1968) Evolution in closely adjacent plant populations. V. Evolution of self-fertility. *Heredity*, 23, 219–238.

Antonovics, J. & Levin, D. A. (1980) The ecological and genetic consequences of density-dependent regulation in plants. *Annual Review of Ecology and Systematics*, 11, 411–452.

Antonovics, J. & Primack, R. B. (1982) Experimental ecological genetics in *Plantago* VI. The demography of seedling transplants of *P. Lanceolata*. *Journal of Ecology*, 70, 55–75.

Antonovics, J. & van Tienderen, P. H. (1991) Ontoecogenophyloconstraints? The chaos of constraint terminology. *Trends in Ecology and Evolution*, 6, 166–167.

The Arabidopsis Initiative (2000) Analysis of the genome sequence of the flowering plant *Arabidopsis thaliana*. *Nature*, 408, 796–815.

Argyres, A. Z. & Schmitt, J. (1991) Microgeographic genetic structure of morphological and life history traits in a natural population of *Impatiens capensis*. *Evolution*, 45, 178–189.

Arizaga, S. & Ezcurra, E. (1995) Insurance against reproductive failure in a semelparous plant: Bulbil formation in *Agave macroacantha* flowering stalks. *Oecologia*, 101, 329–334.

Armstrong, D. P. & Westoby, M. (1993) Seedlings from large seeds tolerate defoliation better: a test using phylogenetically independent contrasts. *Ecology*, 74, 1092–1100.

Arriaga, L., Maya, Y., Diaz, S. & Cancino, J. (1993) Association between cacti and nurse perennials in a heterogeneous tropical dry forest in northwestern Mexico. *Journal of Vegetation Science*, 4, 349–356.

Arthur, A. E., Gale, J. S. & Lawrence, K. J. (1973) Variation in wild populations of Papaver dubium: VII. Germination time. *Heredity*, 30, 189–197.

Ashman, T.-L. (1992) The relative importance of inbreeding and maternal sex in determining fitness in Sidalcea oregana ssp. spicata, a gynodioecious plant. *Evolution*, 46, 1862–1874.

Ashman, T. L. (1994) Reproductive allocation in hermaphrodite and female plants of Sidalcea oregana ssp spicata (Malvaceae) using four currencies. *American Journal of Botany*, 81, 433–438.

Ashton, P. S., Givnish, T. J. & Appanah, S. (1988) Staggered flowering in the Dipterocarpaceae: New insights into floral induction and the evolution of mast fruiting in the aseasonal tropics. *American Naturalist*, 132, 44–66.

Atkinson, W. D. & Shorrocks, B. (1981) Competition on a divided and ephemeral resource: a simulation model. *Journal of Animal Ecology*, 50, 461–471.

Atsatt, P. R. & O'Dowd, D. J. (1976) Plant defense guilds. *Science*, 193, 24–29.

Augspurger, C. K. (1983) Seed dispersal of the tropical tree Platypodium elegans, and the escape of its seedlings from fungal pathogens. *Journal of Ecology*, 71, 759–771.

Augspurger, C. K. (1984) Seedling survival of tropical trees: interactions of dispersal distance, light gaps, and pathogens. *Ecology*, 65, 1705–1712.

Auld, T. D. (1986a) Population dynamics of the shrub Acacia suaveolens (Sm.) Willd. Dispersal and the dynamics of the soil seed-bank. *Australian Journal of Ecology*, 11, 235–254.

Auld, T. D. (1986b) Population dynamics of the shrub Acacia suaveolens (Sm.) Willd. Fire and the transition to seedlings. *Australian Journal of Ecology*, 11, 373–385.

Auld, T. D. (1987) Population dynamics of the shrub Acacia suaveolens (Sm.) Willd: survivorship throughout the life cycle, a synthesis. *Australian Journal of Ecology*, 12, 139–151.

Auld, T. D. & Myerscough, P. J. (1986) Population dynamics of the shrub Acacia suaveolens (Sm.) Willd. Seed production and predispersal seed predation. *Australian Journal of Ecology*, 11, 219–234.

Austerlitz, F., Mariette, S., Machon, N., Gouyon, P.-H. & Godelle, B. (2000) Effects of colonization processes on genetic diversity: differences between annual plants and tree species. *Genetics*, 154, 1309–1321.

Austin, M. P. (1985) Continuum concept, ordination methods and niche theory. *Annual Review of Ecology and Systematics*, 16, 39–61.

Avise, J. C. (2000) *Phylogeography*. Harvard University Press, Cambridge, MA.

Ayliffe, M. A., Scott, N. S. & Timmis, J. N. (1997) Analysis of plastid DNA-like sequences within the nuclear genomes of higher plants. *Molecular Biology and Evolution*, 15, 738–745.

Baker, H. G. (1955) Self-compatibility and establishment after 'long-distance' dispersal. *Evolution*, 9, 347–348.

Baker, H. G. (1959) Reproductive methods as a factor in speciation in flowering plants. *Science*, 24, 9–24.

Baker, H. G. (1963) Evolutionary mechanisms in pollination biology. *Science*, 139, 877–883.

Baker, H. G. (1972) Seed weight in relation to environmental conditions in California. *Ecology*, 53, 997–1010.

Baker, H. G. (1989) Some aspects of the natural history of seed banks. *Ecology of Soil Seed Banks* (eds M. A. Leck, V. T. Parker & R. L. Simpson), pp. 9–21. Academic Press, San Diego, CA.

Baker, G. A. & O'Dowd, D. J. (1982) Effects of parent density on the production of achene type in the annual Hypochoeris glabra. *Journal of Ecology*, 70, 201–215.

Ballaré, C. L., Scopel, A. L., Roush, M. L. & Radosevich, S. R. (1995) How plants find light in patchy canopies – a comparison between wild-type and phytochrome-b-deficient mutant plants of cucumber. *Functional Ecology*, 9, 859–868.

Ballaré, C. L., Scopel, A. L. & Sanchez, R. A. (1990) Far-red radiation reflected from adjacent leaves: An early signal of competition in plant canopies. *Science*, 247, 329–332.

Barnes, P. W. & Archer, S. (1999) Tree–shrub interactions in a subtropical savanna parkland: competition or facilitation? *Journal of Vegetation Science*, 10, 525–536.

Barone, J. A. (1998) Host-specificity of folivorous insects in a moist tropical forest. *Journal of Animal Ecology*, 67, 400–409.

Barrett, S. C. H. (1992a) Gender variation in Wurmbea dioica (Liliaceae) and the evolution of dioecy. *Journal of Evolutionary Biology*, 5, 423–444.

Barrett, S. C. H. (1992b) Heterostylous genetic polymorphisms: model systems for evolutionary analysis. *Evolution and Function of Heterostyly* (ed. S. C. H. Barrett), pp. 1–29. Springer, Heidelberg.

Barrett, S. C. H., Harder, L. D. & Worley, A. C. (1996) Comparative biology of plant reproductive traits. *Philosophical Transactions of the Royal Society, London Series B*, 351, 1272–1280.

Barrett, S. C. H., Harder, L. D. & Worley, A. C. (1997) The comparative biology of pollination and mating in flowering plants. *Plant Life Histories: Ecology, Phylogeny and Evolution* (eds J. Silvertown, M. Franco & J. L. Harper), pp. 57–76. Cambridge University Press, Cambridge.

Barrett, S. C. H. & Kohn, J. R. (1991) Genetic and evolutionary consequences of small population size in plants: implications for conservation. *Genetics and Conservation of Rare Plants* (eds D. A. Falk & K. E. Holsinger), pp. 3–30. Oxford University Press, New York.

Barrett, S. C. H., Morgan, M. T. & Husband, B. C. (1989) The dissolution of a complex genetic polymorphism: the evolution of self-fertilization in trisylous Eichhornia paniculata (Pontederiaceae). *Evolution*, 43, 1398–1416.

Barrett, S. C. H. & Shore, J. (1987) Variation and evolution of breeding systems in the Turnera ulmifolia L. complex (Turneraceae). *Evolution*, 41, 340–354.

Bartholomew, B. (1970) Bare zone between California shrub and grassland communities: the role of animals. *Science*, 170, 1210–1212.

Barton, N. H. & Charlesworth, B. (1998) Why sex and recombination? *Science*, 281, 1986–1990.

REFERENCES

Baskin, J. M. & Baskin, C. C. (1980) Ecophysiology of secondary dormancy in seeds of *Ambrosia artemisifolia*. *Ecology*, 61, 475–480.

Baskin, J. M. & Baskin, C. C. (1983) Seasonal changes in the germination response of buried seeds of *Arabidopsis thaliana* and ecological interpretation. *Botanical Gazette*, 144, 540–543.

Baskin, C. C. & Baskin, J. M. (1998) *Seeds*. Academic Press, San Diego, CA.

Baskin, J. M. & Baskin, C. C. (1972) Influence of germination date on survival and seed production in a natural population of Leavenworthia stylosa. *American Midland Naturalist*, 88, 318–323

Baskin, J. M. & Baskin, C. C. (1985) The annual dormancy cycle in buried weeds seeds: a continuum. *Bioscience*, 35, 492–498.

Baskin, J. M. & Baskin, C. C. (1987) Environmentally induced changes in the dormancy states of buried weed seeds. *British Crop Protection Conference*, 7C-2, 695–706.

Bawa, K. S. (1980) Evolution of dioecy in flowering plants. *Annual Review of Ecology and Systematics*, 11, 15–39.

Bazzaz, F. A., Levin, D. A. & Schmierbach, M. R. (1982) Differential survival of genetic variants in crowded populations of *Phlox*. *Journal of Applied Ecology*, 19, 891–900.

Beatley, J. C. (1970) Perennation in *Astragalus lentiginosus* and *Tridens pulchellus* in relation to rainfall. *Madroño*, 20, 326–332.

Beattie, A. J. (1985) *The Evolutionary Ecology of Ant-Plant Mutualisms*. Cambridge University Press, Cambridge.

Beattie, A. J. & Culver, D. C. (1979) Neighbourhood size in *Viola*. *Evolution*, 33, 1226–1229.

Becker, P. & Wong, M. (1985) Seed dispersal, seed predation, and juvenile mortality of *Aglaia* sp. (Meliaceae) in lowland dipterocarp rainforest. *Biotropica*, 17, 230–237.

Bekker, R. M., Bakker, J. P., Grandin, U. *et al.* (1998) Seed size, shape and vertical distribution in the soil: indicators of seed longevity. *Functional Ecology*, 12, 834–842.

Belhassen, E., B. Dommée, A., Atlan, P.-H. *et al.* (1991) Complex determination of male sterility in *Thymus vulgaris*: genetic and molecular analysis. *Theoretical and Applied Genetics*, 82, 137–143.

Bell, G. (1980) The costs of reproduction and their consequences. *American Naturalist*, 116, 45–76.

Bell, G. (1996) *Selection. The Mechanism of Evolution*. Chapman & Hall, New York.

Benjamin, L. R. (1984) Role of foliage habit in the competition between differently sized plants in carrot crops. *Annals of Botany*, 53, 549–557.

Bennett, K. D. (1983) Postglacial population expansion of forest trees in Norfolk, U.K. *Nature*, 303, 164–167.

Bennington, C. C., McGraw, M. C. & Vavrek, M. C. (1991) Ecological genetic variation in seed banks. II. Phenotypic and genetic differences between young and old subpopulations of *Luzula parviflora*. *Journal of Ecology*, 79, 627–643.

Bennington, C. C. & Stratton, D. A. (1998) Field tests of density- and frequency-dependent selection in *Erigeron annuus* (Compositae). *American Journal of Botany*, 85, 540–545.

Benton, T. G. & Grant, A. (1996) How to keep fit in the real world: Elasticity analyses and selection pressures on life histories in a variable environment. *American Naturalist*, 147, 115–139.

Berendse, F. (1981) Competition between plant populations with different rooting depths II. Pot experiments. *Oecologia*, 48, 334–341.

Berendse, F. (1982) Competition between plant populations with different rooting depths III. Field experiments. *Oecologia*, 53, 50–55.

Bergelson, J. (1990) Life after death: site pre-emption by the remains of *Poa annua*. *Ecology*, 71, 2157–2165.

Bergelson, J. (1993) Details of local dispersion improve the fit of neighborhood competition models. *Oecologia*, 95, 299–302.

Bergelson, J. M. & Crawley, M. J. (1988) Mycorrhizal infection and plant species diversity. *Nature*, 334, 202.

Bergelson, J. & Purrington, C. B. (1996) Surveying patterns in the cost of resistance in plants. *American Naturalist*, 148, 536–558.

Berrigan, D. & Koella, J. C. (1994) The Evolution of Reaction Norms – Simple Models for Age and Size at Maturity. *Journal of Evolutionary Biology*, 7, 549–566.

Bever, J. D., Westover, K. M. & Antonovics, J. (1997) Incorporating the soil community into plant population dynamics: the utility of the feedback approach. *Journal of Ecology*, 85, 561–573.

Bhattacharya, M. K., Smith, A. M., Ellis, T. H. N., Hedley, C. & Martin, C. (1990) The wrinkled-seed character of pea described by Mendel is caused by a transposon-like insertion in a gene encoding starch-branching enzyme. *Cell*, 60, 115–122.

Bierzychudek, P. (1987a) Patterns in plant parthenogenesis. *The Evolution of Sex and its Consequences* (ed. S. C. Stearns), pp. 197–217. Birkhaüser, Basel.

Bierzychudek, P. (1987b) Resolving the paradox of sexual reproduction: a review of experimental tests. *The Evolution of Sex and its Consequence* (ed. S. Stearns), pp. 163–174. Birkhaüser, Basel.

Bierzychudek, P. (1999) Looking backwards: Assessing the projections of a transition matrix model. *Ecological Applications*, 9, 1278–1287.

Bigwood, D. W. & Inouye, D. W. (1988) Spatial pattern analysis of seed banks: An improved method and optimized sampling. *Ecology*, 69, 497–507.

Birks, H. J. B. (1989) Holocene Isochrone maps and patterns of tree-spreading in the British Isles. *Journal of Biogeography*, 16, 503–540.

Bishop, G. F., Davy, A. J. & Jeffries, R. L. (1978) Demography of *Hieracium pilosella* in a Breck grassland. *Journal of Ecology*, 66, 615–629.

Black, J. N. (1958) Competition between plants of different initial seed sizes in swards of subterranean clover (*Trifolium subterraneum* L.) with particular reference to leaf area and the microclimate. *Australian Journal of Agricultural Research*, 9, 299–318.

Black, J. N. (1959) Seed size in herbage legumes. *Herbage Abstracts*, 29, 235–241.

Bøcher, T. W. (1949) Racial divergences in *Prunella vulgaris* in relation to habitat and climate. *New Phytologist*, 48, 285–314.

Bøcher, T. W. & Larsen, K. (1958) Geographical distribution of initiation of flowering, growth habit and other factors in *Holcus lanatus*. *Botaniska Notiser*, 3, 289–300.

Bolker, B. M. & Pacala, S. W. (1999) Spatial moment equations for plant competition: Understanding spatial strategies and the advantages of short dispersal. *American Naturalist*, 153, 575–602.

Bonan, G. B. (1988) The size structure of theoretical plant populations: spatial patterns and neighbourhood effects. *Ecology*, 69, 1721–1730.

Bonan, G. B. (1993) Analysis of neighborhood competition among annual plants – implications of a plant growth model. *Ecological Modelling*, 65, 123–136.

Bond, W. J. (1985) Canopy-stored seed reserves (serotiny) in Cape Proteaceae. *South African Journal of Botany*, 51, 181–186.

Bond, W. J., Maze, K. & Desmet, P. (1995) Fire life histories and the seeds of chaos. *Ecoscience*, 2, 252–260.

Borchert, M. I. (1985) Serotiny and cone-habit variation in populations of *Pinus coulteri* (Pinaceae) in the southern coastal ranges of California. *Madroño*, 32, 29–48.

ter Borg, S. J. (1979) Some topics in plant population biology. *The study of vegetation*. (ed M. J. A. Werger), pp. 13–55. Junk, The Hague.

Bossema, I. (1979) Jays & oaks: an eco-ethological study of a symbiosis. *Behaviour*, 70, 1–117.

Boyce, M. S. (1984) Restitution of r-selection and k-selection as a model of density-dependent natural-selection. *Annual Review of Ecology and Systematics*, 15, 427–447.

Boyd, M., Silvertown, J. & Tucker, C. (1990) Population ecology of heterostyle and homostyle *Primula vulgaris*: Growth, survival and reproduction in field populations. *Journal of Ecology*, 78, 799–813.

Braakhekke, W. G. (1980) *On Coexistence: a Causal Approach to Diversity and Stability in Grassland Vegetation*. Centre for Agricultural Publishing and Documentation, Wageningen.

Braakhekke, W. G. & Hooftman, D. A. P. (1999) The resource balance hypothesis of plant species diversity in grassland. *Journal of Vegetation Science*, 10, 187–200.

Bradshaw, A. D. (1965) Evolutionary significance of phenotypic plasticity in plants. *Advances in Genetics*, 13, 115–155.

Bradshaw, A. D. & McNeilly, T. (1981) *Evolution and Pollution*. Edward Arnold, London.

Bradstock, R. A. & Myerscough, P. J. (1981) Fire effects on seed release and the emergence and establishment of seedlings of *Banksia ericifolia* L.f. *Australian Journal of Botany*, 29, 521–532.

Brand, D. G. & Magnusson, S. (1988) Asymmetric, two-sided competition in even-aged monocultures of red pine. *Canadian Journal of Forestry Research*, 18, 901–910.

Bräuer, I., Maibom, W., Matthies, D. & Tscharntke, T. (1999) Populationsgrösse und Aussterberisiko gefährdeter Pflanzenarten in Niedersachsen. *Verhandlungen der Gesellschaft für Ökologie*, 29, 505–510.

Brokaw, N. V. L. (1982) The definition of treefall gap and its effect on measures of forest dynamics. *Biotropica*, 14, 158–160.

Brokaw, N. V. L. (1985) Gap-phase regeneration in a tropical forest. *Ecology*, 66, 682–687.

Brook, B. W., O'Grady, J. J., Chapman, A. P. *et al.* (2000) Predictive accuracy of population viability analysis in conservation biology. *Nature*, 404, 385–387.

Brown, A. H. D. (1979) Enzyme polymorphism in plant populations. *Theoretical Population Biology*, 15, 1–42.

Brown, A. H. D. & Clegg, M. T. (1984) Influence of flower colour polymorphism on genetive transmission in a natural population of the common morning glory *Ipomoea purpurea*. *Evolution*, 38, 796–803.

Brown, J. H., Reichman, O. J. & Davidson, D. W. (1979) Granivory in desert ecosystems. *Annual Review of Ecology and Systematics*, 10, 201–227.

Broyles, S. B. (1998) Post-glacial migration and the loss of allozyme variation in *Asclepias exaltata* (Asclepiadaceae). *American Journal of Botany*, 85, 1091–1097.

Broyles, S. B. & Wyatt, R. (1991) Effective pollen dispersal in a natural population of *Asclepias exaltata*: the influence of pollinator behavior, genetic similarity, and mating success. *American Naturalist*, 138, 1239–1249.

Brunsfeld, S. J., Sullivan, J., Soltis, D. E. & Soltis, P. S. (2001) Comparative phylogeography of Northwestern North America: A synthesis. *Integrating Ecology and Evolution in a Spatial Context* (eds J. Silvertown & J. Antonovics). Blackwell Science, Oxford.

Buchanan, G. A., Crowley, R. H., Street, J. E. & McGuire, J. A. (1980) Competition of sicklepod (*Cassia obtusifolia*) and redroot pigweed (*Amaranthus retroflexus*) with cotton (*Gossypium hirsutum*). *Weed Science*, 28, 258–262.

Bullock, J. M., Hill, B. C., Dale, M. P. & Silvertown, J. (1994) An experimental study of the effects of sheep grazing on vegetation change in a species-poor grassland and the role of seedlings recruitment into gaps. *Journal of Applied Ecology*, 31, 493–507.

Bullock, J. M., Hill, B. C. & Silvertown, J. (1994) Demography of *Cirsium vulgare* in a grazing experiment. *Journal of Ecology*, 82, 101–111.

Bullock, J. M., Hill, B. C., Silvertown, J. & Sutton, M. (1995) Gap colonization as a source of grassland community change – effects of gap size and grazing on the rate and mode of colonization by different species. *Oikos*, 72, 273–282.

Burczyk, J. (1996) Variance effective population size based on multilocus gamete frequencies in coniferous populations: An example of a scots pine clonal seed orchard. *Heredity*, 77, 74–82.

Burd, M. (1994) Bateman's principle and plant reproduction – the role of pollen limitation in fruit and seed set. *Botanical Review*, 60, 83–139.

Burdon, J. J. (1987) *Diseases and Plant Population Biology*. Cambridge University Press, Cambridge.

Burdon, J. J. & Chilvers, J. A. (1975) Epidemiology of damping-off disease (*Pythium irregulare*) in relation to density of *Lepidium sativum* seedlings. *Annals of Applied Biology*, 81, 135–143.

Burdon, J. J. & Chilvers, J. A. (1982) Host density as a factor in plant disease ecology. *Annual Review of Phytopathology*, 20, 143–166.

Burdon, J. J., Groves, R. H., Kaye, P. E. & Speer, S. S. (1984) Competition in mixtures of susceptible and resistant genotypes of *Chondrilla juncea* differentially infected with rust. *Oecologia*, 64, 199–203.

Burdon, J. J. & Jarosz, A. M. (1988) The ecological genetics of plant–pathogen interactions in natural communities. *Philosophical Transactions of the Royal Society of London, B*, 321, 349–363.

Burdon, J. J. & Shattock, R. C. (1980) Disease in plant communities. *Applied Biology*, 5, 145–220.

Burger, J. C. & Louda, S. M. (1994) Indirect versus direct effects of grasses on growth of a cactus (*Opuntia fragilis*): Insect herbivory versus competition. *Oecologia*, 99, 79–87.

Burgman, M. A. & Lamont, B. B. (1992) A stochastic model for the viability of *Banksia cuneata* populations – environmental, demographic and genetic effects. *Journal of Applied Ecology*, 29, 719–727.

Caballero, A. (1994) Developments in the prediction of effective population size. *Heredity*, 73, 657–679.

Cabin, R. J. (1996) Genetic comparisons of seed bank and seedling populations of a perennial desert mustard, *Lesquerella fendleri*. *Evolution*, 50, 1830–1841.

Cabin, R. J., Mitchell, R. J. & Marshall, D. L. (1998) Do surface plant and soil seed bank populations differ genetically? A multipopulation study of the desert mustard *Lesquerella fendleri* (Brassicaceae). *American Journal of Botany*, 85, 1098–1109.

Cain, M. L., Dudle, D. A. & Evans, J. P. (1996) Spatial models of foraging in clonal plant species. *American Journal of Botany*, 83, 76–85.

Caldwell, M. M., Dawson, T. E. & Richards, J. H. (1998) Hydraulic lift: Consequences of water efflux from the roots of plants. *Oecologia*, 113, 151–161.

Caldwell, M. M. & Richards, J. H. (1989) Hydraulic lift – water efflux from upper roots improves effectiveness of water-uptake by deep roots. *Oecologia*, 79, 1–5.

Callaway, R. M. (1995) Positive interactions among plants. *Botanical Review*, 61, 306–349.

Callaway, R. M. & Walker, L. R. (1997) Competition and facilitation: a synthetic approach to interactions in plant communities. *Ecology*, 78, 1958–1965.

Campbell, R. K. (1974) A provenance-transfer model for boreal regions. *Medd. Nor. Inst. Skoforsk*, 31, 542–566.

Campbell, J. J. N. (1985) Bamboo flowering patterns: a global view with special reference to East Asia. *Journal of the American Bamboo Society*, 6, 17–35.

Canham, C. D. (1985) Suppression and release during canopy recruitment in *Acer saccharum*. *Bulletin of Torrey Botanical Club*, 112, 134–145.

Cannell, M. G. R., Rothery, P. & Ford, E. D. (1984) Competition within stands of *Picea sitchensis* and *Pinus contorta*. *Annals of Botany*, 53, 349–362.

Caraco, T. & Kelly, C. K. (1991) On the adaptive value of physiological integration in clonal plants. *Ecology*, 72, 81–93.

Carlsson, B. A., Jónsdóttir, B. M., Svensson, B. M. & Callaghan, T. V. (1990) Aspects of clonality in the arctic: a comparison between *Lycopodium antoninum* and *Carex bigelowii*. *Clonal Growth in Plants* (eds J. van Groenendael & H. de Kroon), pp. 131–151. SPB Academic Publishing, The Hague.

Carter, R. N. & Prince, S. D. (1981) Epidemic models used to explain biogeographical distribution limits. *Nature*, 293, 644–645.

Carter, R. N. & Prince, S. D. (1988) Distribution limits from a demographic viewpoint. *Plant Population Ecology* (eds A. J. Davy, M. J. Hutchings & A. R. Watkinson), pp. 165–184. Blackwell Scientific Publications, Oxford.

Casper, B. B. (1990) Seedling establishment from one- and two-seeded fruits of *Cryptantha flava*: a test of parent-offspring conflict. *American Naturalist*, 136, 167–177.

Casper, B. B. & Charnov, E. L. (1982) Sex allocation in heterostylous plants. *Journal of Theoretical Biology*, 96, 143–149.

Casper, B. B., Heard, S. B. & Apanius, V. (1992) Ecological correlates of single-seededness in a woody tropical flora. *Oecologia*, 90, 212–217.

Casper, B. B. & Jackson, R. B. (1997) Plant competition underground. *Annual Review of Ecology and Systematics*, 28, 545–570.

Caswell, H. (1989) *Matrix population models: Construction, analysis, and interpretation*. First Edition. Sinauer Associates, Sunderland, MA.

Caswell, H. (2000a) *Matrix Population Models: Construction, Analysis, and Interpretation*. Sinauer Associates, Sunderland, MA.

Caswell, H. (2000b) Prospective and retrospective perturbation analyses: Their roles in conservation biology. *Ecology*, 81, 619–627.

Cavers, P. B. & Benoit, D. L. (1989) *Management and Soil Seed Banks*. In: *Ecology of Soil Seed Banks* (ed M. A. Leck). Academic Press, San Diego, CA.

Cavers, P. B. & Steele, M. G. (1984) Patterns of change in seed weight over time on individual plants. *American Naturalist*, 124, 324–335.

Century, K. S., Holub, E. B. & Staskawicz, B. J. (1995) NDR1, a locus of *Arabidopsis thaliana* that is required for disease resistance to both a bacterial and a fungal pathogen. *Proceedings of the National Academy of Science of the USA*, 92, 6597–6601.

Chaboudez, P. & Burdon, J. (1995) Frequency-dependent selection in a wild plant-pathogen system. *Oecologia*, 102, 490–493.

Channell, R. & Lomolino, M. V. (2000) Dynamic biogeography and conservation of endangered species. *Nature*, 403, 84–86.

Chanway, C. P., Holl, F. B. & Turkington, R. (1989) Effect of *Rhizobium leguminosarum* biovar *trifolii* genotype on specificity between *Trifolium repens* and *Lolium perenne*. *Journal of Ecology*, 77, 1150–1160.

Charlesworth, B. (1980a) The cost of sex in relation to mating system. *Journal of Theoretical Biology*, 84, 655–671.

Charlesworth, B. (1980b) *Evolution in Age-Structured Populations*. Cambridge University Press, Cambridge.

Charlesworth, D. (1984) Androdioecy and the evolution of dioecy. *Biological Journal of the Linnean Society*, 23, 333–348.

Charlesworth, D. (1985) Distribution of dioecy and self-incompatibility in angiosperms. *Evolution – Essays in Honour of John Maynard Smith* (eds P. J. Greenwood & M. Slatkin), pp. 237–268. Cambridge University Press, Cambridge.

Charlesworth, D. (1988) Evolution of homomorphic sporophytic self-incompatibility. *Heredity*, 60, 445–453.

Charlesworth, B. (1991) The evolution of sex chromosomes. *Science*, 251, 1030–1033.

Charlesworth, B. (1992) Evolutionary rates in partially self-fertilizing species. *American Naturalist*, 140, 126–148.

Charlesworth, B. (1994) *Evolution in Age-Structured Populations*, 2nd edn. Cambridge University Press, Cambridge.

Charlesworth, B. & Charlesworth, D. (1978) A model for the evolution of dioecy and gynodioecy. *American Naturalist*, 112, 975–997.

Charlesworth, D. & Charlesworth, B. (1978) Population genetics of partial male-sterility and the evolution of monoecy and dioecy. *Heredity*, 41, 137–153.

Charlesworth, D. & Charlesworth, B. (1979) The evolution and breakdown of S-allele systems. *Heredity*, 43, 41–55.

Charlesworth, D. & Charlesworth, B. (1981) Allocation of resources to male and female functions in hermaphrodites. *Biological Journal of the Linnean Society*, 15, 57–74.

Charlesworth, D. & Charlesworth, B. (1990) Inbreeding depression with heterozygote advantage and its effect on selection for modifiers changing the outcrossing rate. *Evolution*, 44, 870–888.

Charlesworth, D. & Charlesworth, B. (1995) Quantitative genetics in plants: The effect of the breeding system on genetic variability. *Evolution*, 49, 911–920.

Charlesworth, B., Charlesworth, D. & Morgan, M. T. (1990) Genetic loads and estimates of mutation rates in hghly inbred plant populations. *Nature*, 347, 380–382.

Charlesworth, D. & Laporte, V. (1998) The male-sterility polymorphism of *Silene vulgaris*. Analysis of genetic data from two populations, and comparison with *Thymus vulgaris*. *Genetics*, 150, 1267–1282.

Charlesworth, D. & Morgan, M. T. (1991) Allocation of resources to sex functions in flowering plants. *Philosophical Transactions of the Royal Society of London, B*, 332, 91–102.

Charlesworth, B., Nordborg, M. & Charlesworth, D. (1997) The effects of local selection, balanced polymorphism and background selection on equilibrium patterns of genetic diversity in subdivided populations. *Genetical Research*, 70, 155–174.

Charnov, E. L. (1982) *The Theory of Sex Allocation*. Princeton University Press, Princeton, NJ.

Charnov, E. L. & Berrigan, D. (1990) Dimensionless numbers and life history evolution: Age of maturity versus the adult lifespan. *Evolutionary Ecology*, 4, 273–275.

Charnov, E. L., Smith, J. M. & Bull, J. J. (1976) Why be an hermaphrodite? *Nature, London*, 263, 125–126.

Cheliak, W. M., Dancik, B. P., Morgan, K., Yeh, F. C. H. & Strobeck, C. (1985) Temporal variation of the mating system in a natural population of jack pine. *Genetics*, 109, 569–584.

Cheplick, G. P. (1983) Differences between plants arising from aerial and subterranean seeds in the amphicarpic annual *Cardamine chenopodifolia* (Cruciferae). *Bulletin of Torrey Botanical Club*, 110, 442–448.

Cheplick, G. P. (1987) The ecology of amphicarpic flowers. *Trends in Ecology and Evolution*, 2, 97–101.

Cheplick, G. P. (1992) Sibling competition in plants. *Journal of Ecology*, 80, 567–575.

Cheplick, G. P. (1996) Do seed germination patterns in cleistogamous annual grasses reduce the risk of sibling competition? *Journal of Ecology*, 84, 247–255.

Cheplick, G. P. & Quinn, J. A. (1982) Amphicarpum purshii and the 'pessimistic strategy' in amphicarpic annuals with subterranean fruit. *Oecologia*, 52, 327–332.

Christy, E. J. & Mack, R. N. (1984) Variation in demography of juvenile *Tsuga heterophylla* across the substratum mosaic. *Journal of Ecology*, 72, 75–91.

Cipollini, M. L., Wallace Senft, D. A. & Whigham, D. F. (1994) A model of patch dynamics, seed dispersal, and sex ratio in the dioecious shrub *Lindera benzoin* (Lauraceae). *Journal of Ecology*, 82, 621–633.

Cipollini, M. L., Whigham, D. F. & O'Neill, J. (1993) Population growth, structure and seed dispersal in the understorey herb *Cynoglossum virginianum* L. *Plant Species Biology*, 8, 1117–1129.

Clapham, D. H., Dormling, I., Ekberg, I., Eriksson, G., Qamaruddin, M. & VincePrue, D. (1998) Latitudinal cline of requirement for far-red light for the photoperiodic control of budset and extension growth in *Picea abies* (Norway spruce). *Physiologia Plantarum*, 102, 71–78.

Clark, J. S. (1998) Why trees migrate so fast: Confronting theory with dispersal biology and the paleorecord. *American Naturalist*, 152, 204–224.

Clark, J. S., Beckage, B., Camill, P. *et al.* (1999) Interpreting recruitment limitation in forests. *American Journal of Botany*, 86, 1–16.

Clark, D. A. & Clark, D. B. (1984) Spacing dynamics of a tropical forest tree: evaluation of the Janzen-Connell model. *American Naturalist*, 124, 769–788.

Clark, J. S., Silman, M., Kern, R., Macklin, E. & Hille Ris Lambers, J. (1999) Seed dispersal near and far: patterns across temperate and tropical forests. *Ecology*, 80, 1475–1494.

Clausen, J., Keck, D. D. & Hiesey, W. M. (1948) Experimental studies on the nature of species. III. Environmental responses of climatic races of *Achillea*. *Publications of the Carnegie Institution of Washington*, 581, 114–133.

Clay, K. (1990) Fungal endophytes of grasses. *Annual Review of Ecology and Systematics*, 21, 275–297.

Clay, K. & Holah, J. (1999) Fungal endophyte symbiosis and plant diversity in successional fields. *Science*, 285, 1742–1744.

REFERENCES

Cody, M. L. & Overton, J. M. (1996) Short-term evolution of reduced dispersal in island plant populations. *Journal of Ecology*, 84, 53–61.

Cohen, D. (1966) Optimising reproduction in a randomly varying environment. *Journal of Theoretical Biology*, 12, 119–129.

Cohen, D. (1967) Optimizing reproduction in a randomly varying environment when a correlation may exist between the conditions at the time a choice has to be made and the subsequent outcome. *Journal of Theoretical Biology*, 16, 1–14.

Cole, L. C. (1954) The population consequences of life history phenomena. *Quarterly Review of Biology*, 29, 103–137.

Coley, P. D. & Barone, J. A. (1996) Herbivory and plant defenses in tropical forests. *Annual Review of Ecology and Systematics*, 27, 305–335.

Comes, H. P. & Abbott, R. J. (1999) Population genetic structure and gene flow across arid versus mesic environments: a comparative study of two parapatric *Senecio* species from the Near East. *Evolution*, 53, 36–54.

Condit, R., Hubbell, S. P. & Foster, R. B. (1992) Recruitment near conspecific adults and the maintenance of tree and shrub diversity in a neotropical forest. *American Naturalist*, 140, 261–286.

Condit, R., Hubbell, S. P. & Foster, R. B. (1995) Mortality-rates of 205 neotropical tree and shrub species and the impact of a severe drought. *Ecological Monographs*, 65, 419–439.

Condit, R., Hubbell, S. P. & Foster, R. B. (1996) Assessing the response of plant functional types to climatic change in tropical forests. *Journal of Vegetation Science*, 7, 405–416.

Connell, J. H. (1971) On the role of natural enemies in preventing competitive exclusion in some marine animals and in rain forests. *Dynamics of Populations* (eds P. J. den Boer & G. R. Gradwell), pp. 298–310. Centre for Agricultural Publishing and Documentation. (PUDOC), Wageningen.

Connell, J. H. (1978) Diversity in tropical rainforests and coral reefs. *Science*, 199, 1302–1310.

Connell, J. H. (1990) Apparent versus 'real' competition in plants. *Perspectives on Plant Competition* (eds J. B. Grace & D. Tilman), pp. 9–26. Academic Press, San Diego, CA.

Connell, H. H., Tracey, J. G. & Webb, L. J. (1984) Compensatory recruitment, growth, and mortality as factors maintaining rain forest tree diversity. *Ecological Monographs*, 54, 141–164.

Connolly, J. (1986) On difficulties with replacement series methodology in mixture experiments. *Journal of Applied Ecology*, 23, 125–137.

Connolly, J. (1987) On the use of response models in mixture experiments. *Oecologia*, 72, 95–103.

Connolly, J., Wayne, P. & Murray, R. (1990) Time course of plant–plant interactions in experimental mixtures of annuals: Density, frequency, and nutrient effects. *Oecologia*, 82, 513–526.

Cook, R. E. (1980) Germination and size-dependent mortality in *Viola blanda*. *Oecologia*, 47, 115–117.

Cook, R. E. (1983) Clonal plant populations. *American Scientist*, 71, 244–253.

Coomes, D. A. & Grubb, P. J. (2000) Impacts of root competition in forests and woodlands: A theoretical framework and review of experiments. *Ecological Monographs*, 70, 171–207.

Cottam, D. A., Whittaker, J. B. & Malloch, A. J. C. (1986) The effects of chrysomelid beetle grazing and plant competition on the growth of *Rumex obtusifolius*. *Oecologia*, 70, 452–456.

Courtney, A. D. (1968) Seed dormancy and field emergence in *Polygonum aviculare*. *Journal of Applied Ecology*, 5, 675–684.

Cousens, R. (1985) A simple model relating yield loss to weed density. *Annals of Applied Biology*, 73, 925–933.

Crawford, T. J. (1984) What is a population? *Evolutionary Ecology* (ed. B. Shorrocks), pp. 135–173. Blackwell Scientific Publications, Oxford.

Crawley, M. J. (1987) What makes a community invasible? *Colonization, Succession and Stability* (eds A. J. Gray, M. J. Crawley & P. J. Edwards), pp. 429–454. Blackwell Scientific Publications, Oxford.

Crawley, M. J. & Long, C. R. (1995) Alternate bearing, predator satiation and seedling recruitment in *Quercus robur* L. *Journal of Ecology*, 83, 683–696.

Croat, T. B. (1978) *Flora of Barro Colorado Island*. Stanford University Press, Stanford, CA.

Crone, E. E. (1997) Parental environmental effects and cyclical dynamics in plant populations. *American Naturalist*, 150, 708–729.

Crone, E. E. & Taylor, D. R. (1996) Complex dynamics in experimental populations of an annual plant, *Cardamine pensylvanica*. *Ecology*, 77, 289–299.

Crosby, J. L. (1949) Selection of an unfavourable gene-complex. *Evolution*, 3, 212–230.

Cruden, R. W. (1977) Pollen-ovule ratios: a conservative index of breeding systems in flowering plants. *Evolution*, 31, 32–46.

Cummings, M. P. & Clegg, M. T. (1998) Nucleotide sequence diversity at the alcohol dehydrogenase 1 locus in wild barley (*Hordeum vulgare* ssp. *spontaneum*): An evaluation of the background selection hypothesis. *Proceedings of the National Academy of Sciences of the USA*, 95, 5637–5642.

Curran, L. M., Caniago, I., Paoli, G. D. *et al.* (1999) Impact of El Nino and logging on canopy tree recruitment in Borneo. *Science*, 286, 2184–2188.

Cwynar, L. C. & MacDonald, G. M. (1987) Geographical variation of lodgepole pine in relation to population history. *American Naturalist*, 129, 463–469.

D'Hertefeldt, T. & Jónsdóttir, I. (1999) Extensive physiological integration in intact clonal systems of *Carex arenaria*. *Journal of Ecology*, 87, 258–264.

Damme, J. M. M. V. (1983) Gynodioecy in *Plantago lanceolata* L. II. Inheritance of three male sterility types. *Heredity*, 50, 253–273.

Damuth, J. D. (1981) Population-density and body size in mammals. *Nature*, 290, 699–700.

Damuth, J. D. (1998) Population ecology – Common rules for animals and plants. *Nature*, 395, 115–116.

Darwin, C. R. (1859) *The Origin of Species*. John Murray, London.

Darwin, C. R. (1862) *The Various Contrivances by which Orchids are Fertilised by Insects*. John Murray, London.

Darwin, C. R. (1868) *Variation of Animals and Plants Under Domestication*. John Murray, London.

Darwin, C. R. (1876) *The Effects of Cross and Self Fertilization in the Vegetable Kingdom*. John Murray, London.

Darwin, C. R. (1877) *The Different Forms of Flowers on Plants of the Same Species*. John Murray, London.

Davidson, D. W., Samson, D. A. & Inouye, R. S. (1985) Granivory in the Chihuahua desert: interactions within and between trophic levels. *Ecology*, 66, 486–502.

Davies, S. J., Palmiotto, P. A., Ashton, P. S., Lee, H. S. & Lafrankie, J. V. (1998) Comparative ecology of 11 sympatric species of *Macaranga* in Borneo: tree distribution in relation to horizontal and vertical resource heterogeneity. *Journal of Ecology*, 86, 662–673.

Davy, A. J. & Smith, H. (1985) Population differentiation in the life-history characteristics of salt-marsh annuals. *Vegetatio*, 61, 117–125.

Davy, A. J. & Smith, H. (1988) Life-history variation and environment. In: *Plant Population Ecology* (eds A. J. Davy, M. J. Hutchings & A. R. Watkinson), pp. 1–22. Blackwell Scientific Publications, Oxford.

De Vries, H. (1905) *Species and Varieties, Their Origin by Mutation*. Open Court Publishing, Chicago

Debaeke, P. (1988) Population dynamics of some broad-leaved weeds in cereal. I Relation between standing vegetation and soil seed bank [fr]. *Weed Research*, 28, 251–263.

Del Castillo, R. F. (1994) Factors influencing the genetic structure of *Phacelia dubia*, a species with a seed bank and large fluctuations in population size. *Heredity*, 72, 446–458.

Delph, L. F. (1990) Sex-ratio variation in the gynodioecious shrub *Hebe strictissima* (Scrophulariaceae). *Evolution*, 44, 134–142.

Delph, L. F. & Lloyd, D. G. (1996) Inbreeding depression in the gynodioecious shrub *Hebe subalpina* (Scrophulariaceae). *New Zealand Journal of Botany*, 34, 241–247.

Denslow, J. S. (1980) Gap partitioning among tropical rainforest trees. *Biotropica*, 12 (Suppl.), 47–55.

Denslow, J. S. (1987) Fruit removal rate from aggregated and isolated bushes of the red elderberry, *Sambucus pubens*. *Canadian Journal of Botany*, 65, 1229–1235.

Desfeux, C., Maurice, S., Henry, J. P., Lejeune, B. & Gouyon, P. H. (1996) Evolution of reproductive systems in the genus Silene. *Proceedings of the Royal Society of London, B*, 263, 409–414.

Dieckman, U., Law, R. & Metz, J. A. J. (eds) (2000) *The Geometry of Ecological Interactions: Simplifying Spatial Complexity*. Cambridge University Press, Cambridge.

van Dijk, H., Boudry, P., McCombie, H. & Vernet, P. (1997) Flowering time in wild beet (*Beta vulgaris* ssp. *maritima*) along a latitudinal cline. *Acta Oecologica*, 18, 47–60.

van-Dijk, H., Wolff, K. & Fries, A. D. (1988) Genetic variability in *Plantago* species in relation to their ecology. 3. genetic structure of populations of *Plantago major*, *P. lanceolata* and *P. coronopus*. *Theoretical and Applied Genetics*, 75, 518–528.

Dodd, M. E. & Silvertown, J. (2000) Size-specific fecundity and the influence of lifetime size variation upon effective population size in *Abies balsamea*. *Heredity*, 85, 604–609.

Dole, J. A. (1992) Reproductive assurance mechanisms in three taxa of the *Mimulus guttatus* complex (Scrophulariaceae). *American Journal of Botany*, 79, 650–659.

Dole, J. A. & Ritland, K. (1993) Inbreeding depression in two *Mimulus* taxa measured by multigenerational changes in the inbreeding coefficient. *Evolution*, 47, 361–373.

Dow, B. D. & Ashley, M. V. (1996) Microsatellite analysis of seed dispersal and parentage of saplings in bur oak, *Quercus macrocarpa*. *Molecular Ecology*, 5, 615–627.

Dow, B. D. & Ashley, M. V. (1998) High levels of gene flow in bur oak revealed by paternity analysis using microsatellites. *Journal of Heredity*, 89, 62–70.

Dressler, R. L. (1990) *The Orchids*. Harvard University Press, Cambridge, MA.

Dudash, M. R. & Fenster, C. B. (1997) Multiyear study of pollen limitation and cost of reproduction in the iteroparous *Silene virginica*. *Ecology*, 78, 484–493.

Dumolin-Lapègue, S., Demesure, B., Fineschi, S., LeCorre, V. & Petit, R. J. (1997) Phylogeographic structure of white oaks throughout the European continent. *Genetics*, 146, 1475–1487.

Dumolin-Lapegue, S., Kremer, A. & Petit, R. J. (1999) Are chloroplast and mitochondrial DNA variation species independent in oaks? *Evolution*, 53, 1406–1413.

Dumolin-Lapegue, S., Pemonge, M. H. & Petit, R. J. (1998) Association between chloroplast and mitochondrial lineages in oaks. *Molecular Biology and Evolution*, 15, 1321–1331.

Dybdahl, M. F. & Lively, C. M. (1998) Host-parasite coevolution: Evidence for rare advantage and time-lagged selection in a natural population. *Evolution*, 52, 1057–1066.

Ebert, T. A. (1999) *Plant and Animal Populations: Methods in Demography*. Academic Press, San Diego, CA.

Eckert, C. G., Manicacci, D. & Barrett, S. C. H. (1996) Genetic drift and founder effect in native versus introduced populations of an invading plant *Lythrum salicaria* (Lythraceae). *Evolution*, 50, 1512–1519.

Edwards, M. (1980) Aspects of the population ecology of charlock. *Journal of Applied Ecology*, 17, 151–171.

Edwards, J. (1984) Spatial pattern and clone structure of the perennial herb, *Aralia nudicaulis* L. (Araliaceae). *Bulletin of Torrey Botanical Club*, 111, 28–33.

Eguiarte, L. E., Burquez, A., Rodriguez, J., Martinez Ramos, M., Sarukhan, J. & Pinero, D. (1993) Direct and indirect estimates of neighborhood and effective population size in a tropical palm, *Astrocaryum mexicanum*. *Evolution*, 47, 75–87.

Ehrlén, J. & van Groenendael, J. (1998) Direct perturbation analysis for better conservation. *Conservation Biology*, 12, 470–474.

Eissenntat, D. M. & Caldwell, M. M. (1988) Competitive ability is linked to rates of water extraction. A field study of two aridland tussock grasses. *Oecologia*, 75, 1–7.

Ekstam, B. (1995) Ramet size equalization in a clonal plant, *Phragmites australis. Oecologia*, 104, 440–446.

Ellenberg, H. (1953) Physiologisches und ökologisches verhalten derselben Pflanzenarten. *Berichter der Deutschen Botanischen Gesellschaft*, 65, 350–361.

Ellison, A. M. (1987) Density-dependent dynamics of *Salicornia europaea* monocultures. *Ecology*, 68, 737–741.

Ellner, S. (1986) Germination dimorphisms and parent-offspring conflict in seed germination. *Journal of Theoretical Biology*, 123, 173–185.

Ellstrand, N. C., Prentice, H. C. & Hancock, J. F. (1999) Gene flow and introgression from domesticated plants into their wild relatives. *Annual Review of Ecology and Systematics*, 30, 539–563.

Emerman, S. H. & Dawson, T. E. (1996) Hydraulic lift and its influence on the water content of the rhizosphere: An example from sugar maple, *Acer saccharum. Oecologia*, 108, 273–278.

Emlen, J. M., Freeman, D. C. & Wagstaff, F. (1989) Interaction assessment: Rationale and a test using desert plants. *Evolutionary Ecology*, 3, 115–149.

Endler, J. A. (1986) *Natural Selection in the Wild*. Princeton University Press, Princeton, NJ.

Ennos, R. A. (1981) Detection of selection in populations of white clover (*Trifolium repens* L). *Biological Journal of the Linnean Society*, 15, 75–82.

Ennos, R. A. (1985) The significance of genetic variation for root growth within a natural population of white clover (*Trifolium repens*). *Journal of Ecology*, 73, 615–624.

Ennos, R. A. (1994) Estimating the relative rates of pollen and seed migration among plant-populations. *Heredity*, 72, 250–259.

Enquist, B. J., Brown, J. H. & West, G. B. (1998) Allometric scaling of plant energetics and population density. *Nature*, 395, 163–165.

Enquist, B. J., West, G. B., Charnov, E. L. & Brown, J. H. (1999) Allometric scaling of production and life-history variation in vascular plants. *Nature*, 401, 907–911.

Enti, A. A. (1968) Distribution and ecology of *Hildegardia barteri* (Mast.) Kosterm. *Bulletin de l'IFAN*, 30, 881–895.

Epling, C., Lewis, H. & Ball, E. M. (1960) The breeding group and seed storage: a study in population dynamics. *Evolution*, 14, 238–255.

Eriksson, O. (1996) Regional dynamics of plants: a review of evidence for remnant, source-sink and metapopulations. *Oikos*, 77, 248–258.

Eriksson, O. & Jerling, L. (1990) Hierarchical selection and risk spreading in clonal plants. *Clonal Growth in Plants* (eds J. van Groenendael & H. de Kroon), pp. 79–94. SPB Academic Publishing, The Hague.

Erickson, R. O. (1945) The *Clematis fremontii* var. *reihlii* population in the Ozarks. *Annals of the Missouri Botanical Garden*, 32, 413–460.

Evans, D. R., Hill, J., Williams, T. A. & Rhodes, I. (1985) Effects of coexistence on the performance of white clover-perennial ryegrass mixtures. *Oecologia*, 66, 536–539.

Ewell, J. J. (1986) *Invasibility: Lessons from South Florida. Ecology of Biological Invasions of North America and Hawaii*. Springer, New York.

Facelli, E., Facelli, J. M., Smith, S. E. & McLaughlin, M. J. (1999) Interactive effects of arbuscular mycorrhizal symbiosis, intraspecific competition and resource availability on *Trifolium subterraneum* cv. Mt. Barker. *New Phytologist*, 141, 535–547.

Faille, A., Lemée, G. & Pontailler, J. Y. (1984) Dynamique des clairières d'une fôret inexploitée (réserves biologiques de la fôret de Fontainbleau) I. Origine et état actuel des ouvertures. *Acta Oecologica Oecologica Generalis*, 5, 35–51.

Falconer, D. S. & Mackay, T. F. C. (1996) *Introduction to Quantitative Genetics*. Longman, Harlow.

Fang, D. Q., Roose, M. L., Krueger, R. R. & Federici, C. T. (1997) Fingerprinting trifoliate orange germ plasm accessions with isozymes, RFLPs, and inter-simple sequence repeat markers. *Theoretical and Applied Genetics*, 95, 211–219.

Federoff, N. (1984) Transposable elements in maize. *Scientific American*, 250, 64–74.

Fenner, M. (1980) Germination tests on thirty-two East African weed species. *Weed Research*, 20, 135–138.

Fenster, C. B. (1991a) Gene flow in *Chamaecrista fasciculata* (Leguminosae). I. Gene dispersal. *Evolution*, 45, 389–409.

Fenster, C. B. (1991b) Gene flow in *Chamaecrista fasciculata* (Leguminosae). II. Gene establishment. *Evolution*, 45, 410–422.

Firbank, L. G. & Watkinson, A. R. (1987) On the analysis of competition at the level of the individual plant. *Oecologia*, 71, 308–317.

Firbank, L. G. & Watkinson, A. R. (1990) On the effects of competition: from monocultures to mixtures. *Perspectives on Plant Competition* (eds J. B. Grace & D. Tilman), pp. 165–192. Academic Press, San Diego, CA.

Fischer, M. & Matthies, D. (1997) Mating structure and inbreeding and outbreeding depression in the rare plant Gentianella germanica (Gentianaceae). *American Journal of Botany*, 84, 1685–1692.

Fischer, M. & Matthies, D. (1998a) Effects of population size on performance in the rare plant *Gentianella germanica. Journal of Ecology*, 86, 195–204.

Fischer, M. & Matthies, D. (1998b) Experimental demography of the rare *Gentianella germanica*: seed bank formation and microsite effects on seedling establishment. *Ecography*, 21, 269–278.

Fischer, M. & Matthies, D. (1998c) RAPD variation in relation to population size and plant fitness in the rare *Gentianella germanica* (Gentianaceae). *American Journal of Botany*, 85, 811–819.

Fischer, M. & Stocklin, J. (1997) Local extinctions of plants in remnants of extensively used calcareous grasslands 1950–85. *Conservation Biology*, 11, 727–737.

Fisher, R. A. (1941) Average excess and average effect of a gene substitution. *Annals of Eugenics*, 11, 53–63.

Flavell, A. J., Pearce, S. R. & Kumar, A. (1994) Plant transposable elements and the genome. *Current Opinions in Genetics and Development*, 4, 838–844.

Fleming, T. H. & Williams, C. F. (1990) Phenology, seed dispersal, and recruitment in *Cecropia peltata* (Moraceae) in Costa Rican tropical dry forest. *Journal of Tropical Ecology*, 6, 163–178.

Flores-Martinez, A., Ezcurra, E. & Sanchezcolon, S. (1994) Effect of *Neobuxbaumia tetetzo* on growth and fecundity of its nurse plant *Mimosa luisana*. *Journal of Ecology*, 82, 325–330.

Forcella, F. & Harvey, S. J. (1988) Patterns of weed migration in northwestern USA. *Weed Science*, 36, 194–201.

Ford, E. D. (1975) Competition and stand structure in some even-aged plant monocultures. *Journal of Ecology*, 63, 311–333.

Forget, P. M. (1996) Removal of seeds of Carapa procera (Meliaceae) by rodents and their fate in rainforest in French Guiana. *Journal of Tropical Ecology*, 12, 751–761.

Foster, S. A. & Janson, C. H. (1985) The relationship between seed size and establishment conditions in tropical woody plants. *Ecology*, 66, 773–780.

Fowells, H. & Schubert, G. H. (1956) Seed crops of forest trees in the pine region of California. *USDA Technical Bulletin*, 1150, 1–48.

Fowler, N. L. (1981) Competition and coexistence in a North carolina grassland II. The effects of the removal of species. *Journal of Ecology*, 69, 843–854.

Fowler, N. L. (1986) The role of competition in plant communities in arid and semiarid regions. *Annual Review of Ecology and Systematics*, 17, 89–110.

Fox, J. F. (1981) Intermediate levels of disturbance maximize alpine plant species diversity. *Nature*, 293, 564–565.

Franco, M. & Harper, J. L. (1988) Competition and the formation of spatial pattern in spacing gradients an example using *Kochia scoparia*. *Journal of Ecology*, 76, 959–974.

Franco, M. & Kelly, C. K. (1998) The interspecific mass-density relationship and plant geometry. *Proceedings of the National Academy of Sciences of the USA*, 95, 7830–7835.

Freckleton, R. P. & Watkinson, A. R. (1998a) How does temporal variability affect predictions of weed population numbers? *Journal of Applied Ecology*, 35, 340–344.

Freckleton, R. P. & Watkinson, A. R. (1998b) Predicting the determinants of weed abundance: a model for the population dynamics of *Chenopodium album* in sugar beet. *Journal of Applied Ecology*, 35, 904–920.

Freckleton, R. P. & Watkinson, A. R. (2000a) Designs for greenhouse studies of interactions between plants: an analytical perspective. *Journal of Ecology*, 88, 386–391.

Freckleton, R. P. & Watkinson, A. R. (2000b) Predicting the outcome of competition in plant mixtures: reciprocity, transitivity and correlations with life-history traits. *Ecology Letters*, 3, 423–432.

Freckleton, R. P. & Watkinson, A. R. (2000c) On detecting and measuring competition in spatially structured plant communities. *Ecology Letters*, 3, 423–432.

Freckleton, R. P., Watkinson, A. R., Dowling, P. M. & Leys, A. R. (2000) Determinants of the abundance of invasive annual weeds: community structure and non-equilibrium dynamics. *Proceedings of the Royal Society of London, B*, 267, 1153–1161.

Freeman, D. C. & Emlen, J. M. (1995) Assessment of interspecific interactions in plant communities – an illustration from the cold desert saltbush grasslands of North America. *Journal of Arid Environments*, 31, 179–198.

Frissell, S. S. (1973) The importance of fire as a natural ecological factor in Itasca State Park. *Quaternary Research*, 3, 397–407.

Fritsch, P. & Rieseberg, L. H. (1992) High outcrossing rates maintain male and hermaphrodite individuals in populations of the flowering plant *Datisca glomerata*. *Nature*, 359, 633–636.

Futuyma, D. J. (1999) *Evolutionary Biology*, 3rd edn. Sinauer Associates, Sunderland, MA.

Gadgil, M. & Prasad, S. N. (1984) Ecological determinants of life history evolution of two Indian bamboo species. *Biotropica*, 16, 161–172.

Gagnon, P. S., Vadas, R. L., Burdick, D. B. & May, B. (1980) Genetic identity of annual and perennial forms of *Zostera marina* L. *Aquatic Botany*, 18, 157–162.

Galen, C. (1997) Rates of floral evolution: adaptation to bumblebee pollination in an Alpine wildflower, *Polemonium viscosum*. *Evolution*, 50, 120–125.

Galen, C., Shore, J. & Deyoe, H. (1991) Ecotypic divergence in alpine *Polemonium viscosum* – genetic-structure, quantitative variation, and local adaptation. *Evolution*, 45, 1218–1228.

Galen, C., Zunmer, K. A. & Newport, M. E. (1987) Pollination in floral scent morphs of *Polemonium viscosum*: a mechanism for disruptive selection on flower size. *Evolution*, 41, 599–606.

Garbutt, K. & Whitcombe, J. (1986) The inheritance of seed dormancy in *Sinapis arvensis* L. *Heredity*, 56, 25–31.

Gardner, S. N. & Mangel, M. (1997) Can a clonal organism escape senescence? *American Naturalist*, 150, 462–490.

Garrett, K. A. & Dixon, P. M. (1998) When does the spatial pattern of weeds matter? Predictions from neighborhood models. *Ecological Applications*, 8, 1250–1259.

Garwood, N. C. (1982) *Seasonal Rhythm of Seed Germination in a Semideciduous Tropical Forest. The Ecology of a Tropical Forest: Seasonal Rhythms and Long-Term Changes.* Smithsonian Institution Press, Washington, DC.

Garwood, N. C. (1983) Seed germination in a seasonal tropical forest in Panama: A community study. *Ecological Monographs*, 53, 159–181.

Gaut, B. S. (1998) Molecular clocks and nucleotoide substitution rates in higher plants. *Evolutionary Biology*, 30, 93–120.

Geber, M. A. (1990) The cost of meristem limitation in *Polygonum arenastrum*: Negative genetic correlations between fecundity and growth. *Evolution*, 44, 799–819.

Gentry, A. H. (1988) Tree species richness of upper Amazonian forests. *Proceedings of the National Academy of Sciences of the USA*, 85, 156–159.

Geritz, S. A. H. (1995) Evolutionarily stable seed polymorphism and small-scale spatial variation in seedling density. *American Naturalist*, 146, 685–707.

Geritz, S. A. H. (1998) Co-evolution of seed size and seed predation. *Evolutionary Ecology*, 12, 891–911.

Geritz, S. A. H., van der Meijden, E. & Metz, J. A. J. (1999) Evolutionary dynamics of seed size and seedling competitive ability. *Theoretical Population Biology*, 55, 324–343.

Ghazoul, J., Liston, K. A. & Boyle, T. J. B. (1998) Disturbance-induced density-dependent seed set in *Shorea siamensis* (Dipterocarpaceae), a tropical forest tree. *Journal of Ecology*, 86, 462–473.

Gibson, D. J., Connolly, J., Hartnett, D. C. & Weidenhamer, J. D. (1999) Designs for greenhouse studies of interactions between plants. *Journal of Ecology*, 87, 1–16.

Gilbert, G. S., Hubbell, S. P. & Foster, R. B. (1994) Density and distance-to-adult effects of a canker disease of trees in a moist tropical forest. *Oecologia*, 98, 100–108.

Gillman, M. & Hails, R. (1997) *Population Models in Ecology*. Blackwell Science, Oxford.

Gilpin, M. E. & Soulé, M. E. (1986) Minimum viable populations: Processes of species extinction. *Conservation Biology* (ed. M. E. Soulé), pp. 19–34. Sinauer, Sunderland, MA.

Gitzendanner, M. A. & Soltis, P. S. (2000) Patterns of variation in rare and widespread plant congeners. *American Journal of Botany*, 87, 783–792.

Givnish, T. J. (1981) Serotiny, geography and fire in the pine barrens of New Jersey. *Evolution*, 35, 101–123.

Gliddon, C. & Saleem, M. (1985) Gene-flow in *Trifolium repens* – an expanding genetic neighbourhood. *Genetic Differentiation and Dispersal in Plants* (eds P. Jacquard, G. Heim & J. Antonovics), pp. 293–309. Springer, Heidelberg.

Glover, D. E. & Barrett, S. C. H. (1986) Variation in the mating system of *Eichhornia paniculata* (Spreng.) Solms. (Pontederiaceae). *Evolution*, 40, 1122–1131.

Goldberg, D. E. (1987) Neighbourhood competition in an old-field plant community. *Ecology*, 68, 1211–1223.

Goldberg, D. E. (1990) Components of resource competition in plant communities. *Perspectives on Plant Competition* (eds J. B. Grace & D. Tilman). Academic Press, San Diego, CA.

Goldberg, D. E. & Barton, A. M. (1992) Patterns and consequences of interspecific competition in natural communities: a review of field experiments with plants. *American Naturalist*, 139, 771–801.

Goldberg, D. E. & Fleetwood, L. (1987) Competitive effect and response in four annual plants. *Journal of Ecology*, 75, 1131–1143.

Goldberg, D. E. & Landa, K. (1991) Competitive effect and response: hierarchies and correlated traits in the early stages of competition. *Journal of Ecology*, 79, 1013–1030.

Goldberg, D. E. & Scheiner, S. M. (1993) ANOVA and ANCOVA: Field competition experiments. *Design and Analysis of Ecological Experiments* (eds S. M. Scheiner & J. Gurevitch), pp. 69–93. Chapman and Hall, New York.

Goldberg, D. E. & Werner, P. A. (1983) Equivalence of competitors in plant comunities: a null hypothesis and a field experimental approach. *American Journal of Botany*, 70, 1098–1104.

Goldstein, D. B., Linares, A. R., Cavalli Sforza, L. L. & Feldman, M. W. (1994) An evaluation of genetic distances for use with microsatellite loci. *Genetics*, 139, 463–471.

Gornall, R. J. (1999) Population genetic structure in agamospermous plants. *Molecular Systemtics and Plant Evolution* (eds P. M. Hollingsworth, R. M. Bateman & R. J. Gornall). Taylor and Francis, London.

Gottlieb, L. D. (1977) Genotypic similarity of large and small individuals in a natural population of the annual plant *Stephanomeria exigua* ssp. *coronaria* (Compositae). *Journal of Ecology*, 65, 127–134.

Grace, J. B. (1985) Juvenile vs adult competitive abilities in plants: Size-dependance in cattails (*Typha*). *Ecology*, 66, 1630–1638.

Grace, J. B. (1995) The adaptive significance of clonal reproduction in angiosperms: an aquatic perspective. *Aquatic Botany*, 44, 159–180.

Grace, J. B., Guntenspergen, G. R. & Keough, J. (1993) The examination of a competition matrix for transitivity and intransitive loops. *Oikos*, 68, 91–98.

Grace, J. B. & Tilman, D. (1990) *Perspectives in Plant Competition*. Academic Press, London.

Graham, F. B. & Bormann, F. H. (1966) Natural root grafts. *Botanical Review*, 32, 255–292.

Grant, J. D. (1983) The activities of earthworms and the fates of seeds. *Earthworm Ecology* (ed. J. E. Satchell), pp. 107–122. Chapman and Hall, London.

Grant, A. & Benton, T. G. (2000) Elasticity analysis for density-dependent populations in stochastic environments. *Ecology*, 81, 680–693.

Grant, M. C., Mitton, J. B. & Linhart, Y. B. (1992) Even larger organisms. *Nature*, 360, 216.

Grime, J. P. (1977) Evidence for the existence of three primary strategies in plants and its relevance to ecological and evolutionary theory. *American Naturalist*, 111, 1169–1194.

Grime, J. P. (1979) *Plant Strategies and Vegetation Processes*. Wiley, Chichester.

Grime, J. P. & Jeffrey, D. W. (1965) Seedling establishment in vertical gradients of sunlight. *Journal of Ecology*, 53, 621–642.

Grime, J. P., Hodgson, J. G., Hunt, R. *et al.* (1997) Integrated screening validates primary axes of specialisation in plants. *Oikos*, 79, 259–281.

Grime, J. P., Mackey, J. M. L., Hillier, S. H. & Read, D. J. (1987) Floristic diversity in a model system using experimental microcosms. *Nature*, 328, 420–422.

van Groenendael, J. (1985) Differences in life histories between two ecotypes of *Plantago lanceolata* L. *Studies on Plant Demography* (ed J. White), Chapter 4, pp. 51–67.

van Groenendael, J. M., Klimes, L., Klimesova, J. & Hendriks, R. J. J. (1997) Comparative ecology of clonal plants. *Plant Life Histories: Ecology, Phylogeny and Evolution* (eds J. Silvertown, M. Franco & J. L. Harper), pp. 191–209. Cambridge University Press, Cambridge.

van Groenendael, J. M. & Slim, P. (1988) The contrasting dynamics of two populations of *Plantago lanceolata* classified by age and size. *Journal of Ecology*, 76, 585–599.

Groom, M. J. (1998) Allee effects limit population viability of an annual plant. *American Naturalist*, 151, 487–496.

316

Gross, K. L. (1981) Predictions of fate from rosette size in four 'Biennial' plant species: *Verbascum thapsus, Oenothera biennis, Daucus carota*, and *Tragopogon dubius. Oecologia*, 48, 209–213.

Groves, R. H. & Williams, J. D. (1975) Growth of skeleton weeds (*Chondrilla juncea* L.) as affected by growth of subterranean clover (*Trifolium subterraneum* L.) and infection by *Puccinia chondrilla* Bubak and Syd. *Australian Journal of Agricultural Research*, 26, 975–983.

Grubb, P. J. (1977) The maintenance of species richness in plant communities: The importance of the regeneration niche. *Biological Reviews*, 52, 107–145.

Grubb, P. J. (1985) Plant populations and vegetation in relation to habitat, disturbance and competition: Problems of generalization. *The Population Structure of Vegetation* (ed. J. White), pp. 595–562. Dr W. Junk, Dordrecht.

Grubb, P. J. (1986) Problems Posed by Sparse and Patchily Distributed Species in Species-Rich Plant Communities. In: *Community Ecology* (eds T. J. case & J. Diamond). Harper and Row, New York.

Grubb, P. J. (1988) The uncoupling of disturbance and recruitment, two kinds of seed bank, and persistence of plant populations at the regional and local scales. *Annales Zoologici Fennici*, 25, 23–36.

Grubb, P. J. (1998) A reassessment of the strategies of plants which cope with shortages of resources. *Perspectives in Plant Ecology, Evolution and Systematics*, 1, 3–31.

Grubb, P. J. & Metcalfe, D. J. (1996) Adaptation and inertia in the Australian tropical lowland rain-forest flora: Contradictory trends in intergeneric and intrageneric comparisons of seed size in relation to light demand. *Functional Ecology*, 10, 512–520.

Gurevitch, J., Morrow, L. L., Wallace, A. & Walsh, J. S. (1992) A Meta-Analysis of competition in field experiments. *American Naturalist*, 140, 539–572.

Gustaffson, A. (1946) Apomixis in higher plants. I. *Lunds Universitet Arsskriften*, 42, 1–67.

Gustaffson, A. (1947) Apomixis in higher plants. II–III. *Lunds Universitet Arsskriften*, 43 (69–179), 183–370.

Hacker, S. D. & Bertness, M. D. (1999) Experimental evidence for factors maintaining plant species diversity in a New England salt marsh. *Ecology*, 80, 2064–2073.

Haig, D. (1996) The pea and the coconut: Seed size in safe sites. *Trends in Ecology & Evolution*, 11, 1–2.

Haldane, J. B. S. (1937) The effect of variation on fitness. *American Naturalist*, 71, 67–100.

Hamilton, W. D. & May, R. M. (1977) Dispersal in stable habitats. *Nature*, 269, 578–581.

Hammond, D. S. & Brown, V. K. (1995) Seed size of woody plants in relation to disturbance, dispersal, soil type in wet neotropical forests. *Ecology*, 76, 2544–2561.

Hamrick, J. L. & Godt, M. J. (1990) Allozyme diversity in plant species. *Plant Population Genetics, Breeding, and Genetic Resources* (eds A. H. D. Brown, M. T. Clegg, A. L. Kahler & B. S. Weir), pp. 43–63. Sinauer Associates, Sunderland, MA.

Hamrick, J. L. & Godt, M. J. W. (1997) Effects of life history traits on genetic diversity in plant species. *Plant Life Histories*

(eds J. Silvertown, M. Franco & J. L. Harper), pp. 102–118. Cambridge University Press, Cambridge.

Handel, S. N. (1976) Dispersal ecology of *Carex pedunculata* (Cyperaceae), a new North American mymecochore. *American Journal of Botany*, 63, 1071–1079.

Hanski, I. (1985) Single-species spatial dynamics may contribute to long-term rarity and commonness. *Ecology*, 66, 335–343.

Hanski, I. (1999) *Metapopulation Ecology*. Oxford University Press, Oxford.

Hanski, I. & Ovaskainen, O. (2000) The metapopulation capacity of a fragmented landscape. *Nature*, 404, 755–758.

Hanzawa, F. M., Beattie, A. J. & Culver, D. C. (1988) Directed dispersal: demographic analysis of an ant-seed mutualism. *American Naturalist*, 131, 1–13.

Hara, T. & Wyszomirski, T. (1994) Competitive asymmetry reduces spatial effects on Size-Structure dynamics in plant populations. *Annals of Botany*, 73, 285–297.

Harada, Y., Kawano, S. & Iwasa, Y. (1997) Probability of clonal identity: inferring the relative success of sexual versus clonal reproduction from spatial genetic patterns. *Journal of Ecology*, 85, 591–600.

Harberd, D. J. (1961) Observations on population structure and longevity of *Festuca rubra. New Phytologist*, 61, 85–100.

Harberd, D. J. (1962) Some observations on natural clones of *Festuca ovina. New Phytologist*, 61, 85–100.

Harlan, J. R. (1976) Diseases as a factor in plant evolution. *Annual Review of Phytopathology*, 14, 31–51.

Harms, K. E., Wright, S. J., Calderon, O., Hernandez, A. & Herre, E. A. (2000) Pervasive density-dependent recruitment enhances seedling diversity in a tropical forest. *Nature*, 404, 493–495.

Harper, J. L. (1977) *Population Biology of Plants*. Academic press, London.

Harper, J. L. (1981) The population biology of modular organisms. *Theoretical Ecology* (ed. R. M. May), pp. 53–77. Sinauer Associates, Sunderland, MA.

Harper, J. L. (1982) After description. *The Plant Community as a Working Mechanism* (ed. E. I. Newman), pp. 11–25.

Harper, J. L., Lovell, P. H. & Moore, K. G. (1970) The shapes and size of seeds. *Annual Review of Ecology and Systematics*, 1, 327–356.

Harper, J. L., Williams, J. T. & Sagar, G. R. (1965) The behaviour of seeds in the soil: I. The heterogeneity of soil surfaces and its role in determining the establishment of plants. *Journal of Ecology*, 53, 273–286.

Harris, M. & Chung, C. S. (1998) Masting enhancement makes pecan nut casebearer pecans ally against pecan weevil. *Journal of Economic Entomology*, 91, 1005–1010.

Hartgerink, A. P. & Bazzaz, F. A. (1984) Seedling-scale environmental heterogeneity influences individual fitness and population structure. *Ecology*, 65, 198–206.

Hartl, D. L. & Clark, A. G. (1997) *Principles of Population Genetics*. Sinauer Associates, Sunderland, MA.

Hartnett, D. C. & Wilson, G. W. T. (1999) Mycorrhizae influence plant community structure and diversity in tallgrass prairie. *Ecology*, 80, 1187–1195.

Harvey, P. H. & Pagel, M. D. (1991) *The Comparative Method in Evolutionary Biology*. Oxford University Press, Oxford.

Hassell, M. P., Lawton, J. H. & May, R. M. (1976) Patterns of dynamical behaviour in single-species populations. *Journal of Animal Ecology*, 45, 471–486.

Hedrick, P. W. (1995) Genetic polymorphism in a temporally varying environment: effects of delayed germination or diapause. *Heredity*, 75, 164–170.

van der Heijden, M. G. A., Boller, T., Wiemken, A. & Sanders, I. R. (1998a) Different arbuscular mycorrhizal fungal species are potential determinants of plant community structure. *Ecology*, 79, 2082–2091.

van der Heijden, M. G. A., Klironomos, J. N., Ursic, M. *et al.* (1998b) Mycorrhizal fungal diversity determines plant biodiversity, ecosystem variability and productivity. *Nature*, 396, 69–72.

Helgason, T., Daniell, T. J., Husband, R., Fitter, A. H. & Young, J. P. W. (1998) Ploughing up the wood-wide web? *Nature*, 394, 431.

Hendrix, S. D. (1984) Variation in seed weight and its effects on germination in *Pastinaca sativa* L. (Umbelliferae). *American Journal of Botany*, 71, 795–802.

Henry, J. D. & Swan, J. M. A. (1974) Reconstructing forest history from live and dead plant material – an approach to the study of forest succession in southwest New Hampshire. *Ecology*, 55, 772–783.

Herben, T., Krahulec, F., Hadincová, V. & Pecháčková, S. (1997) Fine-scale species interactions of clonal plants in a mountain grassland: a removal experiment. *Oikos*, 78, 299–310.

Herrera, C. M. (1998) Population-level estimates of interannual variability in seed production: what do they actually tell us? *Oikos*, 82, 612–616.

Herrera, C. M., Jordano, P., Guitian, J. & Traveset, A. (1998) Annual variability in seed production by woody plants and the masting concept: Reassessment of principles and relationship to pollination and seed dispersal. *American Naturalist*, 152, 576–594.

Heschel, M. S. & Paige, K. N. (1995) Inbreeding depression, environmental stress, and population size variation in scarlet gilia (*Ipomopsis aggregata*). *Conservation Biology*, 9, 126–133.

Hetrick, B. A. D., Wilson, G. W. T. & Hartnett, D. C. (1989) Relationship between mycorrhizal dependence and competitive ability of two tallgrass prarie grasses. *Canadian Journal of Botany*, 67, 2608–2615.

Hett, J. M. (1971) A dynamic analysis of age in sugar maple seedlings. *Ecology*, 52, 1071–1074.

Heuch, I. (1980) Loss of incompatibility types in finite populations of the heterostylous plant *Lythrum salicaria*. *Hereditas*, 92, 53–57.

Hewitt, N. (1998) Seed size and shade-tolerance: a comparative analysis of North American temperate trees. *Oecologia*, 114, 432–440.

Hewitt, G. M. & Ibrahim, K. M. (2001) Inferring glacial refugia and historical migrations with molecular phylogenies. In: *Integrating Ecology and Evolution in a Spatial Context* (eds J. Silvertown & J. Antonovics). Blackwell Science, Oxford.

Heywood, J. S. (1986) The effect of plant size variation on genetic drift in populations of annuals. *American Naturalist*, 127, 851–861.

Hobbs, R. J. & Mooney, H. A. (1985) Community and population dynamics of serpentine grassland annuals in relation to gopher disturbance. *Oecologia*, 67, 342–351.

Hodge, A., Robinson, D., Griffiths, B. S. & Fitter, A. H. (1999) Why plants bother: root proliferation results in increased nitrogen capture from an organic patch when two grasses compete. *Plant, Cell and Environment*, 22, 811–820.

Hodgson, J. G. & Mackey, J. M. L. (1986) The ecological specialization of dicotyledonous families within a local flora: some factors constraining optimization of seed size and their possible evolutionary significance. *New Phytologist*, 104, 497–515.

Hodkinson, D. J., Askew, A. P., Thompson, K. *et al.* (1998) Ecological correlates of seed size in the British flora. *Functional Ecology*, 12, 762–766.

Holm, S., Ghatnekar, L. & Bengtsson, B. O. (1997) Selfing and outcrossing but no apomixis in two natural populations of diploid *Potentilla argentea*. *Journal of Evolutionary Biology*, 10, 343–352.

Holsinger, K. E. (1986) Dispersal and plant mating systems: the evolution of self-fertilization in subdivided populations. *Evolution*, 40, 405–413.

Holsinger, K. E. (1991) Mass action models of plant mating systems: the evolutionary stability of mixed mating systems. *American Naturalist*, 138, 606–622.

Holsinger, K. E. (2000) Reproductive systems and evolution in vascular plants. *Proceedings of the National Academy of Sciences of the USA*, 97, 7037–7042.

Holt, R. D. & Gomulkiewicz, R. (1997) How does immigration influence local adaptation? A reexamination of a familiar paradigm. *American Naturalist*, 149, 563–572.

Holt, R. D. & Lawton, J. H. (1994) The ecological consequences of shared natural enemies. *Annual Review of Ecology and Systematics*, 25, 495–520.

Holtsford, T. P. & Ellstrand, N. C. (1990) Inbreeding depression in *Clarkia tembloriensis* (Onagraceae) populations with different natural outcrossing rates. *Evolution*, 44, 2031–2046.

Horvitz, C. C. & Schemske, D. W. (1986) Seed dispersal and environmental heterogeneity in a neotropical herb: a model of population and patch dynamics. *Frugivores and Seed Dispersal* (eds A. Estrada & T. H. Fleming), pp. 170–186. Dr. W. Junk, Dordrecht.

Horvitz, C. C. & Schemske, D. W. (1995) Spatiotemporal variation in demographic transitions of a tropical understory herb: Projection matrix analysis. *Ecological Monographs*, 65, 155–192.

Houle, D. (1992) Comparing evolvability and variability of quantitative traits. *Genetics*, 130, 195–204.

Howe, H. F. & Smallwood, J. (1982) Ecology of seed dispersal. *Annual Review of Ecology and Systematics*, 13, 201–228.

Hubbell, S. P. (1979) Tree dispersion, abundance, and diversity in a tropical dry forest. *Science*, 203, 1299–1309.

Hubbell, S. P., Condit, R. & Foster, R. B. (1990) Presence and absence of density dependence in a neotropical tree community. *Philosophical Transactions of the Royal Society of London, B*, 330, 269–281.

Hubbell, S. P. & Foster, R. B. (1986a) Biology, chance, and history and the structure of tropical rain forest tree communities. *Community Ecology* (eds J. Diamond & T. J. Case), pp. 314–329. Harper and Row, New York.

Hubbell, S. P. & Foster, R. B. (1986b) Canopy gaps and the dynamics of a neotropical forest. *Plant Ecology* (ed. M. J. Crawley), pp. 77–96. Blackwell Scientific Publicaitons, Oxford.

Hubbell, S. P. & Foster, R. B. (1990a) Structure, dynamics, and equilibrium status of old-growth forest on Barro Colorado Island. *Four Neotropical Forests* (ed. A. H. Gentry), pp. 522–541. Yale UP, Newhaven.

Hubbell, S. P. & Foster, R. B. (1990b) The fate of juvenile trees in a neotropical forest: implications for the natural maintenance of tropical tree diversity. *Reproductive Ecology of Tropical Forest Plants* (eds K. S. Bawa & M. Hadley), pp. 325–349. UNESCO/IUBS, Paris.

Hubbell, S. P., Foster, R. B., O'brien, S. T. *et al.* (1999) Light-gap disturbances, recruitment limitation, and tree diversity in a neotropical forest. *Science*, 283, 554–557.

Hubbell, S. P. & Werner, P. A. (1979) On measuring the intrinsic rate of increase of populations with heterogeneous life histories. *American Naturalist*, 113, 277–293.

Huff, D. R. & Wu, L. (1992) Distribution and inheritance of inconstant sex forms in natural populations of dioecious buffalograss (*Buchloe dactyloides*). *American Journal of Botany*, 79, 207–215.

Hunter, A. F. A. & Aarssen, L. W. (1988) Plants helping plants. New evidence indicates that beneficence is important in vegetation. *Bioscience*, 38, 34–40.

Huntly, N. (1991) Herbivores and the dynamics of communities and ecosystems. *Annual Review of Ecology and Systematics*, 22, 477–503.

Husband, B. C. & Barrett, S. C. H. (1992) Effective population size and genetic drift in tristylous *Eichhornia paniculata* (Pontederiaceae). *Evolution*, 46, 1875–1890.

Husband, B. C. & Barrett, S. C. H. (1992) Genetic drift and the maintenance of the style length polymorphism in tristylous populations of *Eichhornia paniculata* (Pontederiaceae). *Heredity*, 69, 440–449.

Husband, B. C. & Barrett, S. C. H. (1993) Multiple origins of self-fertilization in tristylous *Eichhornia paniculata* (Pontederiaceae): inferences from style morph and isozyme variation. *Journal of Evolutionary Biology*, 6, 591–608.

Husband, B. C. & Barrett, S. C. H. (1998) Spatial and temporal variation in population size of *Eichhornia paniculata* in ephemeral habitats: implications for metapopulation dynamics. *Journal of Ecology*, 86, 1021–1031.

Husband, B. C. & Schemske, D. W. (1995) Evolution of the magnitude and timing of inbreeding depression in plants. *Evolution*, 50, 54–70.

Huston, M. (1979) A general hypothesis of species diversity. *American Naturalist*, 113, 81–101.

Hutchings, M. J. (1979) Weight-density relationships in ramet populations of clonal perennial herbs, with special reference to the −3/2 power law. *Journal of Ecology*, 67, 21–33.

Hutchings, M. J. & Bradbury, I. K. (1986) Ecological perspectives on clonal perennial herbs. *Bioscience*, 36, 178–182.

Hutchings, M. J. & Wijesinghe, D. K. (1997) Patchy habitats, division of labour and growth dividends in clonal plants. *Trends in Ecology and Evolution*, 12, 390–394.

Hutchinson, G. E. (1957) The multivariate niche. *Cold Spring Harbour Symposia in Quantitative Biology*, 22, 415–421.

Hutchinson, G. E. (1978) *An Introduction to Population Ecology.* Yale University Press, New Haven, CT.

Huxley, J. S. (1942) *Evolution: the Modern Synthesis.* Allen and Unwin, London.

Ims, R. A. (1990) On the adaptive value of reproductive synchrony as a predator-swamping strategy. *American Naturalist*, 136, 485–498.

Inghe, O. & Tamm, C. O. (1985) Survival and flowering of perennial herbs IV. The behaviour of *Hepatica nobilis* and *Sanicula europaea* on permanent plots during 1943–81. *Oikos*, 45, 400–420.

Inghe, O. & Tamm, C. O. (1988) Survival and flowering of perennial herbs V. patterns of flowering. *Oikos*, 51, 203–219.

Inouye, R. S. & Schaffer, W. M. (1981) On the ecological meaning of ratio (de Wit) diagrams in plant ecology. *Ecology*, 62, 1679–1681.

Isagi, Y., Sugimura, K., Sumida, A. & Ito, H. (1997) How does masting happen and synchronize? *Journal of Theoretical Biology*, 187, 231–239.

Isagi, Y., Kanazashi, T., Suzuki, W., Tanaka, H. & Abe, T. (2000) Microsatellite analysis of the regeneration process of *Magnolia obovata* Thunb. *Heredity*, 84, 143–151.

Ives, A. R. (1995) Measuring competition in a spatially heterogeneous environment. *American Naturalist*, 146, 911–936.

Izmailow, R. (1996) Reproductive strategy in the *Ranunculus auricomus* complex (Ranunculaceae). *Acta Societatis Botanicorum Poloniae*, 65, 167–170.

Jackson, L. E. (1985) Ecological origins of California's mediterranean grasses. *Journal of Biogeography*, 12, 349–361.

Jackson, H. D., Steane, D. A., Potts, B. M. & Vaillancourt, R. E. (1999) Chloroplast DNA evidence for reticulate evolution in *Eucalyptus* (Myrtaceae). *Molecular Ecology*, 8, 739–751.

Jalloq, M. (1975) The invasion of molehills by weeds as a possible factor in the degeneration of reseeded pasture. *Journal of Applied Ecology*, 12, 643–657.

Janson, C. H. (1992) Measuring evolutionary constraints: a markov model for phylogenetic transitions among seed dispersal syndromes. *Evolution*, 46, 136–158.

Janzen, D. H. (1969) Seed-eaters versus seed size, number, toxicity and dispersal. *Evolution*, 23, 1–27.

Janzen, D. H. (1970) Herbivores and the number of tree species in tropical forest. *American Naturalist*, 104, 501–528.

Janzen, D. H. (1975) Behavior of *Hymanaea coubaril* when its predispersal seed predator is absent. *Science*, 189, 145–147.

Janzen, D. H. (1976) Why bamboos wait so long to flower. *Annual Review of Ecology and Systematics*, 7, 347–391.

Janzen, D. H. (1980) Specificity of seed-attacking beetles in a Costa Rican deciduous forest. *Journal of Ecology*, 68, 929–952.

Jarne, P. (1995) Mating system, bottlenecks and genetic polymorphism in hermaphroditic animals. *Genetic Research (Cambridge)*, 65, 193–207.

Jarne, P. & Charlesworth, D. (1993) The evolution of the selfing rate in functionally hermaphrodite plants and animals. *Annual Review of Ecology and Systematics*, 23, 441–466.

Jefferies, R. L., Davy, A. J. & Rudmik, T. (1981) Population biology of the salt marsh annual *Sailcornia europaea* agg. *Journal of Ecology*, 69, 17–31.

Johnson, M. L. & Gaines, M. S. (1990) Evolution of dispersal: theoretical models and empirical tests using birds and mammals. *Annual Review of Ecology and Systematics*, 21, 449–480.

Johnson, W. C. & Webb, T. I. (1989) The role of blue jays (*Cyanocitta cristata* L.) in the postglacial dispersal of fagaceous trees in eastern North America. *Journal of Biogeography*, 16, 561–571.

Johnston, M. O. & Schoen, D. J. (1995) Mutation rates and dominance levels of genes affecting total fitness in two angiosperm species. *Science*, 267, 226–229.

Jolliffe, P. A. (2000) The replacement series. *Journal of Ecology*, 88, 371–385.

Jones, C. G., Ostfeld, R. S., Richard, M. P., Schauber, E. M. & Wolff, J. O. (1998) Chain reactions linking acorns to gypsy moth outbreaks and Lyme disease risk. *Science*, 279, 1023–1026.

de Jong, T. J., Klinkhamer, P. G. L. & Metz, J. A. J. (1987) Selection for biennial life histories in plants. *Vegetatio*, 70, 149–156.

Jónsdóttir, I. S. & Callaghan, T. V. (1990) Intraclonal translocation of ammonium and nitrate nitrogen in *Carex bigelowii* Torr. ex Schwein. using ^{15}N and nitrate reductase assays. *New Phytologist*, 114, 419–428.

Jónsdóttir, I. S. & Watson, M. A. (1997) Extensive physiological integration: An adaptive trait in resource-poor environments? *The Ecology and Evolution of Clonal Plants* (eds H. de Kroon & J. van Groenendael), pp. 109–136. Backhuys, Leiden.

Kachi, N. & Hirose, T. (1985) Population dynamics of *Oenothera glazioviana* in a sand-dune system with special reference to the adaptive significance of size-dependent reproduction. *Journal of Ecology*, 73, 887–901.

Kadereit, J. W. & Briggs, D. (1985) Speed of development of radiate and non-radiate plants of *Senecio vulgaris* L. from habitats subject to different degrees of weeding pressure. *New Phytologist*, 99, 155–169.

Kadmon, R. & Shmida, A. (1990) Spatiotemporal demographic processes in plant populations: an approach and a case study. *American Naturalist*, 135, 382–397.

Kadmon, R. & Tielborger, K. (1999) Testing for source-sink population dynamics: an experimental approach exemplified with desert annuals. *Oikos*, 86, 417–429.

Kalisz, S. & McPeek, M. A. (1992) The demography of an age-structured annual: resampled projection matrices, elasticity analysis and seed bank effects. *Ecology*, 73, 1082–1094.

Kay, Q. O. N. (1978) The rôle of preferential and assortative pollination in the maintenance of flower colour polymorphisms. *The Pollination of Flowers by Insects* (ed. A. J. Richards), pp. 175–190. Linnean Society, London.

Kays, S. & Harper, J. L. (1974) The regulation of plant and tiller density in a grass sward. *Journal of Ecology*, 62, 97–105.

Keddy, P. A. (1981) Experimental demography of the sand-dune annual, *Cakile edentula*, growing along an environmental gradient in Nova Scotia. *Journal of Ecology*, 69, 615–630.

Keddy, P. A. (1982) Population ecology on an environmental gradient: *Cakile edentula* on a sand dune. *Oecologia*, 52, 348–355.

Keddy, P. A. & Constabel, P. (1986) Germination of ten shoreline plants in relation to seed size, soil particle size and water level: an experimental study. *Journal of Ecology*, 74, 133.

Keeler, K. (1978) Intra-population differentiation in annual plants II. Electrophoretic variation in *Veronica peregrina*. *Evolution*, 32, 638–645.

Keeley, J. E. & Bond, W. J. (1999) Mast flowering and semelparity in bamboos: The bamboo fire cycle hypothesis. *American Naturalist*, 154, 383–391.

Kelly, D. (1994) The evolutionary ecology of mast seeding. *Trends in Ecology and Evolution*, 9, 465–470.

Kelly, C. K. (1995) Seed size in tropical trees: a comparative study of factors affecting seed size in Peruvian angiosperms. *Oecologia*, 102, 377–388.

Kelly, C. K. & Purvis, A. (1993) Seed size and establishment conditions in tropical trees – on the use of taxonomic relatedness in determining ecological patterns. *Oecologia*, 94, 356–360.

Kelly, D. & Sullivan, J. J. (1997) Quantifying the benefits of mast seeding on predator satiation and wind pollination in *Chionochloa pallens* (Poaceae). *Oikos*, 78, 143–150.

Kelly, A. J. & Willis, J. H. (1998) Polymorphic microsatellite loci in *Mimulus guttatus* and related species. *Molecular Ecology*, 7, 769–774.

Kemp, P. R. (1989) *Seed Banks and Vegetation Processes in Desert*. Ecology of soil seed banks. Academic Press, San Diego, CA.

Kendall, B. E. (1998) Estimating the magnitude of environmental stochasticity in survivorship data. *Ecological Applications*, 8, 184–193.

Kenkel, N. C. (1988) Pattern of self-thinning in Jack pine: testing the random mortality hypothesis. *Ecology*, 69, 1017–1024.

Kenkel, N. C., Hendrie, M. L. & Bella, I. E. (1997) A long-term study of *Pinus banksiana* population dynamics. *Journal of Vegetation Science*, 8, 241–254.

Kimura, M. (1983) *The Neutral Theory of Molecular Evolution*. Cambridge University Press, Cambridge.

Kimura, M. & Ohta, T. (1971) *Theoretical Topics in Population Genetics*. Princeton University Press, Princeton, NJ.

King, T. J. (1977) The plant ecology of ant-hills in calcareous grasslands I. patterns of species in relation to ant-hills in southern England. *Journal of Ecology*, 65, 235–256.

King, R. A. & Ferris, C. (2000) Chloroplast DNA and nuclear DNA variation in the sympatric alder species, *Alnus cordata* (Lois.) Duby and *A. glutinosa* (L.) Gaertn. *Biological Journal of the Linnean Society*, 70, 147–160.

Kirkpatrick, M. & Barton, N. H. (1997) Evolution of a species' range. *American Naturalist*, 150, 1–23.

Kitajima, K. & Augspurger, C. K. (1989) Seed and seedling ecology of a monocarpic tropical tree, *Tachigalia versicolor*. *Ecology*, 70, 1102–1114.

Kleijn, D. & van Groenendael, J. M. (1999) The exploitation of heterogeneity by a clonal plant in habitats with contrasting productivity levels. *Journal of Ecology*, 87, 873–884.

Klekowski, E. J. & Godfrey, P. J. (1989) Aging and mutation in plants: a comparison of woody mangroves and herbaceous annuals. *Nature*, 340, 389–391.

Klekowski, E. J. & Lloyd, R. M. (1968) Reproductive biology of the Pteridophyta. 1. General considerations and a study of *Onoclea sensibilis* L. *Botanical Journal of the Linnean Society*, 60, 315–324.

Kobe, R. K. (1999) Light gradient partitioning among tropical tree species through differential seedling mortality and growth. *Ecology*, 80, 187–201.

Kobe, R. K., Pacala, S. W., Silander, J. A. & Canham, C. D. (1995) Juvenile tree survivorship as a component of shade tolerance. *Ecological Applications*, 5, 517–532.

Koenig, W. D. & Knops, J. M. H. (1998) Scale of mast-seeding and tree-ring growth. *Nature*, 396, 225–226.

Koenig, W. D., Knops, J. M. H., Carmen, W. J., Stanback, M. T. & Mumme, R. L. (1996) Acorn production by oaks in central coastal California: Influence of weather at three levels. *Canadian Journal of Forest Research*, 26, 1677–1683.

Koenig, W. D., Mumme, R. L., Carmen, W. J. & Stanback, M. T. (1994) Acorn production by oaks in central coastal California – variation within and among years. *Ecology*, 75, 99–109.

Kohn, J. (1988) Why be female? *Nature*, 335, 431–433.

Konieczny, A. & Ausubel, F. M. (1993) A procedure for mapping *Arabidopsis* mutations using co-dominant ecotype-specific PCR-based markers. *Plant Journal*, 4, 403–410.

de Kroon, H. & Hutchings, M. J. (1995) Morphological Plasticity in Clonal Plants – The Foraging Concept Reconsidered. *Journal of Ecology*, 83, 143–152.

de Kroon, H., Plaisier, A. & van Groenendael, J. M. (1986) Elasticity: the relative contribution of demographic parameters to population growth rate. *Ecology*, 67, 1427–1431.

de Kroon, H., van Groenendael, J. & Ehrlén, J. (2000) Elasticities: a review of methods and model limitations. *Ecology*, 81, 607–618.

Kudoh, H., Ishiguri, Y. & Kawano, S. (1996) Phenotypic plasticity in age and size at maturity and its effects on the integrated phenotypic expressions of life history traits of *Cardamine flexuosa* (Cruciferae). *Journal of Evolutionary Biology*, 9, 541–570.

Kumar, A. & Bennetzen, J. L. (1999) Plant retrotransposons. *Annual Review of Genetics*, 33, 479–532.

Kumar, V., Spangenberg, O. & Konrad, M. (2000) Cloning of the guanylate kinase homologues ACK-1 and AGK-2 from *Arabidopsis thaliana* and characterization of AGK-1. *European Journal of Biochemistry*, 207, 606–615.

Lacey, E. P. (1986) The genetic and environmental control of reproductive timing in a short-lived monocarpic species *Daucus carota* (Umbelliferae). *Journal of Ecology*, 74, 73.

Lacey, E. P. (1988) Latitudinal variation in reproductive timing of a short-lived monocarp, *Daucus carota* (Apiaceae). *Ecology*, 69, 220–232.

Lack, A. J. & Kay, Q. O. N. (1988) Allele frequencies, genetic relationships and heterozygosity in *Polygala vulgaris* populations from contrasting habitats in southern Britain. *Botanical Journal of the Linnean Society*, 34, 119–147.

Lagercrantz, U. & Ryman, N. (1990) Genetic structure of Norway spruce (*Picea abies*): concordance of morphological and allozyme variation. *Evolution*, 44, 38–53.

Lalonde, R. G. & Roitberg, B. D. (1989) Resource limitation and offspring size and number trade-offs in *Cirsium arvense* (Asteraceae). *American Journal of Botany*, 76, 1107–1113.

Lalonde, R. G. & Roitberg, B. D. (1992) On the evolution of masting behaviour in trees: predation or weather. *American Naturalist*, 139, 1293–1304.

Lamont, B. B., Klinkhamer, P. G. L. & Witkowski, E. T. F. (1993) Population fragmentation may reduce fertility to zero in *Banksia goodii* – a demonstration of the Allee effect. *Oecologia*, 94, 446–450.

Lande, R. (1988) Genetics and demography in biological conservation. *Science*, 241, 1455–1460.

Lande, R. & Arnold, S. J. (1983) The measurement of selection on correlated characters. *Evolution*, 37, 1210–1226.

Lande, R. & Schemske, D. W. (1985) The evolution of self-fertilization and inbreeding depression in plants. I. Genetic models. *Evolution*, 39, 24–40.

Larson, M. M. & Schubert, G. H. (1970) Cone crops of ponderosa pine in central Arizona, including the influence of Abert squirrels. *USDA Forest Service Research Paper*, RM58, 1–15.

Law, R. (1979) The cost of reproduction in annual meadow grass. *American Naturalist*, 113, 3–16.

Law, R. (1983) A model for the dynamics of a plant population containing individuals classified by age and size. *Ecology*, 64, 224–230.

Law, R., Bradshaw, A. D. & Putwain, P. D. (1977) Life history variation in *Poa annua*. *Evolution*, 31, 233–246.

Law, R., Herben, T. & Dieckmann, U. (1997) Non-manipulative estimates of competition coefficients in a montane grassland community. *Journal of Ecology*, 85, 505–518.

Law, R. & Watkinson, A. R. (1987) Response-surface analysis of two-species competition: an experiment on *Phleum arenarium* and *Vulpia fasciculata*. *Journal of Ecology*, 75, 871–886.

Lawrence, M. J. (1984) The genetical analysis of ecological traits. *Evolutionary Ecology* (ed. B. Shorrocks), pp. 27–63. Blackwell Scientific Publications, Oxford.

Lawrence, M. J., Lane, M. D., O'Donnell, S. & Franklin-Tong, V. E. (1993) The population genetics of the self-incompatibility polymorphism in *Papaver rhoeas*. V. Cross-classification of the S-alleles from three natural populations. *Heredity*, 71, 581–590.

Leishman, M. R. & Westoby, M. (1994a) The role of seed size in seedling establishment in dry soil conditions – Experimental evidence from semi-arid species. *Journal of Ecology*, 82, 249–258.

Leishman, M. R. & Westoby, M. (1994b) The role of large seed size in shaded conditions – experimental evidence. *Functional Ecology*, 8, 205–214.

Leishman, M. R. & Westoby, M. (1998) Seed size and shape are not related to persistence in soil in Australia in the same way as in Britain. *Functional Ecology*, 12, 480–485.

Leishman, M. R., Westoby, M. & Jurado, E. (1995) Correlates of seed size variation: a comparison among five temperate floras. *Journal of Ecology*, 83, 517–529.

Lennon, J. J., Turner, J. R. G. & Connell, D. (1997) A metapopulation model of species boundaries. *Oikos*, 78, 486–502.

Levin, D. A. & Clay, K. (1984) Dynamics of synthetic *Phlox drummondii* populations at the species margin. *American Journal of Botany*, 71, 1040–1050.

Levin, S. A., Cohen, D. & Hastings, A. (1984) Dispersal strategies in patchy environments. *Theoretical Population Biology*, 26, 165–191.

Levin, D. A. & Kerster, H. W. (1974) Gene flow in seed plants. *Evolutionary Biology*, 7, 139–220.

Levin, D. A. & Wilson, J. B. (1978) The genetic implications of ecological adaptations in plants. *Structure and Functioning of Plant Communities* (eds A. H. J. Freysen & J. W. Woldendorp), pp. 75–100. PUDOC, Wageningen.

Levings, C. S., Kim, B. D., Pring, D. R., Conde, M. F., Mans, R. J., Laughnan, J. R. & Gabay-Laughnan, S. J. (1980) Cytoplasmic reversion of *cms-S* in maize: association with a transpositional event. *Science*, 209, 1021–1023.

Levins, R. (1970) Extinction. *Lectures on Mathematics in the Life Science 2* (ed. M. Gerstenhaber), pp. 77–107. American Mathematical Society, Providence, RI.

Levins, R. & Culver, D. (1971) Regional coexistence of species and competition between rare species. *Proceedings of the National Academy of Science of the USA*, 68, 1246–1248.

Lewis, D. (1941) Male sterility in natural populations of hermaphrodite plants. *New Phytologist*, 40, 56–63.

Lewontin, R. C. (1974) *The Genetic Basis of Evolutionary Change*. Columbia University Press, New York.

Li, W.-H. (1997) *Principles of Molecular Evolution*. Sinauer Associates, Sunderland, MA.

Lieberman, D. & Lieberman, M. (1987) Forest tree growth and dynamics at La Selva, Costa Rica (1969–82). *Journal of Tropical Ecology*, 3, 347–358.

Lieberman, D., Lieberman, M., Peralta, R. & Hartshorn, G. S. (1985a) Mortality patterns and stand turnover rates in a wet tropical forest in Costa Rica. *Journal of Ecology*, 73, 915–924.

Lieberman, D., Lieberman, M., Hartshorn, G. & Peralta, R. (1985b) Growth rates and age-size relationships of tropical wet forest trees in Costa Rica. *Journal of Tropical Ecology*, 1, 97–109.

Lieberman, M., Lieberman, D., Peralta, R. & Hartshorn, G. S. (1995) Canopy closure and the distribution of tropical forest tree species at La Selva, Costa Rica. *Journal of Tropical Ecology*, 11, 161–178.

Lindquist, J. L., Rhode, D., Puettmann, K. J. & Maxwell, B. D. (1994) The influence of plant population spatial arrangement on individual plant yield. *Ecological Applications*, 4, 518–524.

Linhart, Y. B. (1988) Intrapopulation differentiation in annual plants. 3. The contrasting effects of intraspecific and interspecific competition. *Evolution*, 42, 1047–1064.

Linhart, Y. B. & Baker, I. (1973) Intra-population differentiation in response to flooding in a population of *Veronica peregrina* L. *Nature*, 242, 275–276.

Linhart, Y. B. & Grant, M. C. (1996) Evolutionary significance of local genetic differentiation in plants. *Annual Review of Ecology and Systematics*, 27, 237–277.

Lloyd, D. G. (1965) Evolution of self-compatibility and racial differentiation in *Leavenworthia* (Cruciferae). *Contributions to the Gray Herbarium, Harvard University*, 195, 3–134.

Lloyd, D. G. (1975) The maintenance of gynodioecy and androdioecy in angiosperms. *Genetica*, 45, 325–339.

Lloyd, D. G. (1979a) Evolution towards dioecy in heterostylous plants. *Plant Systematics and Evolution*, 131, 71–80.

Lloyd, D. G. (1979b) Some reproductive factors affecting the selection of self-fertilization in plants. *American Naturalist*, 113, 67–79.

Lloyd, D. G. (1980a) Demographic factors and mating patterns in angiosperms. *Demography and Evolution in Plant Populations* (ed. O. T. Solbrig), pp. 67–88. Blackwell Scientific Publications, Oxford.

Lloyd, D. G. (1980b) The distributions of gender in four angiosperm species illustrating two evolutionary pathways to dioecy. *Evolution*, 34, 123–134.

Lloyd, D. G. (1984) Gender allocations in outcrossing cosexual plants. *Perspectives on Plant Population Ecology* (eds R. Dirzo & J. Sarukhan), pp. 277–300. Sinauer Associates, Sunderland, MA.

Lloyd, D. G. (1987a) Allocations to pollen, seeds and pollination mechanisms in self-fertilizing plants. *Functional Ecology*, 1, 83–89.

Lloyd, D. G. (1987b) Selection of offspring size at independence and other size-versus-number strategies. *American Naturalist*, 129, 800–817.

Loehle, C. (1988) Tree life histories: the role of defences. *Canadian Journal of Forest Research*, 18, 209–222.

Lonsdale, W. M. (1990) The self-thinning rule: dead or alive? *Ecology*, 71, 1373–1388.

Lord, E. M. (1981) Cleistogamy: a tool for the study of floral morphogenesis, function and evolution. *Botanical Review*, 47, 421–449.

Lord, J., Egan, J., Clifford, T. *et al.* (1997) Larger seeds in tropical floras: Consistent patterns independent of growth form and dispersal mode. *Journal of Biogeography*, 24, 205–211.

Lord, J., Westoby, M. & Leishman, M. (1995) Seed size and phylogeny in 6 temperate floras – constraints, niche conservatism, and adaptation. *American Naturalist*, 146, 349–364.

Lotz, L. A. P. (1990) The relation between age and size at first flowering of *Plantago major* in various habitats. *Journal of Ecology*, 78, 757–771.

Loveless, M. D., Hamrick, J. L. & Foster, R. B. (1998) Population structure and mating system in *Tachigali versicolor*, a monocarpic neotropical tree. *Heredity*, 81, 134–143.

Lovett Doust, L. (1981) Population dynamics and local specialization in a clonal perennial (*Ranunculus repens*). I. The dynamics of ramets in contrasting habitats. *Journal of Ecology*, 69, 743–755.

MacArthur, R. H. (1972) *Geographical Ecology*. Princeton University Press, Princeton, NJ.

MacArthur, R. H. & Wilson, E. O. (1967) *The Theory of Island Biogeography*. Princeton University Press, Princeton, NJ.

MacDonald, S. E. & Chinnappa, C. C. (1989) Population differentiation for phenotypic plasticity in the *Stellaria longipes* complex. *American Journal of Botany*, 76, 1627–1637.

MacDonald, G. M. & Cwynar, L. C. (1991) Post-glacial population growth rates of *Pinus contorta* ssp. *latifolia* in western Canada. *Journal of Ecology*, 79, 417–429.

Mack, R. N. (1981) Invasion of *Bromus tectorum* L. into western North America: An ecological chronicle. *Agroecosystems*, 7, 145–165.

Mack, A. L. (1998) An advantage of large seed size: Tolerating rather than succumbing to seed predators. *Biotropica*, 30, 604–608.

Mack, R. N. & Pyke, D. A. (1983) The demography of *Bromus tectorum*: variation in time and space. *Journal of Ecology*, 71, 69–93.

Macnair, M. R. & Cumbes, Q. J. (1990) The genetic architecture of interspecific variation in *Mimulus*. *Genetics*, 122, 211–222.

Mahdi, R., Law, R. & Willis, A. J. (1989) Large niche overlaps among coexisting plant species in a limestone grassland community. *Journal of Ecology*, 77, 386–400.

Mahy, G., Vekemans, X. & Jacquemart, A. L. (1999) Patterns of allozymic variation within Calluna vulgaris populations at seed bank and adult stages. *Heredity*, 82, 432–440.

Marks, M. & Prince, S. (1981) Influence of germination date on survival and fecundity in wild lettuce *Lactuca serriola*. *Oikos*, 36, 326–330.

Marshall, C. (1990) *Source-Sink Relations of Interconnected Ramets. Clonal growth in plants*. SPB Academic, The Hague.

Martin, M. P. L. D. & Field, R. J. (1984) The nature of competition between perennial ryegrass and white clover. *Grass and Forage Science*, 39, 247–253.

Martinez-Ramos, M., Alvarez-Buylla, E. & Pinero, D. (1988) Treefall age determination and gap dynamics in a tropical forest. *Journal of Ecology*, 76, 700–716.

Martínez-Ramos, M., Alvarez-Buylla, E. & Sarukhán, J. (1989) Tree demography and gap dynamics in a tropical rain forest. *Ecology*, 70, 555–558.

Matheson, A. C. & Raymond, C. A. (1986) A review of provenance–environment interaction: its practical importance and use with particular reference to the tropics. *Commonwealth Forestry Review*, 65, 283–302.

Mathews, S., Lavin, M. & Sharrock, R. A. (1995) Evolution of the phytochrome gene family and its utility for phylogenetic analyses of angiosperms. *Annals of Missouri Botanical Garden*, 82, 296–321.

Mathews, S. & Sharrock, R. A. (1997) Phytochrome gene diversity. *Plant, Cell and Environment*, 20, 666–671.

de Matos, M. B. & Silva Matos, D. M. (1998) Mathematical constraints on transition matrix elasticity analysis. *Journal of Ecology*, 86, 706–716.

Mátyás, C. (1994) Modeling climate change effects with provenance test data. *Tree Physiology*, 14, 797–804.

Mátyás, C. & Yeatman, C. W. (1992) Effect of geographical transfer on growth and survival of jack pine (*Pinus banksiana* Lamb) populations. *Silvae Genetica*, 41, 370–376.

Mayer, S. S. & Charlesworth, D. (1991) Cryptic dioecy in flowering plants: its occurrence and significance. *Trends in Ecology and Evolution*, 6, 320–325.

Mayer, S. S. & Charlesworth, D. (1992) Genetic evidence for multiple origins of dioecy in the Hawaiian shrub *Wikstroemia*. *Evolution*, 46, 207–215.

Maynard Smith, J. (1978) *The Evolution of Sex*. Cambridge University Press, Cambridge.

Maynard Smith, J. (1993) *The Theory of Evolution*. Cambridge University Press, Cambridge.

Mazer, S. J. (1987) Parental effects on seed development and seed yield in *Raphanus raphanistrum*: Implications for natural and sexual selection. *Evolution*, 41, 340–354.

Mazer, S. (1989) Ecological, taxonomic and life history correlates of seed mass among Indiana dune angiosperms. *Ecological Monographs*, 59, 153–175.

Mazer, S. J. (1990) Seed mass of Indiana Dune genera and families: taxonomic and ecological correlates. *Evolutionary Ecology*, 4, 326–357.

Mazer, S. J., Delesalle, V. A. & Neal, P. R. (1999) Responses of floral traits to selection on primary sexual investment in *Spergularia marina*: The battle between the sexes. *Evolution*, 53, 717–731.

McAuliffe, J. R. (1984) Saguaro–nurse tree associations in the Sonoran Desert; competitive effects of saguaros. *Oecologia*, 64, 319–321.

McCauley, D. E. (1994) Contrasting the distribution of chloroplast DNA and allozyme polymorphism among local populations of *Silene alba*: implications for studies of gene flow in plants. *Proceedings of the National Academy of Sciences of the USA*, 91, 8127–8131.

REFERENCES

McCauley, D. E. (1997) The relative contributions of seed and pollen movement to the local genetic structure of *Silene alba*. *Journal of Heredity*, 88, 257–263.

McCauley, D. E., Raveill, J. & Antonovics, J. (1995) Local founding events as determinants of genetic structure in a plant metapopulation. *Heredity*, 75, 630–636.

McCauley, D. E., Richards, C. M., Emery, S. N., Smith, R. A. & McGlothlin, J. W. (2001) The interaction of genetic and demographic processes in plant metapopulations: A case study of *Silene alba*. *Integrating Ecology and in a Spatial Context* (eds J. Silvertown & J. Antonovics). Blackwell Science, Oxford.

McCue, K. A. & Holtsford, T. P. (1998) Seed bank influences on genetic diversity in the rare annual *Clarkia springvillensis* (Onagraceae). *American Journal of Botany*, 85, 30–36.

McGinley, M. A., Smith, C. C., Elliott, P. F. & Higgins, J. J. (1990) Morphological constraints on seed mass in lodgepole pine. *Functional Ecology*, 4, 183–192.

McGraw, J. B. & Antonovics, J. (1983) Experimental ecology of *Dryas octopetala* ecotypes I. Ecotypic differentiation and life-cycle stages of selection. *Journal of Ecology*, 71, 879–897.

McMaster, G. S. & Zedler, P. H. (1981) Delayed seed dispersal in *Pinus torreyana* (Torrey pine). *Oecologia*, 51, 62–66.

Meagher, T. R. (1986) Analysis of paternity within a single natural population of *Chamaelirium luteum*. 1. Identification of most-likely male parents. *American Naturalist*, 128, 199–215.

Meagher, T. R. & Thompson, E. (1987) Analysis of parentage for naturally established seedlings of *Chamaelirium luteum* (Liliaceae) *Ecology*, 68, 803–812.

van der Meijden, E., Klinkhamer, P. G. L., De Jong, T. J. & Vanwijk, C. A. M. (1992) Meta-Population dynamics of biennial plants – how to exploit temporary habitats. *Acta Botanica Neerlandica*, 41, 249–270.

Menges, E. S. (2000) Population viability analyses in plants: challenges and opportunities. *Trends in Ecology and Evolution*, 15, 51–56.

Metcalfe, D. J. & Grubb, P. J. (1995) Seed mass and light requirement for regeneration in South-East Asian rain forest. *Canadian Journal of Botany*, 73, 817–826.

Meyer, S. E. & Allen, P. S. (1999a) Ecological genetics of seed germination regulation in *Bromus tectorum* L. I. Phenotypic variance among and within populations. *Oecologia*, 120, 27–34.

Meyer, S. E. & Allen, P. S. (1999b) Ecological genetics of seed germination regulation in *Bromus tectorum* L. II. Reaction norms in response to a water stress gradient imposed during seed maturation. *Oecologia*, 120, 35–43.

Michaels, H. J., Benner, B., Hartgerink, A. P. *et al.* (1988) Seed size variation: magnitude, distribution, and ecological correlates. *Evolutionary Ecology*, 2, 157–166.

Michalakis, Y. & Excoffier, L. (1996) A generic estimation of population subdivision using distance between alleles with special reference for microsatellite loci. *Genetics*, 142, 1061–1064.

Michelmore, R. W. & Meyers, B. C. (1998) Clusters of resistance genes in plants evolve by divergent selection and a birth-and-death process. *Genome Research*, 8, 1113–1130.

Milberg, P. & Lamont, B. B. (1997) Seed/cotyledon size and nutrient content play a major role in early performance of species on nutrient-poor soils. *New Phytologist*, 137, 665–672.

Milberg, P., Perez Fernandez, M. A. & Lamont, B. B. (1998) Seedling growth response to added nutrients depends on seed size in three woody genera. *Journal of Ecology*, 86, 624–632.

Miller, T. E. (1987) Effects of emergence time on survival and growth in an early old-field plant community. *Oecologia*, 72, 272–278.

Miller, T. E. & Werner, P. A. (1987) Competitive effects and responses between plant species in a first-year old-field community. *Ecology*, 68, 1201–1210.

Milligan, B. G. (1992) Is organelle DNA strictly maternally inherited? Power analysis of a binomial distribution. *American Journal of Botany*, 79, 1325–1328.

Mills, K. E. & Bever, J. D. (1998) Maintenance of diversity within plant communities: Soil pathogens as agents of negative feedback. *Ecology*, 79, 1595–1601.

Mitchell-Olds, T. & Shaw, R. G. (1987) Regression analysis of natural selection: statistical inference and biological interpretation. *Evolution*, 41, 1149–1161.

Mithen, R., Harper, J. L. & Weiner, J. (1984) Growth and mortality of individual plants as a function of 'available area'. *Oecologia*, 57, 57–60.

Miyashita, N. T., Kawabe, A., Innan, H. & Terauchi, R. (1998) Intra- and interspecific DNA variation and codon bias of the alcohol dehydrogenase (*Adh*) locus in *Arabis* and *Arabidopsis* species. *Molecular Biology and Evolution*, 15, 1420–1429.

Moegenburg, S. M. (1996) Sabal palmetto seed size: Causes of variation, choices of predators, and consequences for seedlings. *Oecologia*, 106, 539–543.

Mogie, M. & Hutchings, M. J. (1990) Phylogeny, ontogeny and clonal growth in vascular plants. *Clonal Growth in Plants* (eds J. van Groenendael & H. de Kroon), pp. 3–22. SPB Academic, The Hague.

Mohler, C. L. (1990) Co-occurrence of oak sub-genera: Implications for niche differentiation. *Bulletin of Torrey Botanical Club*, 117, 247–255.

Mohler, C. L., Marks, P. L. & Sprugel, D. G. (1978) Stand structure and allometry of trees during self-thinning of pure stands. *Journal of Ecology*, 66, 599–614.

Mojonnier, L. (1998) Natural selection on two seed-size traits in the common morning glory *Ipomoea purpurea* (Convolvulaceae): Patterns and evolutionary consequences. *American Naturalist*, 152, 188–203.

Moloney, K. A. (1988) Fine-scale spatial and temporal variation in the demography of a perennial bunchgrass. *Ecology*, 69, 1588–1598.

Moloney, K. A. (1990) Shifting demographic control of a perennial bunchgrass along a natural habitat gradient. *Ecology*, 71, 1133–1143.

Moody, M. E. & Mack, R. N. (1988) Controlling the spread of plant invasions: the importance of nascent foci. *Journal of Applied Ecology*, 25, 1009–1021.

Moora, M. & Zobel, M. (1998) Can arbuscular mycorrhiza change the effect of root competition between conspecific plants of different ages? *Canadian Journal of Botany*, 76, 613–619.

del Moral, R. (1983) Competition as a control mechanism in subalpine meadows. *American Journal of Botany*, 70, 232–245.

Morgenstern, E. K. & Mullin, T. J. (1990) Growth and survival of black spruce in the range-wide provenance study. *Canadian Journal of Botany*, 20, 130–143.

Mortimer, A. M. (1984) Population ecology and weed science. *Perspectives on Plant Population Ecology* (eds R. Dirzo & J. Sarukhán), pp. 363–388. Sinauer Associates, Sunderland, MA.

Muenchow, G. A. (1987) Is dioecy associated with fleshy fruit?. *American Journal of Botany*, 74, 287–293.

Muir, P. S. & Lotan, J. E. (1984) Serotiny and life history of *Pinus contorta* var. latifolia. *Canadian Journal of Botany*, 63, 938–945.

Muir, P. S. & Lotan, J. E. (1985) Disturbance history and serotiny of *Pinus contorta* in Western Montana. *Ecology*, 66, 1658–1668.

Nason, J. D. & Hamrick, J. L. (1997) Reproductive and genetic consequences of forest fragmentation: Two case studies of neotropical canopy trees. *Journal of Heredity*, 88, 264–276.

Nasrallah, M. E. & Nasrallah, J. B. (1986) Molecular biology of self-incompatibility in plants. *Trends in Genetics*, 2, 239–244.

Neet, C. R. (1989) Niche overlap measures and hypothesis testing: a review with particular reference to empirical applications. *Coenoses*, 4, 137–144.

Nei, M. (1987) *Molecular Evolutionary Genetics*. Columbia University Press, New York.

Nei, M., Maruyama, T. & Chakraborty, R. (1975) The bottleneck effect and genetic variability in populations. *Evolution*, 29, 1–10.

Nevo, E., Beiles, A. & Krugman, T. (1988) Natural selection of allozyme polymorphisms: a microgeographic climatic differentiation in wild emmer wheat (*Triticum dicoccoides*). *Theoretical and Applied Genetics*, 75, 529–538.

New, J. (1958) A population study of *Spergula arvensis* I. Two clines and their significance. *Annals of Botany*, 22, 457–477.

New, J. K. (1978) Change and stability of clines in *Spergula arvensis* L. (corn spurrey) after 20 years. *Watsonia*, 12, 137–143.

New, J. K. & Herriott, J. C. (1981) Moisture for germination as a factor affecting the distribution of the seedcoat morphs of *Spergula arvensis* L. *Watsonia*, 13, 323–324.

Newman, D. & Pilson, D. (1997) Increased probability of extinction due to decreased genetic effective population size: experimental populations of *Clarkia pulchella*. *Evolution*, 51, 354–362.

Ng, F. S. P. (1983) Ecological principles of tropical lowland rain forest conservation. *Tropical Rain Forest Ecology and Management* (eds S. L. Sutton, T. C. Whitmore & A. C. Chadwick), pp. 359–375. Blackwell Scientific Publications, Oxford.

Nilsson, P., Fagerstrom, T., Tuomi, J. & Astrom, M. (1994) Does seed dormancy benefit the mother plant by reducing sib competition? *Evolutionary Ecology*, 8, 422–430.

Nilsson, L. A., Rabakonandrianina, E. & Petersson, B. (1992) Exact tracking of pollen transfer and mating in plants. *Nature*, 360, 666–668.

Nilsson, S. G. & Wästljung, U. (1987) Seed predation and cross-pollination in mast-seeding beech (*Fagus sylvatica*) patches. *Ecology*, 68, 260–265.

Nordborg, M. & Bergelson, J. (1999) The effect of seed and rosette cold treatment on germination and flowering time in some *Arabidopsis thaliana* (Brassicaceae) ecotypes. *American Journal of Botany*, 86, 470–475.

Novoplansky, A., Cohen, D. & Sachs, T. (1990) How Portulaca seedlings avoid their neighbours. *Oecologia*, 82, 490–493.

Nuñez-Farfan, J. & Dirzo, R. (1988) Within-gap spatial heterogeneity and seedling performance in a Mexican tropical rainforest. *Oikos*, 51, 274–284.

O'Connor, T. G. (1993) The influence of rainfall and grazing on the demography of some African savanna grasses: a matrix modelling approach. *Journal of Applied Ecology*, 30, 119–132.

O'Donnell, S. & Lawrence, M. J. (1984) The population genetics of the self-incompatibility polymorphism of *Papaver rhoeas*. IV. The estimation of numbers of alleles in a population. *Heredity*, 53, 495–507.

Oborny, B. (1994) Growth rules in clonal plants and environmental predictability – A simulation study. *Journal of Ecology*, 82, 341–351.

Olivieri, I. & Gouyon, P. H. (1985) Seed dimorphism for dispersal: theory and observations. *Structure and Functioning of Plant Populations* (eds J. Haeck & J. W. Woldendorp), pp. 77–90. North Holland Press, Amsterdam.

Olivieri, I., Michalakis, Y. & Gouyon, P. H. (1995) Metapopulation genetics and the evolution of dispersal. *American Naturalist*, 146, 202–228.

Oostermeijer, G. (1996) Population viability of the rare *Gentiana pneumonanthe*. The relative importance of demography, genetics and reproductive biology. Amsterdam.

Oostermeijer, J. G. B. (2000) Population viability analysis of the rare *Gentiana pneumonanthe*: importance of demography, genetics and reproductive biology. In: *Genetics, Demography and Viability of Fragmented Populations* (eds A. Young & G. Clarke), pp. 313–334. Cambridge University Press, Cambridge.

Oostermeijer, J. G. B., van Eijck, M. W. & Dennijs, J. C. M. (1994) Offspring fitness in relation to population size and genetic variation in the rare perennial plant species *Gentiana pneumonanthe* (Gentianaceae). *Oecologia*, 97, 289–296.

Orive, M. E. (1995) Senescence in organisms with clonal reproduction and complex life-histories. *American Naturalist*, 145, 90–108.

Osawa, A. S. & Sugita, S. (1989) The self-thinning rule: Another interpretation of Weller's results. *Ecology*, 70, 279–283.

REFERENCES

Ouborg, N. J., Piquot, Y. & Groenendael, J. M. V. (1999) Population genetics, molecular markers and the study of dispersal in plants. *Journal of Ecology*, 87, 551–568.

Pacala, S. W. (1997) Dynamics of plant communities. *Plant Ecology* (ed. M. J. Crawley), pp. 532–555. Blackwell Science, Oxford.

Pacala, S. W., Canham, C. D., Saponara, J. *et al.* (1996) Forest models defined by field-measurements – estimation, error analysis and dynamics. *Ecological Monographs*, 66, 1–43.

Pacala, S. W., Canham, C. D. & Silander, J. A. (1993) Forest models defined by field-measurements. 1. The design of a northeastern forest simulator. *Canadian Journal of Forest Research*, 23, 1980–1988.

Pacala, S. W., Canham, C. D., Silander, J. A. & Kobe, R. K. (1994) Sapling growth as a function of resources in a north temperate forest. *Canadian Journal of Forest Research*, 24, 2172–2183.

Pacala, S. W. & Deutschman, D. H. (1995) Details that matter – the spatial-distribution of individual trees maintains forest ecosystem function. *Oikos*, 74, 357–365.

Pacala, S. W. & Silander, J. A. (1985) Neighbourhood models of plant population dynamics 1. Single-species models of annuals. *American Naturalist*, 125, 385–411.

Pacala, S. W. & Silander, J. A. J. (1987) Neighborhood interference among velvet leaf, *Abutilon theophrasti*, and pigweed, *Amaranthus retroflexus*. *Oikos*, 48, 217–224.

Pacala, S. W. & Silander, J. A. J. (1990) Tests of neighbourhood population dynamic models in field communities of two annual weed species. *Ecological Monographs*, 60, 113–134.

Packer, A. & Clay, K. (2000) Soil pathogens and spatial patterns of seedling mortality in a temperate tree. *Nature*, 404, 278–281.

Palmblad, I. G. (1968) Competition studies on experimental populations of weed with emphasis on the regulation of population size. *Ecology*, 49, 26–34.

Pamphilis, C. W. D. & Palmer, J. D. (1990) Loss of photosynthetic and chlororespiratory genes from the plastid genome of a parasitic flowering plant. *Nature*, 348, 337–339.

Pannell, J. (1997a) Mixed genetic and environmental sex determination in an androdioecious population of *Mercurialis annua*. *Heredity*, 78, 50–56.

Pannell, J. (1997b) Widespread functional androdioecy in *Mercurialis annua* L. (Euphorbiaceae). *Biological Journal of the Linnean Society*, 61, 95–116.

Pannell, J. R. & Barrett, S. C. H. (1998) Baker's law revisited: Reproductive assurance in a metapopulation. *Evolution*, 52, 657–668.

Pannell, J. R. & Charlesworth, B. (1999) Neutral genetic diversity in a metapopulation with recurrent local extinction and recolonization. *Evolution*, 53, 664–676.

Pantone, D. J. & Baker, J. B. (1991) Weed-crop competition models and response-surface analysis of red rice competition in cultivated rice - a review. *Crop Science*, 31, 1105–1110.

Parish, R. & Turkington, R. (1990a) The colonization of dung pats and molehills in permanent pastures. *Canadian Journal of Botany*, 68, 1706–1711.

Parish, R. & Turkington, R. (1990b) The influence of dung pats and molehills on pasture composition. *Canadian Journal of Botany*, 68, 1698–1705.

Parker, M. A. (1994) Pathogens and sex in plants. *Evolutionary Ecology*, 8, 560–584.

Pavone, L. V. & Reader, R. J. (1982) The dynamics of seed bank size and seed state of *Medicago lupulina*. *Journal of Ecology*, 70, 537–547.

Peck, J. R., Yearsley, J. M. & Waxman, D. (1998) Explaining the geographic distributions of sexual and asexual populations. *Nature*, 391, 889–892.

Pedersen, B. (1995) An evolutionary theory of clonal senescence. *Theoretical Population Biology*, 47, 292–320.

Peñalosa, J. (1983) Shoot dynamics and adaptive morphology of *Impomoea phillomega* (Vell.) House (Convolulaceae), a tropical rainforest liana. *Annals of Botany*, 52, 737–754.

Peroni, P. A. (1994) Seed size and dispersal potential of *Acer rubrum* (Aceraceae) samaras produced by populations in early and late successional environments. *American Journal of Botany*, 81, 1428–1434.

Peters, N. C. B. (1985) Competitive effects of *Avena fatua* L. plants derived from seeds of different weights. *Weed Research*, 25, 67–77.

Petit, R. J., Bialozyt, R., Brewer, S., Cheddadi, R. & Comps, B. (2001) From spatial patterns of genetic diversity to postglacial migration processes in forest trees. In: *Integrating Ecology and Evolution in a Spatial Context*. (eds J. Silvertown & J. Antonovics). Blackwell Science, Oxford.

Petit, R. J., Pineau, E., Demesure, B. *et al.* (1997) Chloroplast DNA footprints of postglacial recolonization by oaks. *Proceedings of the National Academy of Sciences of the USA*, 94, 9996–10001.

Pfister, C. A. (1998) Patterns of variance in stage-structured populations: Evolutionary predictions and ecological implications. *Proceedings of the National Academy of Sciences of the USA*, 95, 213–218.

Philippi, T. (1993) Bet-Hedging germination of desert annuals – variation among populations and maternal effects in *Lepidium lasiocarpum*. *American Naturalist*, 142, 488–507.

Phillips, D. L. & MacMahon, J. A. (1981) Competition and spacing patterns of desert shrubs. *Journal of Ecology*, 69, 97–115.

Pianka, E. R. (1970) On r- and K-selection. *American Naturalist*, 104, 592–597.

Pickett, S. T. A. & White, P. S. (1985) *The Ecology of Natural Disturbance and Patch Dynamics*. Academic Press, New York.

Pigliucci, M., Cammell, K. & Schmitt, J. (1999) Evolution of phenotypic plasticity a comparative approach in the phylogenetic neighbourhood of *Arabidopsis thaliana*. *Journal of Evolutionary Biology*, 12, 779–791.

Piñero, D., Martínez-Ramos, M. & Sarukhán, J. (1984) A population model of *Astrocaryum mexicanum* and a sensitivity analysis of its finite rate of increase. *Journal of Ecology*, 72, 977–991.

Piper, J., Charlesworth, B. & Charlesworth, D. (1986) Breeding system evolution in *Primula vulgaris* and the role of reproductive assurance. *Heredity*, 56, 207–217.

Pitelka, L. F. (1984) Application of the -3/2 power law to clonal herbs. *American Naturalist*, 123, 442–449.

Platenkamp, G. A. J. (1990) Phenotypic plasticity and genetic differentiation in the demography of the grass *Anthoxanthum odoratum*. *Journal of Ecology*, 78, 772–788.

Platt, W. J. (1975) The colonization and formation of equilibrium plant species associations on badger disturbances in tallgrass prairie. *Ecological Monographs*, 45, 285–305.

Platt, W. J. & Weiss, I. M. (1977) Resource partitioning and competition within a guild of fugitive prairie plants. *American Naturalist*, 111, 479–513.

Platt, W. J. & Weiss, I. M. (1985) An experimental study of competition among fugitive prairie plants. *Ecology*, 66, 708–720.

Portnoy, S. & Willson, M. F. (1993) Seed dispersal curves – behavior of the tail of the distribution. *Evolutionary Ecology*, 7, 25–44.

Prentice, H. C., Lonn, M., Lager, H., Rosen, E. & Van der Maarel, E. (2000) Changes in allozyme frequencies in *Festuca ovina* populations after a 9-year nutrient/water experiment. *Journal of Ecology*, 88, 331–347.

Prentice, H. C., Lonn, M., Lefkovitch, L. P. & Runyeon, H. (1995) Associations between allele frequencies in *Festuca ovina* and habitat variation in the alvar grasslands on the Baltic island of Oland. *Journal of Ecology*, 83, 391–402.

Price, E. A. C., Marshall, C. & Hutchings, M. J. (1992) Studies of growth in the clonal herbs *Glechoma hederaceae*. Patterns of physiological integration. *Journal of Ecology*, 80, 25–38.

Primack, R. B. & Antonovics, J. (1981) Experimental ecological genetics in *Plantago* V. Components of seed yield in the ribwort plantain *Plantago lanceolata* L. *Evolution*, 35, 1069–1079.

Proctor, M. C. F. & Yeo, P. F. (1973) *The Pollination of Flowers*. Collins, London.

Promislow, D. E. L. & Harvey, P. H. (1990) Living fast and dying young: a comparative analysis of life-history variation among mammals. *Journal of Zoology*, 220, 417–437.

Proulx, M. & Mazumder, A. (1998) Reversal of grazing impact on plant species richness in nutrient-poor vs. nutrient-rich ecosystems. *Ecology*, 79, 2581–2592.

Pryor, T. (1987) The origin and structure of fungal disease resistance in plants. *Trends in Genetics*, 3, 157–161.

Pulliam, H. R. (1989) Sources, sinks and population regulation. *American Naturalist*, 132, 652–661.

Putz, F. E. (1983) Treefall pits and mounds, buried seeds, and the importance of soil disturbance to pioneer trees on Barro Colorado Island, Panama. *Ecology*, 64, 1069–1074.

Quattro, J. M., Avise, J. C. & Vrijenhoek, R. C. (1991) Molecular evidence for multiple origins of hybridogenetic fish clones (Poeciliidae: *Poeciliopsis*). *Genetics*, 127, 391–398.

Rabinowitz, D. (1981) Seven forms of rarity. *The Biological Aspects of Rare Plant Conservation* (ed. H. Synge), pp. 205–217. Wiley, Chichester.

Rabinowitz, D., Rapp, J. K., Cairns, S. & Mayer, M. (1989) The persistence of rare prairie grasses in Missouri: Environmental variation buffered by reproductive output of sparse species. *American Naturalist*, 134, 525–544.

Radosevich, S. R. & Rousch, M. L. (1990) The role of competition in agriculture. *Perspectives on Plant Competition* (eds J. B. Grace & D. Tilman), pp. 341–363. Academic Press, San Diego, CA.

Raijmann, L. E. L., Leeuwen, N. C. V., Kersten, R. *et al.* (1994) Genetic variation and outcrossing rate in relation to population size in *Gentiana pneumonanthe* L. *Conservation Biology*, 8, 1014–1026.

Reader, R. J. (1992) Herbivory as a confounding factor in an experiment measuring competition among plants. *Ecology*, 73, 373–376.

Reboud, X. & Zeyl, C. (1994) Organelle inheritance in plants. *Heredity*, 72, 132–140.

Rees, M. (1993) Trade-offs among dispersal strategies in British plants. *Nature*, 366, 150–152.

Rees, M. (1994) Delayed germination of seeds: a look at the effects of adult longevity, the timing of reproduction, and population age/stage structure. *American Naturalist*, 144, 43–64.

Rees, M. (1995) Community structure in sand dune annuals: Is seed weight a key quantity? *Journal of Ecology*, 83, 857–863.

Rees, M. (1997) Evolutionary ecology of seed dormancy and seed size. *Plant Life Histories: Ecology, Phylogeny and Evolution* (eds J. Silvertown, M. Franco & J. L. Harper), pp. 121–142. Cambridge University Press, Cambridge.

Rees, M., Grubb, P. J. & Kelly, D. (1996) Quantifying the impact of competition and spatial heterogeneity on the structure and dynamics of a four-species guild of winter annuals. *American Naturalist*, 147, 1–32.

Rees, M. & Long, M. J. (1993) The analysis and interpretation of seedling recruitment curves. *American Naturalist*, 141, 233–262.

Rees, M. & Westoby, M. (1997) Game-theoretical evolution of seed mass in multi-species ecological models. *Oikos*, 78, 116–126.

van der Reest, P. J. & Rogaar, H. (1988) The effect of earthworm activity on the vertical distribution of plant seeds in newly reclaimed polder soils in the Netherlands. *Pedobiologia*, 31, 211–218.

Rehfeldt, G. E., Ying, C. C., Spittlehouse, D. L. & Hamilton, D. A. (1999) Genetic responses to climate in *Pinus contorta*: Niche breadth, climate change, and reforestation. *Ecological Monographs*, 69, 375–407.

Reichman, O. J. (1984) Spatial and temporal variation of seed distributions in Sonoran desert soils. *Journal of Biogeography*, 11, 1–11.

Reinartz, J. A. (1984) Life history variation of common mullein (*Verbascum thapsus*). I. Latitudinal differences in population dynamics and timing of reproduction. *Journal of Ecology*, 72, 897–912.

Rejmánek, M., Robinson, G. R. & Rejmánková, E. (1989) Weed-crop competition – experimental-designs and models for data-analysis. *Weed Science*, 37, 276–284.

Renner, S. S. & Ricklefs, R. E. (1995) Dioecy and its correlates in the flowering plants. *American Journal of Botany*, 82, 596–606.

REFERENCES

Rice, K. T. (1987) Evidence for the retention of genetic variation in Erodium seed dormancy by variable rainfall. *Oecologia*, 72, 589–596.

Rice, K. J. (1989) *Impacts of Seed Banks on Grassland Community Structure and Population Dynamics*. Ecology of Soil Seed Banks. Academic Press, San Diego, CA.

Rice, K. J. & Mack, R. N. (1991) Ecological genetics of *Bromus tectorum* III. The demography of repricrocally sown populations. *Oecologia*, 88, 91–101.

Rice, B. & Westoby, M. (1982) Heteroecious rusts as agents of interference competition. *Evolutionary Theory*, 6, 43–52.

Richards, A. J. (1997) *Plant Breeding Systems*, 2nd edn. Stanley Thorne, Cheltenham.

Richards, C. M. (2000) Inbreeding depression and genetic rescue in a plant metapopulation. *American Naturalist*, 155, 383–394.

Richards, C. M., Church, S. & McCauley, D. E. (1999) The influence of population size and isolation on gene flow by pollen in *Silene alba*. *Evolution*, 53, 63–73.

Ricklefs, R. E. (1977) Environmental heterogeneity and plant species diversity: a hypothesis. *American Naturalist*, 111, 376–381.

Riera, B. (1985) Importance des buttes de deracinement dans le regeneration forestiere en Guyane Français. *Revue Ecologie (Terre Vie)*, 40, 321–329.

Riley, J. (1984) A general form of the 'Land Equivalent Ratio'. *Experimental Agriculture*, 20, 19–29.

Ritland, K. (1990a) A series of FORTRAN computer programs for estimating plant mating systems. *Journal of Heredity*, 81, 235–237.

Ritland, K. (1990b) Gene identity and the genetic demography of plant populations. *Plant Population Genetics, Breeding, and Genetic Resources* (eds A. H. D. Brown, M. T. Clegg, A. L. Kahler & B. S. Weir), pp. 181–199. Sinauer Associates, Sunderland, MA.

Ritland, K. & Jain, S. K. (1981) A model for the estimation of outcrossing rate and gene frequencies using *n* independent loci. *Heredity*, 47, 35–52.

Roach, D. A. (1987) Variation in seed and seedling size in *Anthoxanthum odoratum*. *American Midland Naturalist*, 117, 258–264.

Roberts, H. A. (1970) Viable weed seeds in cultivated soils. *Reports of the National Vegetation Research Station*, 1969, 23–28.

Roberts, H. A. (1981) Seed banks in the soil. *Advances in Applied Biology*, 6, 1–55.

Roberts, H. A. & Feast, P. M. (1973) Emergence and longevity of seeds of annual weeds in cultivated and undisturbed soil. *Journal of Applied Ecology*, 10, 133–143.

Roberts, H. A. (1986) Seed persistence in soils and seasonal emergence in plant species from different habitats. *Journal of Applied Ecology*, 23, 639–656.

Roberts, H. A. & Neilson, J. E. (1981) Seed survival and periodicity of seedling emergence in twelve weedy species of Compositae. *Annals of Applied Biology*, 97, 325–334.

Robinson, D. & Fitter, A. (1999) The magnitude and control of carbon transfer between plants linked by a common mycorrhizal network. *Journal of Experimental Botany*, 50, 9–13.

Robinson, D., Hodge, A., Griffiths, B. S. & Fitter, A. H. (1999) Plant root proliferation in nitrogen-rich patches confers competitive advantage. *Proceedings of the Royal Society of London, B*, 266, 431–435.

Rockwood, L. L. (1985) Seed weight as a function of life form, elevation and life zone in neotropical forests. *Biotropica*, 17, 32–39.

Roff, D. A. (1992) *The Evolution of Life Histories*. Chapman & Hall, London.

Rose, M. R. (1991) *Evolutionary Biology of Aging*. Oxford University Press, New York.

Rosewell, J., Shorrocks, B. & Edwards, K. (1990) Competition on a divided and ephemeral resource: Testing the assumptions. I. Aggregation. *Journal of Animal Ecology*, 59, 977–1001.

Ross, M. A. & Harper, J. L. (1972) Occupation of biological space during seedling establishment. *Journal of Ecology*, 60, 77–88.

Roy, B. A. (1995) The Breeding Systems of 6 Species of *Arabis* (Brassicaceae). *American Journal of Botany*, 82, 869–877.

Roy, B. A. & Kirchner, J. W. (2000) Evolutionary dynamics of pathogen resistance and tolerance. *Evolution*, 54, 51–63.

Runkle, J. R. (1982) Patterns of disturbance in some old-growth mesic forests of Eastern North America. *Ecology*, 63, 1533–1546.

Runkle, J. R. (1998) Changes in southern Appalachian canopy tree gaps sampled thrice. *Ecology*, 79, 1768–1780.

Runkle, J. R. & Yetter, T. C. (1987) Treefalls revisited: gap dynamics in the Southern Appalachians. *Ecology*, 68, 417–424.

Salisbury, E. J. (1942) *The Reproductive Capacity of Plants*. Bell and Sons, London.

Salisbury, E. J. (1974) Seed size and mass in relation to environment. *Proceedings of the Royal Society of London, B*, 186, 83–88.

Samson, D. A. & Werk, K. S. (1986) Size-dependent effects in the analysis of reproductive effort in plants. *American Naturalist*, 127, 667–680.

Sano, Y., Morishima, H. & Oka, H.-I. (1980) Intermediate perennial-annual populations of *Oryza perennis* found in Thailand and their evolutionary significance. *Botanical Magazine, Tokyo*, 93, 291–305.

Sarukhán, J. (1974) Studies on plant demography: *Ranunculus repens* L., *R. bulbosus* L. & *R. acris* L: II. Reproductive strategies and seed population dynamics. *Journal of Ecology*, 62, 151–177.

Sarukhán, J. & Harper, J. L. (1973) Studies on plant demography: *Ranunculus repens* L., *R. Bulbosus* L. and *R. acris* L. I. Population flux and survivorship. *Journal of Ecology*, 61, 675–716

Sarukhán, J., Piñero, D. & Martínez-Ramos, M. (1985) Plant demography: a community-level interpretation. *Studies in Plant Demography* (ed. J. White), pp. 17–31. Academic Press, London.

Sarvas, R. (1968) Investigations on the flowering and seed crop of *Picea abies. Communicationes Instituti Forestalis Fenniae*, 67, 1–69.

Scaife, M. A. & Jones, D. (1976) The relationship between crop yield (or mean plant weight) of lettuce and plant density, length of growing period, and initial plant weight. *Journal of Agricultural Science*, 86, 83–91.

Schaal, B. A. (1980) Reproductive capacity and seed size in *Lupinus texensis. American Journal of Botany*, 67, 703–709.

Schaal, B. A. & Olsen, K. M. (2000) Gene genealogies and population variation in plants. *Proceedings of the National Academy of Sciences of the USA*, 97, 7024.

Schaffer, W. M. (1974) Selection for optimal life histories: The effects of age structure. *Ecology*, 55, 291–303.

Schaffer, W. M. & Gadgil, M. D. (1975) Selection for optimal life histories in plants. *Ecology and Evolution of Communities* (eds M. L. Cody & J. Diamond), pp. 142–156. Belknap Press, Cambridge, MA.

Schaffer, W. M. & Schaffer, M. D. (1977) The adaptive significance of variations in reproductive habit in the Agavaceae. *Evolutionary Ecology* (eds B. Stonehouse & C. M. Perrins), pp. 261–276. Macmillan, London.

Schaffer, W. M. & Schaffer, M. D. (1979) The adaptive significance of variations in reproductive habit in the Agavaceae. II. Pollinator foraging behaviour and selection for increased reproductive expenditure. *Ecology*, 60, 1051–1069.

Scheiner, S. M. & Goodnight, C. J. (1984) The comparison of phenotypic and genotic variation in populations of the grass *Danthonia spicata. Evolution*, 38, 817–832.

Schemske, D. W. (1978) Evolution of reproductive characteristics in *Impatiens* (Balsaminanceae): The significance of cleistogamy and chasmogamy. *Ecology*, 59, 596–613.

Schemske, D. W. (1984) Population structure and local selection in *Impatiens pallida* (Balsaminaceae), a selfing annual. *Evolution*, 38, 817–832.

Schlichting, C. D. & Levin, D. A. (1986) Phenotopic plasticity: an evolving plant character. *Biological Journal of the Linnean Society*, 29, 37–47.

Schmidt, K. P. & Levin, D. A. (1985) The comparative demography of reciprocally sown populations of *Phlox drummondii* Hook I. Survivorship, fecundities and finite rates of increase. *Evolution*, 39, 396–404.

Schmitt, J., Dudley, S. A. & Pigliucci, M. (1999) Manipulative approaches to testing adaptive plasticity: Phytochrome-mediated shade-avoidance responses in plants. *American Naturalist*, 154, S43–S54.

Schnable, P. S. & Wise, R. P. (1998) The molecular basis of cytoplasmic male sterility and fertility restoration. *Trends in Plant Science*, 3, 175–180.

Schoen, D. J. (1982a) The breeding system of *Gilia achilleifolia*: variation in floral characteristics and outcrossing rate. *Evolution*, 36, 352–360.

Schoen, D. J. (1982b) Male reproductive effort and breeding system in an hermaphroditic plant. *Oecologia (Berlin)*, 53, 255–257.

Schoen, D. J. & Clegg, M. T. (1984) Estimation of mating system parameters when outcrossing events are correlated. *Proceedings of the National Academy of Sciences of the USA*, 81, 5258–5262.

Schoen, D. J. & Clegg, M. T. (1985) The influence of flower color on outcrossing rate on male reproductive success in *Ipomoea Purpurea. Evolution*, 39, 1242–1249.

Schoen, D. J., L'Heureux, A.-M., Marsolais, J. & Johnston, M. O. (1997) Evolutionary history of the mating system in *Amsinckia* (Boraginaceae). *Evolution*, 51, 1090–1099.

Schoen, D. J. & Lloyd, D. G. (1984) The selection of cleistogamy and heteromorphic diaspores. *Botanical Journal of the Linnean Society*, 23, 303–322.

Schoen, D. J. & Lloyd, D. G. (1992) Self- and cross-fertilization in plants. III. Methods for studying modes and functional aspects of self-fertilization. *International Journal of Plant Science*, 153, 381–393.

Schoen, D. J., Morgan, M. T. & Bataillon, T. (1997) How does self-pollination evolve? Inferences from floral ecology and molecular genetic variation. *Plant Life Histories* (eds J. Silvertown, M. Franco & J. L. Harper), pp. 102–118. Cambridge University Press, Cambridge.

Schwinning, S. (1996) Decomposition analysis of competitive symmetry and size structure dynamics. *Annals of Botany*, 77, 47–57.

Schwinning, S. & Weiner, J. (1998) Mechanisms determining the degree of size asymmetry in competition among plants. *Oecologia*, 113, 447–455.

Seger, J. & Eckhart, V. M. (1996) Evolution of sexual systems and sex allocation in annual plants when growth and reproduction overlap. *Proceedings of the Royal Society of London, B*, 263, 833–841.

Selman, M. (1970) The population dynamics of *Avena fatua* (wild oats) in continuous spring barley: desirable frequency of spraying with tri-allate. *Proceedings of the 10th British Weed Control Conference*, 1176–1188.

Sharitz, R. R. & McCormick, J. F. (1973) Population dynamics of two competing annual plant species. *Ecology*, 54, 723–740.

Shaw, R. G. (1987) Density dependence in Salvia lyrata: an experimental alteration of densities of established plants. *Journal of Ecology* 75, 1049.

Shipley, B. (1993) A null model for competitive hierarchies in competition matrices. *Ecology*, 74, 1693–1699.

Shmida, A. & Ellner, S. P. (1984) Coexistence of plants with similar niches. *Vegetatio*, 58, 29–55.

Shumway, D. L. & Koide, R. T. (1995) Size and reproductive inequality in mycorrhizal and nonmycorrhizal populations of abutilon-theophrasti. *Journal of Ecology*, 83, 613–620.

Silander, J. A. & Pacala, S. W. (1985) Neighborhood predictors of plant performance. *Oecologia*, 66, 256–263.

Silva Matos, D. M., Freckleton, R. P. & Watkinson, A. R. (1999) The role of density dependence in the population dynamics of a tropical palm. *Ecology*, 80, 2635–2650.

Silvertown, J. (1980a) Leaf-canopy induced seed dormancy in a grassland flora. *New Phytologist*, 85, 109–118.

Silvertown, J. (1980b) The evolutionary ecology of mast seeding in trees. *Biological Journal of the Linnean Society*, 14, 235–250.

Silvertown, J. (1984) Phenotypic variety in seed germination behaviour: the ontogeny and evolution of somatic polymorphism in seeds. *American Naturalist*, 124, 1–16.

Silvertown, J. (1985) When plants play the field. *Evolution: Essays in Honour of John Maynard Smith* (eds P. Greenwood, P. H. Harvey & M. Slatkin), pp. 143–153. Cambridge University Press, Cambridge.

Silvertown, J. (1987) *Introduction to Plant Population Ecology.* Longman, Harlow.

Silvertown, J. (1988) The demographic and evolutionary consequences of seed dormancy. *Plant Population Ecology* (eds A. J. Davy, M. J. Hutchings & A. R. Watkinson), pp. 205–219. Blackwell Scientific Publications, Oxford.

Silvertown, J. (1989) A binary classification of plant life histories and some possibilities for its evolutionary application. *Evolutionary Trends in Plants*, 3, 87–90.

Silvertown, J. (1991) Modularity, reproductive thresholds and plant population dynamics. *Functional Ecology*, 5, 577–582.

Silvertown, J. & Dodd, M. (1997) Comparing plants and connecting traits. *Plant Life Histories: Ecology, Phylogeny and Evolution* (eds J. Silvertown, M. Franco & J. L. Harper), pp. 3–16. Cambridge University Press, Cambridge.

Silvertown, J., Dodd, M. E., Gowing, D. & Mountford, O. (1999) Hydrologically-defined niches reveal a basis for species-richness in plant communities. *Nature*, 400, 61–63.

Silvertown, J. & Franco, M. (1993) Plant demography and habitat: a comparative approach. *Plant Species Biology*, 8, 67–73.

Silvertown, J., Franco, M. & Menges, E. (1996) Interpretation of elasticity matrices as an aid to the management of plant populations for conservation. *Conservation Biology*, 10, 591–597.

Silvertown, J., Franco, M. & Perez-Ishiwara, R. (2001) Evolution of senescence in iteroparous perennial plants. *Evolutionary Ecology Research*, 3, 1–20.

Silvertown, J., Franco, M., Pisanty, I. & Mendoza, A. (1993) Comparative plant demography – relative importance of life-cycle components to the finite rate of increase in woody and herbaceous perennials. *Journal of Ecology*, 81,465–476.

Silvertown, J. & Gordon, D. (1989) A framework for plant behaviour. *Annual Review of Ecology and Systematics*, 20, 349–366.

Silvertown, J. & Law, R. (1987) Do plants need niches? *Trends in Ecology and Evolution*, 2, 24–26.

Silvertown, J. & Wilson, J. B. (2000) Spatial interactions among grassland plant populations. *The Geometry of Ecological Interactions* (eds U. Dieckman, R. Law & H. Metz), pp. 28–47. Cambridge University Press, Cambridge.

Simard, S. W., Perry, D. A., Jones, M. D. *et al.* (1997) Net transfer of carbon between ectomycorrhizal tree species in the field. *Nature*, 388, 579–582.

Simonich, M. T. & Morgan, M. D. (1994) Allozymic uniformity in *Iris lacustris* (dwarf lake iris) in Wisconsin. *Canadian Journal of Botany*, 72, 1720–1722.

Skalova, H., Pechackova, S., Suzuki, J. *et al.* (1997) Within population genetic differentiation in traits affecting clonal growth: *Festuca rubra* in a mountain grassland. *Journal of Evolutionary Biology*, 10, 383–406.

Skellam, J. G. (1951) Random dispersal in theoretical populations. *Biometrika*, 38, 196–218.

Slade, A. J. & Hutchings, M. J. (1987a) Clonal integration and plasticity in foraging behaviour in *Glechoma hederacea. Journal of Ecology*, 75, 1023.

Slade, A. J. & Hutchings, M. J. (1987b) The effects of nutrient availability on foraging in the clonal herb *Glechoma hederacea. Journal of Ecology*, 75, 95–112.

Slatkin, M. (1981) Estimating levels of gene flow in natural populations. *Genetics*, 99, 323–335.

Slatkin, M. (1993) Isolation by distance in equilibrium and non-equilibrium populations. *Evolution*, 47, 264–279.

Smith, H. (2000) Phytochromes and light signal perception by plants – an emerging synthesis. *Nature*, 407, 585–591.

Smith, C. C. & Fretwell, S. D. (1974) The optimal balance between the size and number of offspring. *American Naturalist*, 108, 499–506.

Smith, C. C., Hamrick, J. L. & Kramer, C. L. (1990) The advantage of mast years for wind pollination. *American Naturalist*, 136, 154–166.

Smith, M. D., Hartnett, D. C. & Wilson, G. W. T. (1999) Interacting influence of mycorrhizal symbiosis and competition on plant diversity in tallgrass prairie. *Oecologia*, 121, 574–582.

Snaydon, R. W. & Davies, T. M. (1982) Rapid divergence of plant populations in response to recent changes in soil conditions. *Evolution*, 36, 289–297.

Snaydon, R. W. & Howe, C. D. (1986) Root and shoot competition between established ryegrass and invading grass seedlings. *Journal of Applied Ecology*, 23, 667–674.

Solbrig, O. T., Sarandon, R. & Bossert, W. (1988) A density-dependent growth model of a perennial herb, *Viola fimbriatula. American Naturalist*, 131, 385–400.

Soltis, D. E., Soltis, P. S. & Milligan, B. G. (1992) Intraspecific chloroplast DNA variation: systematic and phylogenetic implications. *Molecular Systematics of Plants* (eds P. S. Soltis, D. E. Soltis & J. J. Doyle), pp. 116–150. Chapman and Hall, New York.

Sork, V. L. (1993) Evolutionary ecology of Mast-Seeding in temperate and tropical oaks (*Quercus* spp). *Vegetatio*, 108, 133–147.

Southwood, T. R. E. (1988) Tactics, strategies and templets. *Oikos*, 52, 3–18.

Spitters, C. J. T. (1983) An alternative approach to the analysis of mixed cropping experiments 1. Estimation of competition coefficients. *Netherlands Journal of Agricultural Science*, 31, 1–11.

Stadler, L. J. (1930) The frequency of mutation of specific genes in maize (Abstract). *Anatomical Record*, 47, 381.

Stanton, M. L. (1984) Developmental and genetic sources of seed weight variation in *Raphanus raphanistrum* L. (Brassicaceae). *American Journal of Botany*, 71, 1090–1098.

Stearns, S. C. (1992) *The Evolution of Life Histories.* Oxford University Press, Oxford.

Stearns, S. C. & Hoekstra, R. E. (2000) *Evolution. An Introduction.* Oxford University Press, Oxford.

Stebbins, G. L. (1950) *Variation and Evolution in Plants.* Columbia University Press, New York.

Stebbins, G. L. (1957) Self fertilization and population variation in the higher plants. *American Naturalist,* 91, 337–354.

Sterner, R. W., Ribic, C. A. & Schatz, G. E. (1986) Testing for life historical changes in spatial pattern of four tropical tree species. *Journal of Ecology,* 74, 621.

Stevens, M. H. H. & Carson, W. P. (1999) The significance of assemblage-level thinning for species richness. *Journal of Ecology,* 87, 490–502.

Stöcklin, J. & Fischer, M. (1999) Plants with longer-lived seeds have lower local extinction rates in grassland remnants 1950-85. *Oecologia,* 120, 539–543.

Stratton, D. A. (1992) Life-cycle components of selection in *Erigeron annuus:* II. Genetic variation. *Evolution,* 46, 107–120.

Stratton, D. A. (1994) Genotype-by-environment interactions for fitness of *Erigeron annuus* show fine-scale selective heterogeneity. *Evolution,* 48, 1607–1618.

Stratton, D. A. (1995) Spatial scale of variation in fitness of Erigeron annuus. *American Naturalist,* 146, 608–624.

Stratton, D. A. & Bennington, C. C. (1998) Fine-grained spatial and temporal variation in selection does not maintain genetic variation in Erigeron annuus. *Evolution,* 52, 678–691.

Streiff, R., Ducousso, A., Lexer, C. *et al.* (1999) Pollen dispersal inferred from paternity analysis in a mixed oak stand of *Quercus robur* L and *Q. petraea* (Matt.) Liebl. *Molecular Ecology,* 8, 831–841.

Sultan, S. E. (1987) Evolutionary implications of phenotypic plasticity in plants. *Evolutonary Biology,* 21, 127–178.

Sun, M. & Ganders, F. R. (1986) Female frequencies in gynodioecious populations correlated with selfing rates in hermaphrodites. *American Journal of Botany,* 73, 1645–1648.

Sutherland, W. J. & Stillman, R. A. (1988) The foraging tactics of plants. *Oikos,* 52, 239–244.

Sutherland, S. & Vickery, R. K. Jr (1988) Trade-offs between sexual and asexual reproduction in the genus Mimulus. *Oecologia,* 76, 330–335.

Suzan, H., Nabhan, G. P. & Patten, D. T. (1996) The importance of *Olneya tesota* as a nurse plant in the Sonoran Desert. *Journal of Vegetation Science,* 7, 635–644.

Symonides, E. (1977) Mortality of seedlings in natural psammophyte populations. *Ekologia Polska,* 25, 635–651.

Symonides, E. (1983) Population size regulation as a result of intra-population interactions. II. Effect of density on the survival of individuals of *Erophila verna* (L.) C.A.M. *Ekologia Polska,* 31, 839–881.

Symonides, E. (1984) Population size regulation as a result of intra-population interactions. III. Effect of *Erophila verna* (L.) C.A.M. population density on the abundance of seedlings. Summing-up and conclusions. *Ekologia Polska,* 32, 557–580.

Symonides, E. (1988) Population dynamics of annual plants. *Plant Population Ecology* (eds A. J. Davy, M. J. Hutchings & A. R. Watkinson), pp. 221–248. Blackwell Scientific Publications, Oxford.

Symonides, E., Silvertown, J. & Andreasen, V. (1986) Population cycles caused by overcompensating density-dependence in an annual plant. *Oecologia,* 71, 156–158.

Taberlet, P., Fumagalli, L., Wust Saucy, A. G. & Cosson, J. F. (1998) Comparative phylogeography and postglacial colonization routes in Europe. *Molecular Ecology,* 7, 453–464.

Takada, T. & Caswell, H. (1997) Optimal size at maturity in size-structured populations. *Journal of Theoretical Biology,* 187, 81–93.

Taylor, D. R. & Aarssen, L. W. (1989) On the density dependence of replacement-series competition experiments. *Journal of Ecology,* 77, 975–988.

Templeton, A. R. & Levin, D. A. (1979) Evolutionary consequences of seed pools. *American Naturalist,* 114, 232–249.

Thomas, C. D. & Kunin, W. E. (1999) The spatial structure of populations. *Journal of Animal Ecology,* 68, 647–657.

Thompson, K. (1986) Small-scale heterogeneity in the seed bank of an acidic grassland. *Journal of Ecology,* 74, 733–738.

Thompson, K. & Grime, J. P. (1979) Seasonal variation in seed banks of herbaceous species in ten contrasting habitats. *Journal of Ecology,* 67, 893–921.

Thompson, K. & Hodkinson, D. J. (1998) Seed mass, habitat and life history: a re-analysis of Salisbury (1942, 1974). *New Phytologist,* 138, 163–166.

Thompson, K., Bakker, J. P., Bekker, R. M. & Hodgson, J. G. (1998) Ecological correlates of seed persistence in soil in the north-west European flora. *Journal of Ecology,* 86, 163–169.

Thompson, K., Band, S. R. & Hodgson, J. G. (1993) Seed size and shape predict persistence in soil. *Functional Ecology,* 7, 236–241.

Thomson, J. D. & Brunet, J. (1990) Hypotheses for the evolution of dioecy in seed plants. *Trends Ecology and Evolution,* 5, 11–16.

Thrall, P. H., Pacala, S. W. & Silander, J. A. Jr (1989) Oscillatory dynamics in populations of an annual weed species *Abutilon theophrasti. Journal of Ecology,* 77, 1135–1149.

Tilman, D. (1982) *Resource Competition and Community Structure.* Princeton University Press, Princeton, NJ.

Tilman, D. (1994) Competition and biodiversity in spatially structured habitats. *Ecology,* 75, 2–16.

Tilman, D. & Wedin, D. (1991) Dynamics of nitrogen competition between successional grasses. *Ecology,* 72, 1038–1049.

Tissue, D. T. & Nobel, P. S. (1990) Carbon relations of flowering in a semelparous clonal desert perennial. *Ecology,* 71, 273–281.

Todokoro, S., Terauchi, R. & Kawano, S. (1995) Microsatellite polymorphisms in natural populations of *Arabidopsis thaliana* in Japan. *Japanese Journal of Genetics,* 70, 543–554.

Tonsor, S. J. (1990) Spatial patterns of differentiation for gene flow in *Plantago lanceolata. Evolution,* 44, 1373–1378.

Tonsor, S. J., Kalisz, S., Fisher, J. & Holtsford, T. P. (1993) A life-history based study of population genetic structure: seed bank to adults in *Plantago lanceolata. Evolution,* 47, 833–843.

REFERENCES

Torstensson, P. & Telenius, A. (1986) Consequences of differential utilization of meristems in the annual *Spergularia marina* and the perennial *S. media. Holarctic Ecology*, 9, 20–26.

van Treuren, R., Bijlsma, R., Ouborg, N. J. & Kwak, M. M. (1994) Relationships between plant density, outcrossing rates and seed set in natural and experimental populations of *Scabiosa columbaria. Journal of Evolutionary Biology*, 7, 287–301.

Turgeon, J. J., Roques, A. & Degroot, P. (1994) Insect fauna of coniferous seed cones – diversity, host plant interactions, and management. *Annual Review of Entomology*, 39, 179–212.

Turkington, R. (1996) Intergenotypic interactions in plant mixtures. *Euphytica*, 92, 105–119.

Turkington, R. A. & Harper, J. L. (1979) The growth, distribution and neighbour relationships of *Trifolium repens* in a permanent pasture. *Journal of Ecology*, 67, 245–254.

Turnbull, L. A., Rees, M. & Crawley, M. J. (1999) Seed mass and the competition/colonization trade-off: a sowing experiment. *Journal of Ecology*, 87, 899–912.

Turner, M. D. & Rabinowitz, D. (1983) Factors affecting frequency distributions of plant mass: the absence of dominance and suppression in competing monocultures of *Festuca paradoxa. Ecology*, 64, 469–475.

Uhl, C. & Clark, K. (1983) Seed ecology of selected Amazon basin successional species. *Botanical Gazette*, 144, 419–425.

Uyenoyama, M. K. (1986) Inbreeding and the cost of meiosis: the evolution of selfing in populations practicing biparental inbreeding. *Evolution*, 40, 388–404.

Uyenoyama, M. K. (1988) On the evolution of genetic incompatibility systems. II. Initial increase of strong gametophytic self-incompatibility under partial selfing and half-sib mating. *American Naturalist*, 131, 700–722.

Uyenoyama, M. K. (1989) On the evolution of genetic incompatibility systems. V. Origin of sporophytic self-incompatibility in response to overdominance in viability. *Theoretical Population Biology*, 36, 339–365.

Valiente-Banuet, A. & Ezcurra, E. (1991) Shade as a cause of the association between the cactus *Neobuxbaumia tetetzo* and the nurse plant *Minosa luisana* in the Tehuacán valley, Mexico. *Journal of Ecology*, 79, 961–971.

Valverde, T. & Silvertown, J. (1995) Spatial variation in the seed ecology of a woodland herb (*Primula vulgaris*) in relation to light environment. *Functional Ecology*, 9, 942–950.

Valverde, T. & Silvertown, J. (1997a) A metapopulation model for *Primula vulgaris*, a temperate forest understorey herb. *Journal of Ecology*, 85, 193–210.

Valverde, T. & Silvertown, J. (1997b) An integrated model of demography, patch dynamics and seed dispersal in a woodland herb, *Primula vulgaris. Oikos*, 80, 67–77.

Valverde, T. & Silvertown, J. (1997c) Canopy closure rate and forest structure. *Ecology*, 78, 1555–1562.

Valverde, T. & Silvertown, J. (1998) Variation in the demography of a woodland understorey herb (*Primula vulgaris*) along the forest regeneration cycle: projection matrix analysis. *Journal of Ecology*, 86, 545–562.

Vandermeer, J. (1984) Plant competition and the yield-density relationship. *Journal of Theoretical Biology*, 109, 393–399.

Vasek, F. C. (1980) Creosote bush: long-lived clones in the Mohave desert. *American Journal of Botany*, 67, 246–255.

Vavrek, M. C., McGraw, M. C. & Bennington, C. C. (1991) Ecological genetic variation in seed banks. III. Phenotypic and genetic differences between young and old seed populations of *Carex bigelowii. Journal of Ecology*, 79, 627–643.

Vazquez-Yanes, C. & Orozco-Segovia, A. (1982) Seed germination of a tropical rainforest pioneer tree (*Heliocarpus donellsmithii*) in response to diurnal fluctuations of temperature. *Physiologia Plantarum*, 56, 295–298.

Vazquez-Yanes, C. & Smith, H. (1982) Phytochrome control of seed germination in the tropical rain forest pioneer trees *Cecropia obtusifolia* and *Piper auritum* and its ecological significance. *New Phytologist*, 92, 477–485.

Venable, D. (1985) The evolutionary ecology of seed heteromorphism. *American Naturalist*, 126, 577–595.

Venable, D. L. (1989) Modelling the evolutionary ecology of seed banks. *Ecology of Soil Seed Banks* (eds M. A. Leck, V. T. Parker & R. L. Simpson), pp. 67–87. Academic Press, San Diego, CA.

Venable, D. L. (1992) Size-number trade-offs and the variation of seed size with plant resource status. *American Naturalist*, 140, 287–304.

Venable, D. L. & Brown, J. S. (1988) The selective interactions of dispersal, dormancy, and seed size as adaptations for reducing risk in variable environments. *American Naturalist*, 131, 360–384.

Venable, D. L. & Brown, J. S. (1993) The Population-Dynamic functions of seed dispersal. *Vegetatio*, 108, 31–55.

Venable, D. L. & Búrquez, A. (1989) Quantitative genetics of size, shape, life-history, and fruit characteritics of the seed-heteromorphic composite *Heterosperma pinnatum* L. variation within and among populations. *Evolution*, 43, 113–124.

Venable, D. L., Búrquez, A., Coral, G., Morales, E. & Espinosa, F. (1987) The ecology of seed heteromorphism in *Heterosperma pinnatum* in Central Mexico. *Ecology*, 68, 65–76.

Venable, D. L. & Lawlor, L. (1980) Delayed germination and dispersal in desert annuals: escape in time and space. *Oecologia*, 46, 272–282.

Venable, D. L. & Pake, C. E. (1999) Population ecology of Sonoran Desert Annual Plants. *Ecology of Sonoran Desert Plants and Plant Communities* (ed. R. H. Robichaux), Chapter 4. University of Arizona Press, Tucson, AZ.

Vendramin, G. G., Anzidei, M., Madaghiele, A., Sperisen, C. & Bucci, G. (2000) Chloroplast microsatellite analysis reveals the presence of population subdivision in Norway spruce (*Picea abies* K.). *Genome*, 43, 68–78.

Vogel, S. (1978) Evolutionary shifts from reward to deception in pollen flowers. *The Pollination of Flowers by Insects* (ed. A. J. Richards), pp. 89–96. Academic Press, London.

Vos, P., Hogers, R., Bleeker, M. *et al.* (1995) AFLP: a new technique for DNA fingerprinting. *Nucleic Acids Research*, 23, 4407–4414.

Wade, M. J. & Kalisz, S. (1990) The causes of natural selection. *Evolution*, 44, 1947–55.

Wade, M. J. & McCauley, D. E. (1988) Extinction and recolonization: Their effects on the genetic differentiation of local populations. *Evolution*, 42, 995–1005.

Wagner, R. G. & Radosevich, S. R. (1998) Neighborhood approach for quantifying interspecific competition in coastal Oregon forests. *Ecological Applications*, 8, 779–794.

Waldmann, P. & Andersson, S. (1998) Comparison of quantitative genetic variation and allozyme diversity within and between populations of *Scabiosa canescens* and *S. columbaria*. *Heredity*, 81, 79–86.

Walker, J. & Peet, R. K. (1983) Composition and species-diversity of pine-wiregrass savannas of the Green Swamp, North Carolina. *Vegetatio*, 55, 163–179.

Waller, D. M. (1979) Models of mast fruiting in trees. *Journal of Theoretical Biology*, 80, 223–232.

Waller, D. M. (1981) Neighbourhood competition in several violet populations. *Oecologia*, 51, 116–122.

Waller, D. M. (1982) Factors influencing seed weight in impatiens capensis (Balsaminaceae). *American Journal of Botany*, 69, 1470–1475.

Walter, L. E. F., Hartnett, D. C., Hetrick, B. A. D. & Schwab, A. P. (1996) Interspecific nutrient transfer in a tallgrass prairie plant community. *American Journal of Botany*, 83, 180–184.

Warner, R. R. & Chesson, P. L. (1985) Coexistence mediated by recruitment fluctuations: a field guide to the storage effect. *American Naturalist*, 125, 769–787.

Waser, N. M. & Price, M. V. (1989) Optimal outcrossing in *Ipomopsis aggregata*: seed set and offspring fitness. *Evolution*, 43, 1097–1109.

Watkinson, A. R. (1980) Density-dependence in single-species populations of plants. *Journal of Theoretical Biology*, 83, 345–357.

Watkinson, A. R. (1984) Yield-density relationships: the influence of resource availability on growth and self-thinning in populations of *Vulpia fasciculata*. *Annals of Botany*, 53, 469–482.

Watkinson, A. (1985) On the abundance of plants along an environmental gradient. *Journal of Ecology*, 73, 569–578.

Watkinson, A. R. (1986) Plant population dynamics. In: *Plant Ecology* (ed M. J. Crawley), pp. 137–184. Blackwell Scientific Publications, Oxford.

Watkinson, A. R. (1990) The population dynamics of *Vulpia fasciculata*: a nine-year study. *Journal of Ecology*, 78, 196–209.

Watkinson, A. R. & Davy, A. J. (1985) Population biology of salt marsh and sand dune annuals. *Vegetatio*, 62, 487–497.

Watkinson, A. R. & Harper, J. L. (1978) The demography of a sand dune annual *Vulpia fasciculata*: I. The natural regulation of populations. *Journal of Ecology*, 66, 15–33.

Watkinson, A. R., Lonsdale, W. M. & Andrew, M. H. (1989) Modelling the population dynamics of an annual plant *Sorghum intrans* in the wet-dry tropics. *Journal of Ecology*, 77, 162–181.

Watkinson, A. R. & Sutherland, W. J. (1995) Sources, sinks and pseudo-sinks. *Journal of Animal Ecology*, 64, 126–130.

Watt, A. S. (1974) Senescence and rejuvenation in ungrazed chalk grassland (grassland B) in Breckland: the significance of litter and moles. *Journal of Applied Ecology*, 11, 1157–1171.

Wayne, P. M. & Bazzaz, F. A. (1997) Light acquisition and growth by competing individuals in CO_2-enriched atmospheres: Consequences for size structure in regenerating birch stands. *Journal of Ecology*, 85, 29–42.

Weaver, S. E. & Cavers, P. B. (1979) The effects of date of emergence and emergence order on seedling survival rates in *Rumex crispus* and *R. obtusifolius*. *Canadian Journal of Botany*, 57, 730–738.

Webb, C. J. (1979) Breeding systems and the evolution of dioecy in New Zealand apioid Umbelliferae. *Evolution*, 33, 662–672.

Webb, S. L. (1986) Potential role of passenger pigeons and other vertebrates in the rapid holocene migrations of nut trees. *Quaternary Research*, 26, 367–375.

Webb, S. L. (1987) Beech range extension and vegetation history. *Ecology*, 68, 1993–2005.

Wedin, D. & Tilman, D. (1993) Competition among grasses along a nitrogen gradient – initial conditions and mechanisms of competition. *Ecological Monographs*, 63, 199–229.

Weiblen, G. D., Oyama, R. K. & Donoghue, M. J. (2000) Phylogenetic analysis of dioecy in monocotyledons. *American Naturalist*, 155, 46–58.

Weiher, E., van der Werf, A., Thompson, K. *et al.* (1999) Challenging Theophrastus: a common core list of plant traits for functional ecology. *Journal of Vegetation Science*, 10, 609–620.

Weiner, J. (1986) How competition for light and nutrients affects size variability in *Ipomoea tricolor* populations. *Ecology*, 67, 1425–1427.

Weiner, J. (1990) Asymmetric competition in plant populations. *Trends in Ecology and Evolution*, 5, 360–364.

Weiner, J. & Solbrig, O. T. (1984) The meaning and measurement of size hierarchies in plant populations. *Oecologia*, 61, 334–336.

Weiner, J. & Thomas, S. C. (1986) Size variability and competition in plant monocultures. *Oikos*, 47, 211–222.

Weiner, J., Wright, D. B. & Castro, S. (1997) Symmetry of below-ground competition between *Kochia scoparia* individuals. *Oikos*, 79, 85–91.

Welden, C. W., Hewett, S. W., Hubbell, S. P. & Foster, R. B. (1991) Sapling survival, growth, and recruitment: Relationship to canopy height in a neotropical forest. *Ecology*, 72, 35–50.

Weller, D. E. (1989) The interspecific size-density relationship among crowded plant stands and its implications for the -3/2 power rule of self thinning. *American Naturalist*, 133, 20–41.

Weller, S. G., Donoghue, M. J. & Charlesworth, D. (1995) The evolution of self-incompatibility in the flowering plants: a phylogenetic approach. *Experimental and Molecular Approaches to Plant Biosystematics* (eds P. C. Hoch & A. G. Stephenson), pp. 355–382. Missouri Botanic Garden, St. Louis, MO.

Weller, S. G. & Sakai, A. K. (1990) The genetic basis of male sterility in *Schiedia globosa* (Caryophyllaceae) an endemic Hawaiian genus. *Heredity*, 67, 265–273.

Weller, S. G., Wagner, W. L. & Sakai, A. K. (1995) A phylogenetic analysis of *Schiedia* and *Alsinidendron* (Caryophyllaceae: Alsinoideae): implications for the evolution of breeding systems. *Systematic Botany*, 20, 315–337.

Werner, P. A. (1975) Predictions of fate from rosette size in teasel (*Dipsacus fullonum* L.). *Oecologia*, 20, 197–201.

Werner, P. A. & Caswell, H. (1977) Population growth rates and age vs stage-distribution models for teasel (*Dipsacus sylvestris* Huds.). *Ecology*, 58, 1103–1111.

Wesselingh, R. A. & de Jong, T. J. (1995) Bidirectional selection on threshold size for flowering in *Cynoglossum officinale* (hound's tongue). *Heredity*, 74, 415–424.

Wesselingh, R. A., De Jong, T. J., Klinkhamer, P. G. L., Vandijk, M. J. & Schlatmann, E. G. M. (1993) Geographical variation in threshold size for flowering in *Cynoglossum officinale*. *Acta Botanica Neerlandica*, 42, 81–91.

Wesselingh, R. A. & Klinkhamer, P. G. L. (1996) Threshold size for vernalization in *Senecio jacobaea*: Genetic variation and response to artificial selection. *Functional Ecology*, 10, 281–288.

Wesselingh, R. A., Klinkhamer, P. G. L., DeJong, T. J. & Boorman, L. A. (1997) Threshold size for flowering in different habitats: Effects of size-dependent growth and survival. *Ecology*, 78, 2118–2132.

West, G. B., Brown, J. H. & Enquist, B. J. (1997) A general model for the origin of allometric scaling laws in biology. *Science*, 276, 122–126.

Westoby, M. (1998) A leaf-height-seed (LHS) plant ecology strategy scheme. *Plant and Soil*, 199, 213–227.

Westoby, M., Leishman, M. & Lord, J. (1997) Comparative ecology of seed size and dispersal. *Plant Life Histories: Ecology, Phylogeny and Evolution* (eds J. Silvertown, M. Franco & J. L. Harper), pp. 143–162. Cambridge University Press, Cambridge.

White, J. (1985) The thinning rule and its application to mixtures of plant populations. *Studies on Plant Demography* (ed. J. White), pp. 291–309. Academic Press, London.

Whitlock, M. C. & McCauley, D. E. (1990) Some population genetic consequences of colony formation and extinction – genetic correlations within founding groups. *Evolution*, 44, 1717–1724.

Whitlock, M. C. & McCauley, D. E. (1999) Indirect measures of gene flow and migration: F_{ST} does not equal $1/(4N_m + 1)$. *Heredity*, 82, 117–125.

Whitmore, T. C. (1983) Secondary succession from seed in tropical rain forests. *Forestry Abstracts*, 44, 767–779.

Whitmore, T. C. (1988) The Influence of Tree Population Dynamics on Forest Species Composition. In: *Plant Population Ecology* (eds A. J. Davy, M. J. Hutchings & A. R. Watkinson). Blackwell Scientific Publications, Oxford.

Whitmore, T. C. (1989) Canopy gaps and the two major groups of forest trees. *Ecology*, 70, 536–538.

Whittaker, R. H. (1969) Evolution of diversity in ecological systems. *Brookhaven Symposium Biol.*, 22, 178–195.

Widén, B. (1991) Phenotypic selection on flowering phenology in *Senecio integrifolius*, a perennial herb. *Oikos*, 61, 205–215.

Widén, B., Cronberg, N. & Widen, M. (1994) Genotypic diversity, molecular markers and spatial distribution of genets in clonal plants: a literature survey. *Folia Geobotanica and Phytotaxonomica*, 29, 245–263.

Willey, R. W. (1979) Intercropping: its importance and research needs Part 1. Competition and yield advantage. *Field Crop Abstracts*, 32, 1–10.

Willey, R. W. & Heath, S. B. (1969) The quantitative relationships between plant population and crop yield. *Advances in Agronomy*, 21, 281–321.

Williamson, M. (1996) *Biological Invasions*. Chapman and Hall, London.

Wills, C., Condit, R., Foster, R. B. & Hubbell, S. P. (1997) Strong density- and diversity-related effects help to maintain tree species diversity in a neotropical forest. *Proceedings of the National Academy of Sciences of the USA*, 94, 1252–1257.

Willson, M. F. (1993) Dispersal mode, seed shadows, and colonization patterns. *Vegetatio*, 108, 261–280.

Wilson, J. B. (1988) Shoot competition and root competition. *Journal of Appl. Ecol*, 25, 279.

Wilson, C. & Gurevitch, J. (1995) Plant size and spatial pattern in a natural population of *Myosotis micrantha*. *Journal of Vegetation Science*, 6, 847–852.

Wilson, R. G., Kerr, E. D. & Nelson, L. A. (1985) Potential for using weed seed content in the soil to predict future weed problems. *Weed Science*, 33, 171–175.

Wilson, J. B. & Levin, D. A. (1986) Some genetic consequences of skewed fecundity distributions in plants. *Theoretical and Applied Genetics*, 73, 113–121.

Wilson, S. D. & Tilman, D. (1991) Components of plant competition along an experimental gradient of nitrogen availability. *Ecology*, 72, 1050–1065.

de Wit, C. T. (1960) On competition. *Verslang an Landbouwkundige Onderzoekingen*, 66, 1–82.

Wolf, P. G., Murray, R. A. & Sipes, S. D. (1997) Species-independent, geographical structuring of chloroplast DNA haplotypes in a montane herb *Ipomopsis* (Polemoniaceae). *Molecular Ecology*, 6, 283–291.

Wolf, L. L., Reed Hainsworth, F., Mercier, T. & Benjamin, R. (1986) Seed-size variation and pollinator uncertainty in *Ipomopsis aggregata* (Polemoniaceae). *Journal of Ecology*, 74, 361–371.

Wolfe, L. M. (1993) Inbreeding depression in *Hydrophyllum appendiculatum*: role of maternal effects, crowding, and parental mating history. *Evolution*, 47, 374–386.

Wolfe, L. M. (1995) The genetics and ecology of seed size variation in a biennial plant, *Hydrophyllum appendiculatum* (Hydrophyllaceae). *Oecologia*, 101, 343–352.

Wolfe, A. D. & Elisens, W. J. (1995) Evidence of chloroplast capture and pollen-mediated gene flow in *Penstemon* Sect Peltanthera (Scrophulariaceae). *Systematic Botany*, 20, 395–412.

Wolfe, A. D., Xiang, Q.-Y. & Kephart, S. R. (1998) Assessing hybridization in natural populations of *Penstemon* (Scrophulariaceae) using hypervariable intersimple sequence repeat (ISSR) bands. *Molecular Ecology*, 7, 1107–1125.

Woodell, S. J., Mooney, H. A. & Hill, A. J. (1969) The behaviour of *Larrea divaricata* (creosote bush) in response to rainfall in California. *Journal of Ecology*, 57, 37–44.

Wright, S. (1939) The distribution of self-sterility alleles in populations. *Genetics*, 24, 538–552.

Wright, S. (1948) On the roles of directed and random changes in gene frequency in the genetics of populations. *Evolution*, 2, 279–295.

Wright, S. (1977) *Evolution and the genetics of populations, Vol. 3. Experimental Results and Evolutionary Deductions.* University of Chicago Press, Chicago.

Wright, A. J. (1981) The analysis of yield-density relationships in binary mixtures using inverse polynomials. *Journal of Agricultural Science*, 96, 561–567.

Wright, S. J. (1982) Competition, differential mortality, and their effect on the spatial pattern of a desert perennial, *Eriogonum inflatum* Torr. & Frem. (Polygonaceae). *Oecologia*, 54, 266–269.

Wright, S. J. (1983) The dispersion of eggs by a bruchid beetle among *Scheelea* palm seeds and the effect of distance to the parent palm. *Ecology*, 64, 1016–1021.

Wulff, R. D. (1986a) Seed size variation in *Desmodium paniculatum*. I. Factors affecting seed size. *Journal of Ecology*, 74, 87.

Wulff, R. D. (1986b) Seed size variation in Desmodium paniculatum. III. Effects on reproductive yield and competitive ability. *Journal of Ecology*, 74, 115.

Wyatt, R., Odrzykoski, I. J. & Stoneburner, A. (1989) High levels of genetic variability in the haploid moss *Plagiomnium ciliare. Evolution*, 43, 1085–1096.

Yampolsky, E. & Yampolsky, H. (1922) Distribution of sex forms in the phanerogamic flora. *Bibliotheca Genetica*, 3, 1–62.

Yeaton, R. I. (1978) A cyclical relationship between *Larrea tridentata* and *Opuntia leptocaulis* in the northern Chihuahua desert. *Journal of Ecology*, 66, 651–656.

Yoda, K., Kira, T., Ogawa, H. & Hozumi, K. (1963) Self-thinning in overcrowded pure stands under cultivated and natural conditions. *Journal of Biology, Osaka City University*, 14, 107–129.

Young, T. P. (1990) Evolution of semelparity in Mount Kenya lobelias. *Evolutionary Ecology*, 4, 157–171.

Young, T. P. & Augspurger, C. K. (1991) Ecology and evolution of long-lived semelparous plants. *Trends in Ecology and Evolution*, 6, 285–289.

Zammit, C. & Zedler, P. H. (1990) Seed yield, seed size and germination behaviour in the annual *Pogogyne abramsii. Oecologia*, 84, 24–28.

Zhang, J. & Maun, M. A. (1990) Effect of sand burial on seed germination, seedling emergence, survival and growth of Agropyron psammophilum. *Canadian Journal of Botany*, 68, 304–310.

Zhang, D. Y. & Wang, G. (1994) Evolutionarily stable reproductive strategies in sexual organisms – an integrated approach to life-history evolution and sex allocation. *American Naturalist*, 144, 65–75.

Zuberi, M. I. & Gale, J. S. (1976) Variation in wild populations of *Papaver dubium*. X. Genotype–environment interaction associated with differences in soil. *Heredity*, 36, 359–368.

Index